ANNALS OF THE NEW YORK ACADEMY OF SCIENCES
Volume 482

COMPUTER SIMULATION OF CHEMICAL AND BIOMOLECULAR SYSTEMS

Edited by David L. Beveridge and William L. Jorgensen

The New York Academy of Sciences
New York, New York
1986

Cover: Transition state for the S_N2 reaction of Cl^- + CH_3Cl and nearby water molecules. Photograph taken from a Silicon Graphics IRIS 3030 workstation using a program written by Jiali Gao at Purdue University.

Library of Congress Cataloging-in-Publication Data

Computer simulation of chemical and biomolecular systems.

(Annals of the New York Academy of Sciences, ISSN 0077-8923; v. 482)
Papers from a conference held by the New York Academy of Sciences, Oct. 2–4, 1985.
Bibliography: p.
Includes index.
1. Molecular dynamics—Mathematical models—Congresses.
2. Molecular structure—Mathematical models—Congresses.
3. Molecular dynamics—Data processing—Congresses.
4. Molecular structure—Data processing—Congresses.
I. Beveridge, David L., 1938–   II. Jorgensen,
William L.   III. New York Academy of Sciences.
IV. Series.
Q11.N5   vol. 482   500 s   86-33230
[QD461]   [541.2'2'0724]
ISBN 0-89766-359-4
ISBN 0-89766-360-8 (pbk.)

SP
Printed in the United States of America
ISBN 0–089766–359–4 (cloth)
ISBN 0–89766–360–8 (paper)
ISSN 0077-8923

ANNALS OF THE NEW YORK ACADEMY OF SCIENCES

Volume 482
December 31, 1986

COMPUTER SIMULATION OF CHEMICAL AND BIOMOLECULAR SYSTEMS[a]

Editors
DAVID L. BEVERIDGE and WILLIAM L. JORGENSEN

CONTENTS

[a]This volume is the result of a conference entitled Computer Simulation of Chemical and Biomolecular Systems, which was held by the New York Academy of Sciences on October 2–4, 1985 in New York City.

Financial assistance was received from:
- Burroughs Wellcome Co.
- Cemcomco
- E.I. du Pont de Nemours & Company—Biomedical Products Department
- ELXSI
- Gould Electronics—Computer Systems Division
- Hoffmann-LaRoche, Inc.
- Office of Naval Research, Department of the Navy (Grant N00014-85-G-0214)
- Rhône-Poulenc Inc.
- Smith Kline & French Laboratories
- Stuart Pharmaceuticals/Division of ICI Americas

The New York Academy of Sciences believes it has a responsibility to provide an open forum for discussion of scientific questions. The positions taken by the participants in the reported conferences are their own and not necessarily those of the Academy. The Academy has no intent to influence legislation by providing such forums.

Preface

D. L. BEVERIDGE

Department of Chemistry
Wesleyan University
Middletown, Connecticut 06457

W. L. JORGENSEN

Department of Chemistry
Purdue University
West Lafayette, Indiana 47907

With the present generation of digital computers and most recently, supercomputers and parallel processing, numerical calculations based on classical statistical mechanics performed on molecular liquids and on macromolecules have become feasible. Both these areas require consideration of an ensemble of configurations and are dealt with via very similar theoretical formalisms. An instructive presentation of the relationship between this area of research and theoretical and computational investigations of solids and gases was devised by van Gunsteren and Berendsen, two of the pioneers in the field, and is shown here as FIGURE 1. Classical statistical mechanics of liquids and macromolecules is now at the forefront of research initiatives in computational chemistry and biochemistry.

Recent years have seen the emergence of computer simulation studies on liquid water, aqueous solutions, and the initial extensions of these procedures to the study of the aqueous hydration of biomolecules. In principle, calculated values of many useful thermodynamic variables and molecular indices can be obtained from computer simulation of aqueous solutions. However, research methodologies in this field are still very much under development. A full reconciliation of the computer simulation results and experimental diffraction data on liquid water has not yet been achieved. A fully reliable description of the interaction of a single water molecule with other molecular species has not yet been realized, much less that of a hydration complex in aqueous solution. Much remains to be learned about and from computer simulation in structural chemistry and biochemistry, both in the case of methodology (sampling procedures, convergence problems, advantageous methods of analysis and characterization of intermolecular potentials) and in application areas, the description of aqueous hydration, solvent effects on conformational stability, and the influence of solvent and counterion atmosphere on protein and nucleic acid structure and function.

A conference designed to provide a state of the art perspective on the capabilities and limitations of computer simulation and on the present status of applications to chemical and biomolecular systems was held in New York City on October 2–4, 1985, and the papers provided in this volume are a record of the proceedings. The papers are grouped into the following categories: Procedure and Methodology, Aqueous Solutions, Crystal Hydrates, Reactions and Interactions, and Biomacromolecules. A

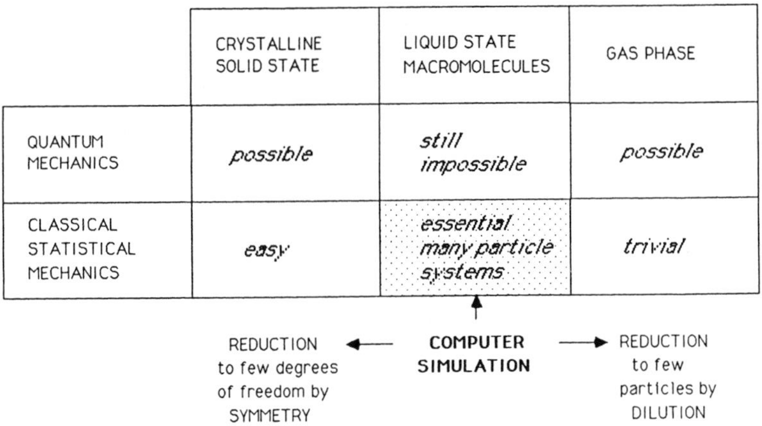

	CRYSTALLINE SOLID STATE	LIQUID STATE MACROMOLECULES	GAS PHASE
QUANTUM MECHANICS	*possible*	*still impossible*	*possible*
CLASSICAL STATISTICAL MECHANICS	*easy*	*essential many particle systems*	*trivial*

REDUCTION   ◄─── **COMPUTER** ───► REDUCTION
to few degrees   **SIMULATION**   to few
of freedom by   particles by
SYMMETRY   DILUTION

FIGURE 1.

number of papers have a purview overlapping two or more of these categories and in these cases the placement was arbitrary.

As organizers we would like to thank the participants for providing a stimulating and provocative conference, and the contributors listed on the last page of the Table of Contents for the critical financial support that made this meeting possible. Dr. Mihaly Mezei of Hunter College deserves a special note of thanks for coordinating the poster session. Ellen Marks of the Conference Department as well as the staff of the Editorial Department of the New York Academy of Sciences also deserve special recognition and thanks from all of us for their organizational and editorial contributions, respectively, to the meeting and the conference proceedings.

Free Energy Simulations[a]

M. MEZEI AND D. L. BEVERIDGE[b]

Chemistry Department
Hunter College of the
City University of New York
New York, New York 10021

INTRODUCTION

Monte Carlo or molecular dynamics simulations involve the numerical determinations of the statistical thermodynamics and related structural, energetic and (in the case of molecular dynamics) dynamic properties of an atomic or molecular assembly on a high-speed digital computer. Applications to molecular systems range from the study of the motions of atoms or groups of atoms of a molecule or macromolecule under the influence of intramolecular energy functions to the exploration of the structure and energetics of condensed fluid phases such as liquid water and aqueous solutions based on intermolecular potentials. The quantities determined in a typical Monte Carlo or molecular dynamics simulation include the average or mean configurational energy (thermodynamic excess internal energy), various spatial distribution functions for equilibrium systems, and time-correlation functions for dynamical systems, along with detailed structural and energetic analyses thereof. Diverse problems in structural and reaction chemistry of molecules in solution, such as solvation potentials, solvent effects on conformational stability and the effect of solvent on chemical reaction kinetics and mechanism via activated complex theory also require a particular knowledge of the configurational free energy, which in principle follows directly from the statistical thermodynamic partition function for the system.

Considerations on free energy in molecular simulations take a distinctly different form for intramolecular and intermolecular degrees of freedom. For the intramolecular case, the problem involves vibrational and librational modes of motion on the intramolecular energy surface. We will discuss briefly at the end of this paper the harmonic and quasiharmonic approximation used to compute vibrational contributions to the free energy, but we will restrict the focus herein to the intermolecular case, where the particles of the system undergo diffusional motion and a harmonic or quasiharmonic treatment breaks down. These considerations apply also in the case of a flexible molecule, where conformational transitions are effectively an intramolecular "diffusional mode."

Conventional Monte Carlo and molecular dynamics procedures for diffusional modes, although firmly grounded in Boltzmann statistical mechanics and dynamics, do not proceed via the direct determination of a partition function because of well-known

[a]This work was supported by Grant GM 24914 from the National Institutes of Health, Grant CHE-8203501 from the National Science Foundation, and a CUNY Faculty Research Award.

[b]Present address: Department of Chemistry, Wesleyan University, Middletown, Connecticut 06457.

1

difficulties in convergence. The Metropolis method used in Monte Carlo procedures is a Markov process designed specifically to avoid the partition function in the calculation of mean energy and related properties in a simulation. In molecular dynamics, the physical nature of the calculated particle trajectories serves this purpose equally well. However, in the absence of a partition function, one is unable to compute the free energy of the system directly and thus under ordinary circumstances molecular simulations lack access to the fundamental index of themodynamic stability of the system.

This article reviews the area of free energy simulation in the forms that are currently being pursued for systems of many molecules. Several new aspects of the methods discussed will also be presented. We begin with an elementary definition of the free energy problem as encountered for this case and discuss two procedures, umbrella sampling and the coupling parameter approach, which are quite useful in this area. Then the diverse individual approaches to free energy simulations are described, considering for each in turn the genesis, methodologic details, advantages and disadvantages, and applications currently reported. We will emphasize applications to molecular liquids and solutions at the expense of the large body of work on simple fluids. These, however, have been reviewed by Barker and Henderson[1] and Levesque and Weiss.[2] The interested reader is also referred to recent reviews related to this subject by Valleau and Torrie,[3] Quirke,[4] Gubbins,[5] and Pohorille and Pratt.[6] A complete simulation of molecular liquids and solutions requires, of course, a consideration of both vibrational and diffusive modes of motion. Some free energy studies where both intramolecular and intermolecular aspects are involved are reviewed at the conclusion of this article.

BACKGROUND

In computer simulations on molecular liquids and solutions, N molecules, typically of $O(10^2)$–(10^3), are configured in an elementary cell subject to periodic boundary conditions, for which various options (simple cubic, face-centered cubic, etc.) are available. The configurational energy E_N[7] of individual complexions of the N-particle system are evaluated from analytical energy functions "parameterized" from experimental data, quantum-mechanical calculations or sometimes a combination of both, truncated at certain predefined limits for computational efficiency. All simulations involve generating large numbers of possible configurations of the N-particle system and forming estimates of ensemble averages such as mean energy. On digital computers circa 1985, it is practical to think in terms of sampling $O(10^6)$ configurations or $O(10)$–$O(10^2)$ picoseconds of atomic and molecular motions. The current generation of supercomputers and/or attached processors make an increase in sampling of an order of magnitude or so immediately feasible.

Important statistical thermodynamic quantities for a discussion of computer simulations are

(*a*) the configurational partition function Z of the system,

$$Z_N = \int \cdots \int \exp\left(-E_N/kT\right) dR^N \tag{1}$$

where k is Boltzmann's constant and T is the absolute temperature, V is the volume of the system, and the integration extends over all space dR^N of the N particle system;

(b) the Boltzmann probability function $P(E_N)$ for a configuration corresponding to a particular E_N,

$$P(E_N) = \exp(-E_N/kT)/Z_N \tag{2}$$

(c) the average or mean energy expression

$$U = \int \cdots \int E_N P(E_N) \, dR^N = \langle E_N \rangle \tag{3}$$

which corresponds to the excess or configurational thermodynamic internal energy of the system; and

(d) the excess or configurational free energy

$$A = -kT \ln(Z_N/V^N) \tag{4}$$

the focus of interest in this article. Our discussion here is presented in terms of the constant volume (T, V, N) canonical ensemble, but the arguments can be extended to the microcanonical or isothermal/isobaric ensemble with no real difficulty. The grand canonical ensemble affords a particular alternative approach to free energy simulation considered briefly at the end of this article.

The essential problem involved in the determination of mean energy via simulation can be appreciated simply from Equations 2 and 3. The determination of mean energy via simulation involves essentially calculating the area under the curve $E_N P(E_N)$ or $E_N \exp(-E_N/kT)$ by generating various configurations of the system and calculating their respective energies and probabilities. Only a narrow range of E_N values results in significant values of the integrand since the region in which E_N and $\exp(-E_N/kT)$ are simultaneously large is relatively limited. A random selection of configurations in a simulation (a crude Monte Carlo method) is thus an inefficient approach to the numerical determination of mean energy since an inordinate amount of time would be spent sampling configurations which make relatively insignificant contributions to the integrand of Equation 3.

This problem in Monte Carlo theory was taken up some years ago by Metropolis et al.,[8] who devised an importance sampling scheme which is now followed in essentially all Monte Carlo studies of fluids. The Metropolis method is a Markov walk through configuration space sampling complexions of the system with a frequency proportional to the corresponding Boltzmann factor. In this realization, the evaluation of mean energy reduces to the simple summation

$$U = \sum_{j=1}^{n} E_N(R_j)/n \quad R_j \in \mathcal{M} \tag{5}$$

where the n configurations R_j are chosen by the Metropolis method ($\mathcal{M}$). Convergence studies indicate that the Metropolis method makes mean energy determinations feasible at presently accessible sampling rates. Other properties of the system computed in parallel with mean energy were each found to have their own profile of convergence, and in general structural properties converge more rapidly than energetic properties, and simple average quantities such as mean energy converge more rapidly than the fluctuation properties such as heat capacity. Higher moments of the energy can be expected to converge even more slowly. Detailed examples of convergence profiles for liquid water simulations are available in recent papers from this laboratory[9-11] and by Rao et al.[12] where these effects are evident. Generally, the mean energy of a pure liquid can be determined to a precision of ± 0.05 kcal/mol or better from a simulation of $O(10^6)$ configurations. However, other energetic quantities such as transfer energies of a solute molecule into solvent water are susceptible to $O(10)$ kcal/mol statistical noise engendered in finding small differences between large, relatively noisy numbers. Free energy simulations, as described later in this account, do not encounter this problem as severely.

A corresponding consideration of free energy follows from the ensemble average expression for the excess free energy,

$$A = kT \ln \langle \exp [+E_N/kT] \rangle \tag{6}$$

a completely equivalent statement of Equation 4. Expanding the exponential in powers of E_N and rearranging terms, it can be shown that

$$A = kT \ln \{[1 + U/kT + (U/2kT)^2 + C_v'/(2k) + 0 (\langle E_N^3 \rangle/(kT)^3) + \cdots]/V^N\} \tag{7}$$

Thus the convergence problem in free energy is equivalent to the problem of determining ensemble averages of the mean energy (first moment) and higher moments of the energy distribution mentioned above. The convergence difficulties in fluctuation properties and, by implication, higher moments of the energy distribution described above indicate that a significant increase in sampling would be required to achieve convergence in free energy relative to mean energy in a simulation.

A broader range of sampling could, or course, be accomplished by a crude Monte Carlo approach, but in practice configuration space for a system of even $O(10^2)$ molecules is so immense that convergence for the partition function integral cannot reasonably be expected. Metropolis sampling, concentrated in regions where $E_N \exp(-E_N/kT)$ is large, is optimal for internal energy, but would not sample broadly enough to provide accurate estimates of the partition function and the free energy. Similar considerations apply in molecular dynamics simulations. Thus we are left with no viable means of the determination of free energy by the usual numerical methods for molecular systems of interest from a mean energy simulation, and with the clear indication that an extent of sampling significantly beyond that suitable for mean energy determinations must be involved in free energy simulations.

In concluding this section, we note that an effective means of extending the range of sampling in a simulation, Monte Carlo or molecular dynamics, has been devised by Valleau, Patey and Torrie.[13,14] The range of sampling in a simulation can be altered in a special form of importance sampling by running the Metropolis procedure or molecular dynamics with modified energy function

$$E_N' = E_N + E_W \tag{8}$$

Here E_W is the modification of the energy that serves to extend the sampling. E_W takes various forms depending upon the particular application. However, introducing E_W carries the simulation to a non-Boltzmann regime (i.e., sampling based on E_N' rather than E_N), and thus the ensemble averages produced would be inappropriate for the determination of statistical thermodynamic quantities. Valleau and coworkers demonstrated that appropriate Boltzmann weighted ensemble averages $\langle Q \rangle$ for any property Q can be extracted from a simulation carried out on the modified energy function E'_N by collecting the additional ensemble average $\langle \exp(E_W/kT) \rangle_W$ and forming the quotient

$$\langle Q \rangle = \langle Q \exp(E_W/kT) \rangle_W / \langle \exp(E_W/kT) \rangle_W \qquad (9)$$

where the subscript W denotes an ensemble average based on the energy function E'_N. Equation 9 can be verified by writing out explicitly the expectation values and recognizing that the $\exp(E_W/kT)$ form Eq. 9 cancels the $\exp(-E_W/kT)$ from the modified Boltzmann factor. This approach is widely known as "umbrella sampling" (US) and, as discussed below, finds wide applicability in free energy simulations.

One obvious limitation of the umbrella sampling procedure is as follows: the more the sampling is to be extended, the larger will be the range of E_W, making the calculation of $\langle \exp(E_W/kT) \rangle_W$ prone to roundoff errors. In fact, if the variations are large, the computed average will be dominated by contributions from only a few configurations, a clearly undesirable effect. Therefore, a limit has to be set for the variations allowed in E_W, which effectively puts a limit to the extension of sampling with this technique. That is, convergence of the E'_N-based ensemble averages still requires careful numerical attention.

The selection of appropriate E_W is generally a nontrivial matter since it essentially presupposes knowledge about the respective probabilities of configurations with different values of the parameters of interest. The more complex these parameters, the more difficult is the determination of an efficient E_W. This, however, suggests an iterative approach to the determination of an effective E_W. Paine and Scheraga[15] have described an adaptive importance sampling technique applied to the conformational study of a free dipeptide molecule and a procedure of this type, called "Adaptive US" has recently been developed and tested on a conformational study of the alanine dipeptide in aqueous solution in this laboratory.[16] Besides yielding possible savings in computational effort, the procedure produces, at least in principle, a uniform distribution of the parameters of interest and is self-checking by construction.

THE COUPLING PARAMETER APPROACH

In molecular systems we are frequently confronted not only with the determination of the absolute excess free energy A, but also with the free energy difference ΔA between two well-defined states. Here for states denoted 0 and 1, ΔA is given in terms of the ratio of the partition functions for the two states, Z_1 and Z_0, as

$$\Delta A = A_1 - A_0 = -kT \ln(Z_1/Z_0) \qquad (10)$$

A straightforward approach to the free energy difference ΔA would require independent determinations of Z_0 and Z_1 based on energy functions E_0 and E_1, which individually are subject to all the numerical difficulties detailed in the preceding section. Several proposed solutions are based on a very useful construct, the "coupling parameter." Notice that here and in the following the subscript N is dropped for brevity. Let us assume that the potential E depends on a continuous parameter λ such that as λ is varied from 0 to 1, $E(\lambda)$ passes smoothly from E_0 to E_1. A free energy function $A(\lambda)$ can be defined as

$$A(\lambda) = - kT \ln Z(\lambda) \tag{11}$$

and calculation of ΔA can be performed by integrating the derivative of $A(\lambda)$ along λ (thermodynamic integration), by designating intermediate states $A(\lambda_i)$ spaced closely enough and using the defining equation of the free energy to compute ΔA in a stepwise manner (perturbation method), or by actually developing $A(\lambda)$ in the [0, 1] interval from simulations where λ is variable (probability ratio method). The function $A(\lambda)$ is equivalent to the potential of mean force in the statistical thermodynamics of fluids.

In a modern sense the coupling parameter approach originates in the derivation of an important integral equation in a liquid state theory by Kirkwood,[17] but the seeds of this idea can be traced to the work of de Donder on chemical affinity and the degree of advancement parameter for a chemical process.[18] The coupling parameter λ is a generalized extent parameter or analytical continuation and defines a physical or sometimes nonphysical path between states 0 and 1. On a physical path, a knowledge of $A(\lambda)$ can be used to determine the free energy of activation $\Delta A^{\ddagger}$ as well as ΔA. A nonphysical path is admissible in the case where the quantity of interest is a state function like ΔA, which is, of course, independent of the path. There is a considerable freedom in the choice of the path, and decisions on the selection of $E(\lambda)$ are usually made by combining physical and numerical requirements.

Some applications of the coupling parameter approach of interest in current computer simulation studies involve topographical changes in the system, and of course we refer this to the molecular topography of the system. We consider here three broad classes of topographical transition coordinates: structural transition coordinates, simple and complex reaction coordinates, and creation/annihilation coordinates (c.f. FIGURE 1).

Structural Transition Coordinates

On a structural transition coordinate, one three-dimensional form of a molecule or macromolecule is carried over into another three-dimensional form.

(a) λ defines a "conformational transition coordinate." The study of molecular conformation in solution requires the free energy of the system as a function of one or more internal coordinates of the system. One of the simplest cases of interest is the torsion angle $\phi(C_1 - C_2 - C_3 - C_4)$ in butane, which can be written in terms of coupling parameter as

$$\phi = (1 - \lambda) \phi_G + \lambda \phi_T \tag{12}$$

where $\phi_T = 180°$ and $\phi_G = 60°$. Here λ can be considered a "conformational transition coordinate," a special case of a structural transition coordinate. This can obviously be generalized to many dimensions.

(b) λ as a "correlated conformational transition coordinate." Our recent studies of solvent effects on the conformational stability of the alanine dipeptide[19] provide a two-dimensional example of this utilization of the coupling parameter approach. In the system, the various conformations of interest differ in values of the Ramachandran torsion angles ψ and ϕ. We mapped the structural change involving the conformational coordinates ψ and ϕ onto a single λ by means of the equation

$$(\psi, \phi) = (1 - \lambda) (\psi_0, \phi_0) + \lambda (\psi_1, \phi_1) \tag{13}$$

where $\lambda = 0$ selects the reference state (ψ_0, ϕ_0) and $\lambda = 1$ selects the state (ψ_1, ϕ_1); here λ is a "correlated conformational transition coordinate." As discussed below, to access computationally tractable sampling procedures in computer simulations it will be frequently desirable to map changes in the structure involving many internal coordinates onto a single λ if possible. The coupling parameter here can be considered

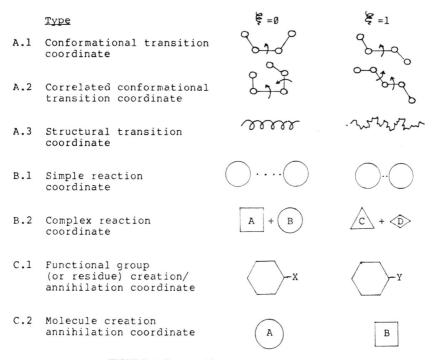

FIGURE 1. Topographical transition coordinates.

essentially a virtual bond coordinate of the sort used extensively in diverse biomolecular conformation problems.

(c) λ as a structural transition coordinate. This use of coupling parameter can be extended to structural changes in bond lengths and bond angles. Combinations thereof with conformational changes, mapped onto a single λ, would produce a "correlated structural transition coordinate." This can be useful for studies of helix-coil transition in polypeptides, protein denaturation, and structural interconversions in nucleic acids involving the A, B, C . . . Z forms.

Reaction Coordinates

A reaction coordinate is a coordinate upon which the evolution of a chemical reaction can be described.

(a) λ as a simple reaction coordinate. Let the simulation be carried out on a segment of coordinate R beginning at R_0 and R_1. Then for any point R between R_0 and R_1

$$R = (1 - \lambda) R_0 + \lambda R_1 \tag{14}$$

defining the coupling parameter such that for $\lambda = 0$, $R = R_0$ and for $\lambda = 1$, $R = R_1$. This is convenient for describing a simple association process in biological systems such as hydrophobic bonding, ion pairing, and hydrogen bonding, and can be easily generalized to many dimensions if necessary.

(b) λ as a complex reaction coordinate. In a complex reaction system, such as a proton transfer studied by Warshel,[20–22] or the reaction of CH_3Cl and Cl^-, treated recently by Jorgensen and coworkers,[23,24] distinct reactant and product species are involved. The use of the coupling parameter in this case is a straightforward generalization of that described above and coincides fully with the definition of reaction coordinate in the transition state theory of chemical reactions.[25] Here, as in the case of a structural transition, it will be useful to map concerted changes onto a single λ and work with a "correlated complex reaction coordinate."

Creation/Annihilation Coordinates

A creation/annihilation coordinate carries the molecular topography of the system smoothly from one structural entity to another. Others in the field refer to this as "Hamiltonian warping"[26] or "mutation."[27]

(a) λ as a functional group, residue or subunit creation/annihilation coordinate. Here the coupling parameter can be used to incorporate one well-defined moiety in the system while simultaneously removing another. This type of creation/annihilation coordinate could be used to pass from one derivative to another in studies of the relative thermodynamics of a homologous series of molecules and is of potential use in the study of biological activity of related sets of molecules via simulation. Possible applications are found in enzyme-inhibitor, drug-receptor, and many other types of problems in biological systems. Recent works by McCammon, Jorgensen and coworkers are in this category.[27–29]

(b) λ as a molecule creation/annihilation coordinate. Here we distinguish the case where λ removes an entire molecule from the system and creates another in its place. Kirkwood's initial use of the coupling parameter can be considered a special case of this category in which only the creation branch is active. A creation/annihilation coordinate can be useful for studying the thermodynamics of hydration of a series of amino acids or nucleotide bases for example, but actually diverse applications are possible. There are, in fact, interesting topographic transition coordinates to define in this class, individually as well as in combination with the other forms defined above. Considering the coupling parameter approach in this generalized form admits considerable imagination into the design of computer simulation studies on chemical and biological systems.

THERMODYNAMIC INTEGRATION

Applying a basic mathematical identity to the free energy function $A(\lambda)$ gives the following (exact) equation:

$$\Delta A = \int_0^1 (\partial A(\lambda)/\partial\lambda) \, d\lambda \tag{15}$$

Substituting Equation 1 into Equation 11 gives

$$\partial A(\lambda)/\partial\lambda = \langle \partial E(\lambda)/\partial\lambda \rangle_\lambda \tag{16}$$

where the subscript λ implies a Boltzmann average based on the function $E(\lambda)$. The direct numerical evaluation of Equation 15, integrating between initial and final states using some thermodynamic relationship, is called "thermodynamic integration" (TI). The simplest application of the coupling parameter approach with λ identified with the volume or with the inverse temperature of the system leads to the virial expression for pressure and the van't Hoff equation, respectively, textbook cases in physical chemistry. The integration variable, however, need not be restricted to thermodynamic quantities, and thus Equation 15 can be used to determine the free energy difference between two states that differ in many conceivable ways, as long as the energy is a smooth function of λ; however, pathways crossing phase transitions have discontinuities that will be problematic.

Linear TI

An important simple particular case is encountered when $E(\lambda)$ is linear in λ:

$$E(\lambda) = (1 - \lambda) \, E_0 + \lambda \, E_1 \tag{17}$$

If the reference state is an ideal gas or otherwise defines the zero of a configurational energy scale,

$$E(\lambda) = \lambda \, E_1 \tag{18}$$

and

$$\Delta A = \int_0^1 U(\lambda) \, d\lambda \tag{19}$$

where $U(\lambda)$ is a mean energy-like quantity developed as a function the coupling parameter, e.g. $\langle E_1 \rangle\lambda$. Implementation of Equation 19 in simulations is quite simple. A series of simulations corresponding to successive discrete values of λ are carried out, giving $U(\lambda_i)$ for $i = 1, 2 \ldots$ The final integration over λ is carried out by an additional numerical procedure. When E_0 is nonzero, Equation 18 becomes

$$E(\lambda) = E_0 + \lambda \, (E_1 - E_0) = E_0 + \lambda\Delta E \tag{20}$$

and the free energy difference is then

$$\Delta A = \int_0^1 \langle \Delta E \rangle_\lambda \, d\lambda \tag{21}$$

The computational procedure, a succession of simulations for discrete values of λ, is essentially similar to that described above.

An important advantage of linear TI is that it only involves computing ensemble averages of energies, not their derivatives, and these are the fastest converging quantities. Also, since the energy calculations are needed for the regular computer simulation runs, the implementation of linear TI into existing programs is a rather simple task. As an additional advantage, it is very easy to demonstrate that for the case of linear λ-dependence the integrand in Equation 15 is a monotonic function of λ.[30] Differentiation gives

$$\frac{\partial}{\partial\lambda}\left\langle \frac{\partial}{\partial\lambda} E(\lambda) \right\rangle_\lambda = -\left[\left\langle \left(\frac{\partial}{\partial\lambda} E(\lambda)\right)^2 \right\rangle_\lambda - \left\langle \frac{\partial}{\partial\lambda} E(\lambda) \right\rangle_\lambda^2 \right] + \left\langle \frac{\partial^2}{\partial\lambda^2} E(\lambda) \right\rangle_\lambda \qquad (22)$$

Obviously, the term containing the second derivative is zero for linear $E(\lambda)$ and the remaining terms in the bracket form a fluctuation expression that is always nonnegative. This result ensures that the interpolation implicit in any numerical quadrature should be reliable for linear TI.

Particular care must be exercised, however, when a creation/annihilation coordinate is involved where the integrand $U(\lambda)$ or $\Delta E(\lambda)$ may diverge at $\lambda = 0$ and/or $\lambda = 1$. This divergence occurs, for example, when a particle is coupled to the system at a location where no particle previously existed. In this event, at $\lambda = 0$ the just-decoupled particle becomes invisible to the generation of new configurations and therefore the rest of the system is free to overlap with it, producing arbitrarily large E_1 or ΔE values. Mruzik et al.[31] remarked that a particle interacting with a system through an r^{-12} potential will give $\langle E_1 \rangle_\lambda \propto \lambda^{-3/4}$ for small values of λ. In general, it can be shown by standard technique that for a particle interacting with a potential of r^{-n} in d dimensions, the limiting behavior of $\langle E_1 \rangle_\lambda$ for $\lambda \to 0$ is proportional to $\lambda^{(d/n)-1}$. This, unfortunately, makes the integral in Equations 19 or 21 an improper one for $d < n$. As a result, direct application of numerical integration may lead to significant error, since no matter how small the smallest λ_0 value used in the numerical quadrature, the contribution from the $[0, \lambda_0]$ interval will depend on the limiting behavior of $\langle E_1 \rangle_\lambda$, about which the numerical quadrature has no information at all. A possible procedure is then to find a λ_0 such that for $\lambda < \lambda_0$ the limiting behavior is observed and use numerical integration only for the range $[\lambda_0, 1]$. The contribution from the interval $[0, \lambda_0]$ should be obtained by exploiting the limiting behavior of $\langle E_1 \rangle_\lambda$. Writing $\langle E_1 \rangle_\lambda = c\lambda_0^{(d/n)-1}$, we obtain

$$\int_0^{\lambda_0} \langle E_1 \rangle_\lambda \, d\lambda = (n/d) \lambda_0 \langle E_1 \rangle_\lambda \qquad (23)$$

A helpful fact in this respect, verifiable by direct substitution, is that $E(\lambda)$ defined in terms of E_0 and E_1 is equivalent to $E(\lambda')$ defined in terms of $E_0' \equiv E(\lambda_0)$ and $E_1' \equiv E(\lambda_1)$ where

$$\lambda' = (\lambda - \lambda_0)/(\lambda_1 - \lambda_0) \qquad (24)$$

Equation 24 allows calculation of ΔA with different choices of λ_0 without having to obtain new ensemble averages.

A different approach, also based on the recognition of the limiting behavior of $\langle E_1 \rangle_\lambda$, was proposed by Mruzik et al.[31] The integration was carried out by introducing a new variable, λ^m:

$$\Delta A = (1/m) \int_0^1 \lambda^{1-m} \langle E_1 \rangle_\lambda \, d(\lambda^m) \tag{25}$$

where the exponent m is chosen so that the λ^{1-m} factor in the integrand makes it finite for all λ. They used $m = 0.25$, appropriate for potentials with an r^{-12} repulsive core. While this transformation makes the integrand finite everywhere, it also puts heavy emphasis on the small λ range. This suggests that the transformation be carried out only in a small interval near the singularity. The λ^m transformation has the further advantage that by choosing different m values a consistency check is obtained on ΔA.

For simultaneous creation and annihilation it is also possible that a similar discontinuity exists at $\lambda = 1$. This case should be handled analogously, either by obtaining the basic integral in the range $[\lambda_0, 1]$ from the limiting behavior of the integrand, as in Equation 23 or by transforming the integration variable, as Mruzik et al. proposed. In this case, the integration has to be broken into two parts and different variables have to be introduced:

$$\int_0^1 \langle E_1 - E_0 \rangle_\lambda = (1/m) \int_0^{0.5^m} d(\lambda^m) \, \lambda^{m-1} \langle E_1 - E_0 \rangle_\lambda +$$

$$(1/m) \int_{0.5^m}^1 d((1 - \lambda)^m) \, (1 - \lambda)^{1-m} \langle E_1 - E_0 \rangle_\lambda \tag{26}$$

Andersen et al.[32] and Wilson et al.[33] in recent research eliminated the singularity problem by truncating the repulsive part of the potential above a few kT, arguing that the strongly repulsive part of the potential is never sampled by the system anyway. However, as demonstrated earlier, the calculation of the free energy requires sampling of a larger than usual region of the configuration space and thus it is not immediately clear that the error caused by the truncation is negligible.

Calculations based essentially on Equation 15 for simple liquids have been reviewed by Barker and Henderson.[1] Mruzik et al.[31] reported a quite early application of this approach to ion hydration energies. The free energy of liquid water was computed using this method by Mezei et al.[34] by integrating between the liquid and an ideal fluid at liquid density on a transcritical tieline. Frischleder computed the free energy of solvation for the dimethyl phosphate anion.[35] Mezei computed the free energy differences between soft spheres and the MCY water and between MCY and ST2 as well as between MCY and SPC waters.[30] Subsequently the free energy difference between the ST2 and SPC waters was determined directly and was found to agree with the indirectly calculated value within the stated error limits.[36] Swope et al. described related procedures for the calculation of formation constants for water clusters.[37] Berens and coworkers described the incorporation of this approach in a molecular dynamics study of liquid water.[38] Postma[39] calculated the free energy of cavity formation in liquid water and obtained the integrand in a molecular dynamics calculation where λ was "forced" to grow from 0 to 1 during the simulation.

Nonlinear TI

Several possible pathways between two states involve an $E(\lambda)$ such that the λ dependence of when $E(\lambda)$ is nonlinear. In these cases TI becomes more complex, since the required derivative will be a function of λ. A simple example of nonlinear TI is the thermodynamic pressure integration mentioned before where the volume of the system is scaled up to infinity. Here we present the formalism for some other cases, and discuss their possible utility.

A general description of altering a molecule in the course of the λ-integration can be obtained with the following energy function:

$$E(\lambda) = E(\mathbf{R}\,(\lambda)),$$
$$\mathbf{R}(\lambda) = \lambda\,\mathbf{R}_1 + (1 - \lambda)\,\mathbf{R}_0 \tag{27}$$

Here $\mathbf{R}$ represents all the parameters of the energy function that change. For this choice, we have for the basic integrand:

$$\frac{\partial}{\partial\lambda}\,E(\lambda) = \sum_k \frac{\partial E}{\partial R_k}\,(R_{1,k} - R_{0,k}) \tag{28}$$

Implementation of Equation 28 into existing programs is more complicated than the implementation of linear TI, but should not lead to significant increase in computational expense for programs that calculate atomic forces anyway (like molecular dynamics or force bias Monte Carlo or variations thereof) since in general most terms going into the R-derivatives are already computed as a partial result.

Equation 28 can describe a conformational transition of a molecule or the change of a molecule into a new one. In the first case, $\mathbf{R}$ describes the atomic coordinates of the molecule to be changed as well as the potential parameters that undergo change during the switch from system 0 to system 1. When applied to the special case of system 1's being an ideal gas particle, a new molecule can be created without the singularity problem.[40] A possible advantage of nonlinear TI over linear TI lies in the fact that the singularity problem is avoided. Also, there is no need to calculate two energy functions, as in linear TI. However, the configurational average to be calculated is not the energy itself, but its derivative, and therefore convergence is likely to be more expensive to achieve and the monotonicity of the integrand is lost.

The nonlinear TI method can be used to change the simulation unit cell size. Conformational transition of large solutes may require a change in the shape of the simulation box, particularly when the periodicity of the solute is built into the periodicity of the solvent system. Changing the unit cell affects directly the energy function since the position of the periodic images will change. Let L describe the three lengths of the rectangular unit cell. The transition from $\mathbf{L}_0$ to $\mathbf{L}_1$ can be described as

$$E(\lambda) = E(L(\lambda))$$
$$L(\lambda) = \lambda\,\mathbf{L}_1 + (1 - \lambda)\,\mathbf{L}_0 \tag{29}$$

The derivative with respect to λ is given (for pairwise additive potentials) as

$$\frac{\partial}{\partial\lambda}\,E(\lambda) = \sum_{i<j}^N \frac{\partial}{\partial\lambda}\,E_{ij}(\lambda) \tag{30}$$

$$\frac{\partial}{\partial\lambda}\,E_{ij}(\lambda) = \sum_{k=1}^3 \frac{\partial}{\partial L_k}\,E_{ij}(\lambda)\,(L_{1,k} - L_{0,k}) \tag{31}$$

$$\frac{\partial}{\partial L_k} E_{ij}(\lambda) = \frac{\partial}{\partial r_{ij,k}} E_{ij}(\lambda) \frac{\partial}{\partial L_k} r_{ij,k} \tag{32}$$

$$\frac{\partial}{\partial L_k} r_{ij,k}(\lambda) = r_{ij,k}\delta(|L_k| - |r_{ij,k}|) + \begin{cases} 1 & \text{if } r_{ij} = r_{ij}^0 + L_k \\ 0 & \text{if } r_{ij} = r_{ij}^0 \\ -1 & \text{if } r_{ij} = r_{ij}^0 - L_k \end{cases} \tag{33}$$

where r_{ij}^0 is the difference between the position vector of i and j and r_{ij} is the difference translated to the nearest image. The first term is a consequence of the possible discontinuity in $E_{ij}(\lambda)$ at $|r_{ij,k}| = L_k$ that arises when the potential cutoff is not less than the simulation cell's inscribed sphere radius. However, its contribution to the TI integrand is zero since the sign of $r_{ij,k}$ is independent of $\partial/\partial r_{ij,k} E(\lambda)$. Therefore, it can be dropped from further consideration.

The implementation of Equations 30–33 requires a similar amount of programming effort as that of Equation 28, but again will not increase the actual computational expenses significantly for gradient bias calculations since all terms in Equations 30–33 are either simple to compute or are already computed during the force calculation.

Another possible use of nonlinear TI comes in the change in the "rate" of coupling during the λ-integration. A simple modification of the linear coupling involves introducing the kth power of the coupling parameter, which can be done in two ways:

$$E(\lambda) = [(1 - \lambda^k) E_0 + \lambda^k E_1] \tag{34}$$

or

$$E(\lambda) = [(1 - \lambda)^k E_0 + \lambda^k E_1] \tag{35}$$

The derivative required is obtained simply as

$$\frac{\partial}{\partial\lambda} E(\lambda) = k \lambda^{k-1} (E_1 - E_0) \tag{36}$$

or

$$\frac{\partial}{\partial\lambda} E(\lambda) = -k(1 - \lambda)^{k-1} E_0 + k\lambda^{k-1} E_1 \tag{37}$$

respectively. For creation/annihilation coupling, both methods will produce an integrand with limiting behavior $\lambda^{(kd/n)-1}$ at $\lambda = 0$. For the limiting behavior at $\lambda = 1$, important when creation and annihilation are done simultaneously, the first gives the same as the linear TI, $(1 - \lambda)^{(d/n)-1}$, and the second version gives the same as the first version at $\lambda = 0$: $(1 - \lambda)^{(kd/n)-1}$. Therefore, for $k \geq n/d$ the singularity is avoided, while the simplicity of the linear TI is maintained. Actually, the first version is a reformulation of the integral transform suggested by Mruzik et al., while the second version can be considered its generalization. The monotonicity of the integrand is, of course, maintained for the first version since it is related to the linear TI by a change of variables via a monotonous function. It is not clear, however, whether the same is true for the second version, and further studies are required at this point. There is an additional point of interest for the second version when applied to changing one large solute into another large but rather different one: It can be expected that in the region $\lambda = 0.5$ the system's pressure will be rather high, since the presence of both solutes is felt by the solvents. This large pressure may slow down convergence. Since $\lambda^k + (1 - \lambda)^k < 1$ for $k > 1$ in the $(0, 1)$ interval, the coupling above would alleviate this problem as well.

Independent Expression of the Entropy in TI

Generally, the entropy change is calculated from ΔA and the internal energy difference $\langle E_1 \rangle_1 - \langle E_0 \rangle_0$. However, it can also be obtained from a TI calculation directly as follows. By taking the derivative of $\langle E(\lambda) \rangle_\lambda$, expressed by Equation 20, we obtain

$$\left\langle \frac{\partial\, E(\lambda)}{\partial\lambda} \right\rangle_\lambda - \frac{\partial\langle E(\lambda) \rangle_\lambda}{\partial\lambda} = \left[\left\langle E(\lambda)\, \frac{\partial E(\lambda)}{\partial\lambda} \right\rangle_\lambda - \langle E(\lambda) \rangle_\lambda \left\langle \frac{\partial E(\lambda)}{\partial\lambda} \right\rangle_\lambda \right] \quad (38)$$

Integrating Equation 38 from 0 to 1 and using Equation 15 gives

$$\Delta A - (\langle E_1 \rangle - \langle E_0 \rangle) = -T\Delta S$$

$$= \int_0^1 \left[\left\langle E(\lambda)\, \frac{\partial E(\lambda)}{\partial\lambda} \right\rangle_\lambda - \langle E(\lambda) \rangle_\lambda \left\langle \frac{\partial\, E(\lambda)}{\partial\lambda} \right\rangle_\lambda \right] d\lambda \quad (39)$$

The ensemble averages in Equation 39 are either already computed during the calculation of ΔA or are trivial to obtain. Notice also that if the internal energy difference is calculated from independent (TVN) ensemble calculations, Equation 39 offers a consistency check.

PERTURBATION METHOD

An alternative expression for ΔA can be obtained by inserting unity into Z_1 of Equation 10 in the form $\exp(E_0/kT)\exp(-E_0/kT)$ leading directly to the equation

$$\Delta A = -kT \ln \langle \exp[-(E_1 - E_0)/kT] \rangle_0 \quad (40)$$

Reversing the role of systems 0 and 1 we obtain the mirror expression of Equation 40:

$$\Delta A = kT \ln \langle \exp[-(E_0 - E_1)/kT] \rangle_1 \quad (41)$$

Use of these equations has been also referred to as perturbation method (PM) since E_1 and E_0 differ by a small "perturbation" in successful application. Bennett[41] recognized that Equations 40 and 41 are equivalent to an infinite order perturbation expansion, that is, they are exact. Therefore, they do not correspond to a "perturbation theory" in the usual sense of the word.

The methodology of the PM involves essentially a simulation carried out by Metropolis-Monte Carlo or molecular dynamics procedures based on the energy function E_0 (or E_1), in which E_1 (or E_0) is also computed at each step and the average of the exponential quantity of Equations 40 or 41 is formed. Successful numerical calculation of these ensemble averages via simulations requires that states 0 and 1 be not too dissimilar. In the event of difficulties with the direct application of Equations 40 and 41, it is possible to use the coupling parameter approach to define a numerically viable path involving intermediate states between states 0 and 1 and to compute the free energy difference as

$$\Delta A = \sum_i \Delta A_i, \tag{42}$$

where

$$\Delta A_i = -kT \ln \langle \exp \left[-(E_{i+1} - E_i) / kT \right] \rangle_i. \tag{43}$$

Here the interval between successive states can in principle be maintained small enough that the similarity condition is always sufficiently satisfied and the ensemble average can be successfully determined; of course, if the states were too different the number of intermediate steps would become prohibitively large.

The two alternative expressions for ΔA, Equations 40 and 41, allow the calculations of ΔA_i and ΔA_{i+1} in the same simulation step. Interestingly, performing a simulation with $E(\lambda_i)$ and computing ΔA_i and ΔA_{i-1} from Equations 40 and 41 is exactly equivalent to an umbrella sampling calculation between $A(\lambda_{i-1})$ and $A(\lambda_{i+1})$ using $E_w = E(\lambda_i) - E(\lambda_{i-1})$ as the non-Boltzmann bias. The combined use of Equations 40 and 41 was actually called "half umbrella sampling" by Scott and Lee.[42] The recognition of this fact, however, also implies that the general umbrella sampling may be superior to half umbrella sampling, particularly if an efficient way of determining the non-Boltzmann bias can be found.

Equations 40 and 41 can not only be used to effect computational savings, but also for a consistency check as an alternative. This check is not too strong, however, since if the two states involved in ΔA_i are too distant, the inadequacy of both Equation 40 and 41 is of the same degree and the true error may not show up. As has been remarked earlier,[30] a stronger test for the adequacy of sampling can obtained by computing in the $(i-1)$th run

$$\langle E_i \rangle_{\lambda_i} = \langle E_i \exp \left[(E_i - E_{i-1})/kT \right] \rangle_{\lambda_{i-1}} / \langle \exp \left[(E_i - E_{i-1})/kT \right] \rangle_{\lambda_{i-1}} \tag{44}$$

and comparing it with the value computed in the ith run directly. An additional point is that the PM is particularly advantageous when the dependence of $E(\lambda)$ on λ is complex since it does not require derivatives of $E(\lambda)$. Care must be taken, however, to make sure that the range of ΔE is limited to a few kT since otherwise the sum of their exponential will be dominated by a few terms, a clearly undesirable effect.

The PM was used very early by Dashevsky et al.[43] to compute ΔA between liquid water and an ideal gas in a single step. Owicki and Scheraga pointed out in an argument similar to that given in the BACKGROUND section based on Equations 6 and 7 that $\langle \exp (+E_N/kT) \rangle$ cannot be expected to converge in a mean energy calculation.[44] Miyazaki et al. used this technique to calculate the surface tension of the Lennard-Jones liquid by separating two slabs of liquid in a stepwise manner.[45] Torrie and Valleau introduced umbrella sampling to enhance the efficiency of Equations 40 and 41) in calculating ΔA between soft spheres and the Lennard-Jones fluid and between Lennard-Jones fluid at various temperatures.[14] Owicki and Scheraga performed calculations on the probability of finding a cavity in liquid water.[46] Scott and Lee calculated the surface tension of the MCY water[42] by combining US with the technique of Miyazaki et al.[45] The difficulty of obtaining efficiently the non-Boltzmann bias was recognized and the half umbrella sampling was proposed. Nakanishi and coworkers[47] computed the free energy of hydration of a methane molecule into liquid water in one step. Umbrella sampling was used and the exponential of the E_W function that these investigators used varied between 1 and 10^{56}. Numerical problems may arise when the range of the weighting function is so large. Recently, Postma, Berendsen, and Haak used this approach to determine the free energy of cavity formation in water.[48] Sussman, Goodfellow, Barnes, and Finney calculated ΔA between liquid water at various temperatures[49] using umbrella sampling. Jorgensen and Ravimohan computed the free energy difference between ethane and ethyl alcohol.[27] McCammon, Tembe, Lybrandt, and Wipf calculated the free energy difference of changing a coenzyme in aqueous solution[28,29] and were able to calculate the free energy of solvation between $[Cl^-]_{aq}$ and $[Br^-]_{aq}$ in one step,[50] but this involves only a relatively small alteration in an ionic radius parameter. Kollman and coworkers[84] very recently report considerable success at computing the relative solvation potential of amino acids using the PM.

In general, successful application of the PM invovled the change in a single molecule only, limiting the range fluctuations in ΔE to a few kT. Changes involving several molecules, however, can not be treated effectively by the PM. Mezei attempted to calculate the ΔA between the ST2 and MCY liquids using US, but repeated attempts at the determination of an effective E_W function failed to give consistent results even when two intermediate states were used, pointing to a serious limitation of the perturbation method.[30]

THE POTENTIAL OF MEAN FORCE

The expression for Helmholtz free energy, Equation 4, for the special case of two of the N particles of the system fixed in space at a distance R takes the form

$$A(R) = -kT \ln \int \cdots \int \exp\left[-E_N(R)/kT\right] dR^{N-2} \tag{45}$$

The radial distribution function for the system is defined as

$$g(R) = [N(N-1)/\rho] \int \cdots \int \exp\left[-E_N(R)/kT\right] dR^{N-2} \tag{46}$$

and thus

$$A(R) = -kT \ln g(R) + \text{constant} \tag{47}$$

The quantity $A(R)$ is the force acting between the fixed particles due to the remaining $N - 2$ other particles of the system[51] and thus $A(R)$, frequently denoted $w(R)$ in the statistical mechanics literature, is known as the potential of mean force. Equation 47 is generally true for any parameter R fixed. The knowledge of $A(R)$ is particularly useful for conformational changes, molecular associations, and chemical reactions.

In general, $g(R)$ is obtained as the ratio of the probability of sampling the coordinate R, obtained from a simulation where R is also allowed to vary, and the volume element of the configuration space corresponding to the coordinate R:

$$g(R) = P(R)/V(R) \qquad (48)$$

When the R coordinate is considered as simply another degree of freedom in an otherwise conventional mean energy calculation, serious sampling problems arise. The simulation, seeking to describe the equilibrium state dictated by the Boltzmann factor, would end up sampling only a small region of the R space rather than the full space. Thus sampling of R requires umbrella sampling techniques to cover the less probable regions of R. The volume element can be interpreted as a quantity proportional to the probability of sampling the parameter R with the potential function set to zero. For example, if R is an intermolecular distance, $V(R) = 4\pi R^2$, and if R is a torsion angle, $V(R) = $ const. Its determination becomes progressively more complex as the dimensionality of R is increased.

Several recent simulations studies to compute A(R) have been reported. Apart from the original study of Patey and Valleau,[13] where a tabulated weighting function was used, all recent works carried out a series of simulations, each constrained to sample the local region about points R_1, R_2, ... respectively. In an individual simulation, a distribution $g(R)_i$ is obtained. A particular simulation is constrained to sample a region about R_i by adding a harmonic constraint to the configurational energy, via

$$E_W(R) = k_H(R - R_i)^2 \qquad (49)$$

as first employed by Pangali Rao and Berne.[52] Successive points R_i are chosen so that $g(R)_i$ are overlapping. Overlapping points in the distribution correspond in principle to the same absolute value, but in practice differ by a normalization constant. Thus the various computed $g(R)_i$ can be arbitrarily shifted up or down, and ultimately matched up to produce a g(R) for the entire range of R. The matching can in principle be carried out for any overlapping points, but in practice one chooses those points with relatively low statistical noise levels. Ideally, the matching should be based on all overlapping points, with higher weight given to points that were sampled more extensively. A formalism, applicable to multidimensional R, has been presented in References 16 and 19.

Other potentials of mean force determinations on R as a reaction coordinate include an early calculation of A(R) for the association of ion pairs by Patey and Valleau[13] and by Berkowitz et al.[53] where successive minima in A(R) correspond to contact and solvent-separated ion pairs. The calculation of A(R) for the association of apolar atoms and molecules has been carried out in studies of the hydrophobic effect by Pangali et al.[52] and by Ravishanker, Mezei and Beveridge.[54,55] Here as well contact and solvent-separated forms were identified with the latter having an unexpectedly high statistical weight, indicating that the hydrophobic effect may act over a longer range of distance than previously suspected by means of solvent-mediated structures. A series of potential of mean torsion studies have been carried out on the n-butane system by Rebertus, Berne and Chandler,[56] Rosenberg, Mikkilineni and Berne,[57] and Jorgensen,[58] as reviewed recently by Jorgensen.[59] Most recently, Chandrashekar, Smith and Jorgensen have determined the potential of mean force on the complex reaction coordinate of the organic SN2 reaction of CH_3Cl and Cl^- in water[23] and DMF.[24] The reaction is predicted to be concerted in water but to proceed via a reaction intermediate in DMF, a previously unanticipated result. Warshel used umbrella sampling to compute the potential of mean force along a proton dissociation coordinate described by a coupling parameter.[20-22] Case and McCammon determined the potential of mean force for a carbon monoxide molecule approaching the myoglobin active site.[60]

PROBABILITY RATIO METHOD

If the free energy function $A(R)$ is evaluated as a function of a coupling parameter λ on the $[0, 1]$ interval, the free energy difference ΔA can be simply obtained as

$$\Delta A = A(1) - A(0) = kT \ln [g(R_0)/g(R_1)] \tag{50}$$

In view of the interpretation of $g(R)$ as a probability per unit volume, the procedure based on Equation 50 will be called probability ratio method (PRM). Also, it can be shown that for the one-dimensional coupling parameter λ,

$$V(\lambda) = \text{constant} \tag{51}$$

and therefore

$$\Delta A = kT \ln P(\lambda_0)/P(\lambda_1) \tag{52}$$

where the probability ratio appears explicitly.

The advantages of the PRM are twofold. First, there is no volume-element ratio to deal with. This appears to contradict the abovementioned necessity of using $V(R) = 4\pi R^2$ for the intersolute distance R. However, the R^2 factor comes in only if the two solutes are allowed to move freely in the three-dimensional space, implying that $R = |\mathbf{R}|$, a parameter depending on three degrees of freedom. If instead the simulation is performed by directly varying the intersolute distance (that is, restricting the movement of the solute to the intersolute line), no volume element correction is needed. This statement can be justified also by recognizing that while the *a priori* probability of moving from R to $R + \Delta R$ and from R to $R - \Delta R$ is the same in the second case, for the first case they are unequal and their ratio is just $(R + \Delta R)^2/(R - \Delta R)^2$. Second, we can apply PRM for computing ΔA between systems described with different potentials (as already pointed out by Bennett earlier[41]). Since it can be generally expected that

$$|\langle E(\lambda + \Delta)\rangle_\lambda - \langle E(\lambda)\rangle_\lambda| > |\langle E(\lambda + \Delta)\rangle_{\lambda+\Delta} - \langle E(\lambda)\rangle_\lambda| \tag{53}$$

it is very likely that the creation of a new particle can be performed efficiently without any sudden increase in the integrand. The above derivation also provides a justification for allowing changes in the energy function when a conformational coordinate is varied. This was the case in Ref. 19 where the atomic charges were also varied with the solute conformation.

A special application of PRM is to systems where the free energy difference between two solute conformations is required. The PRM method only provides the solvent contribution to this free energy difference. The solute contribution has to be computed separately. In that respect it should be pointed out that including the intramolecular energy in the simulation is equivalent to calculating ΔA with the intramolecular energy contribution set at zero during the simulation and correcting the free energy difference with the intramolecular energy difference assuming that there is no coupling between the intermolecular and intramolecular terms.[20] This result can also be obtained if one assumes that the calculation is using the total energy $E + E_{intra}$ but also employs umbrella sampling with $W(R) = -E_{intra}(R)$ where E_{intra} is the intramolecular energy contribution.[19]

In a recent work from this laboratory[19] using the PRM, λ took the form of a correlated conformational transition coordinate defined by Equation 19 as a linear tie-line from the C_7 conformation to a "final" structure α_R or P_{II} of the Ala dipeptide in water. In subsequent studies we calculated the solvent contribution to the free energy difference between various conformations of the dimethyl phosphate anion[61] and between the *cis* and *trans* conformation of N-methyl acetamide[62] in water.

FULL FREE ENERGY SIMULATIONS

We use the terminology "full free energy simulations" or "complete free energy simulations" to refer to theoretical studies where both intramolecular and intermolecular contributions are included. For the case of determining ΔA between two states which correspond to well-defined minima on the intramolecular potential surface, the approximation

$$A = A_{inter} + A_{intra} \qquad (54)$$

can be pursued. The vibrational entropy can be obtained in the harmonic approximation from the calculated vibrational frequencies and vibrational partition function by simple extension of the Einstein oscillator problem,[63] or in the quasiharmonic approximation via an entropy obtained from the covariance matrix of atomic displacements.[64-66] The free energy is then computed from the entropy and the calculated mean of the intramolecular internal energy. As the conformational flexibility of the molecule increases, the likelihood of the system going from the region of one minimum to another is larger and in these instances, harmonic and quasiharmonic methods fail. However, a limited perspective can be still developed from this approach by a detailed study restricted to a few conformations. In these cases, it is reasonable to consider the neighborhood of these conformations independently and to define conformational free energy assuming that only the neighborhood of this selected conformation is accessible to the system, and then to proceed via Equation 56. Limitations arise because of the well-known multiple-minimum problem. At the extreme of complete conformational flexibility, one has to proceed by means of a molecular simulation involving simultaneously both intramolecular and intermolecular degrees of freedom and obtain the free energy of the system via the procedures described in the preceding sections. Ravishanker *et al.*[67] have recently carried out a series of calculations on the intramolecular thermodynamics of the Ala dipeptide in the C_7, C_5, α_R and P_{II} conformations using the quasiharmonic Monte Carlo method[64,68] with energy functions carried over from the CHARMM program,[69] and they have combined these with our estimates of the free energy of hydration as determined with the PRM described in the preceding section. The C_7 form is indicated to be preferentially stabilized in the isolated molecule because of the 7-atom intramolecular ring structure closed with the

NH . . . OC hydrogen bond. The open forms are indicated to be entropically favored, but this contributes little to the intramolecular free energy. In water, the hydration stabilizes the open forms with favorable carbonyl-water hydrogen bonds, mitigated partially by a solvent entropy compensation effect. The net (intramolecular + hydration) free energy of three of the conformational forms turns out to be similar, indicating the molecule to be conformationally flexible in water. This is generally consistent with the experimental results,[70] and suggests that the well-known conformational flexibility of numerous small peptides in water arises as a consequence of hydration competing successfully with intramolecular hydrogen bonds to stabilize open conformational forms.

DISCUSSION

No single method for free energy simulations can be considered as clearly superior to the others and the proper choice depends very much on the system under consideration. At one extreme of the spectrum, where the systems 0 and 1 are very similar, the PM is clearly optimal. However, the number of stages required is roughly proportional to the variation in ΔE. Therefore, when the systems 0 and 1 differ considerably, thermodynamic integration methods are likely to be the method of choice. However, further studies are required for the optimal handling of the singularity problem. At this point little experience is accumulated with the use of the PRM. It has been proposed in this paper that PRM may be rather efficient in "creating" new particles. It is quite possible that for larger systems a combination of the methods would turn out to be optimal—we have in mind a technique where the "seeds" of a new system are introduced with, say, the PRM method (to avoid the singularity problem) and the rest of the system is "grown" with linear TI (to be able to rely on the monotonicity of the integrand). Answers to these questions will come from extensive comparative calculations in future work.

A TI calculation that determines the value of the integrand at one λ value can be considered also as a PM calculation between intermediate states $E_0 = E(\lambda - \Delta)$ and $E_1 = E(\lambda + \Delta)$ for small enough Δ. If the points λ_i in the numerical integration are chosen close enough that a small enough Δ can be found such that $\lambda_{i-1} + \Delta = \lambda_i - \Delta$, then ΔA can be the calculated from the same set of calculations in two different ways (using either Equation 40 or 41), providing a very useful consistency check. However, in most of the cases the integrand is "smooth" enough that the integral can be approximated adequately by evaluating the integrand at only a relatively few points, which demonstrates one of the basic strength of the TI method.

The error characteristics of the different approaches are also different. First, it can be generally said that free energy differences are likely to be more reliable than internal energy differences since the free energy difference calculations involve the energy differences before the averaging while the internal energy differences are usually obtained as a small difference between large quantities with their respective statistical uncertainties. Comparing the three different approaches discussed above, in calculations that require the definition of intermediate states, the errors of individual calculations contribute additively to the error in the final result. By use of thermodynamic integration, however, a weighted sum the error of the individual quadrature points gives the error of the final result. Therefore, if a single calculation has an unusually large error, its effect will be present fully in the final result for the PM or PRM, but in a TI calculation its effect will be scaled down.

The estimation of the error in a free energy simulation has two aspects. First, the error of the individual calculations is to be assessed and second, the propagation of the error to the final results is to be determined. The error of the individual calculations can generally be obtained from the method of batch means.[71,72] Special care must be taken for both the PM and the PRM since these approaches are rather sensitive to the long-range correlations. The propagation of the error is rather straightforward, as discussed above.

An alternative approach to the overall error problem is the use of consistency checks. We showed several examples where the same quantity can be computed in different ways. In fact, the alternative approaches mentioned can also be considered as such. There is also the possibility of computing free energy differences along different paths. A simple example for this is the check employed by Scott and Lee on their PM/US calculations where a calculation between λ_i and λ_{i+1} was checked by computing the free energy difference between systems λ_i and $\lambda_i + (\lambda_{i+1} - \lambda_i)/2$ plus the difference between systems $\lambda_i + (\lambda_{i+1} - \lambda_i)/2$ and λ_{i+1} or the calculations on the ST2, MCY and SPC waters.

There are a number of other free energy techniques that have been developed and tested on relatively simple liquids. The first class of methods is based on a paper by Widom.[73] These methods require the addition or deletion of a particle from the system.[74-77] Simulations in the grand-canonical ensemble also fall into this class.[78-80] These methods, however, can only be applied for small particles or low densities. A different class of methods, originated by Bennett,[41] are based on Equation 52. They do not actually change the coupling parameter, but rely instead on the comparison of energy distributions. The Virtual Overlap and the Overlap Ratio methods of Quirke and Jacucci also fall into this category.[81,82] Their main drawback is that they require accurate estimate of the tail of the energy distributions, which are known to be particularly sensitive to the small but well-documented long-range correlations that exist in simulations.[9,10] Voter[83] published an interesting variation of the original Bennet method but it is only applicable to systems with a small number of extrema. It is unlikely that any of these methods can be applied efficiently to systems consisting of large molecules.

CONCLUSIONS

The results of research investigations described in this article clearly indicate that the time is now at hand for the calculation of free energies of molecular systems via computer simulation. We expect with the advent of supercomputers that the sampling problems inherent in the numerical determination of free energy can be overcome and the goal of producing a full thermodynamic description of molecular assemblies in condensed phase system can be more fully realized.

NOTES AND REFERENCES

1. BARKER, J. A. & D. HENDERSON. 1976. Rev. Mod. Phys. **48**: 587.
2. LEVESQUE, J. J. WEISS & J. P. HANSEN. 1984. *In* Monte Carlo Methods in Statistical Physics II. K. Binder, Ed. Springer Verlag. Berlin.

3. VALLEAU, J. P. & G. M. TORRIE. 1977. *In* Modern Theoretical Chemistry, Vol. 5: Statistical Mechanics. Part A, Equilibrium Techniques. B. J. Berne, Ed. Plenum Press. New York, NY.
4. QUIRKE, N. 1980. Proceedings of the NATO Summerschool on Superionic Conductors, Odense, Denmark. Plenum Press. New York, NY.
5. SHING, K. S. & K. E. GUBBINS. 1983. Advances in Chemistry, Vol. **204:** Molecular-Based Study of Fluids. J. M. Haile & G. A. Mansoori, Eds. American Chemical Society. Washington DC.
6. POHORILLE, A. & L. R. PRATT. 1986. Methods in Enzymology, Vol. **127:** Biomembranes. L. Packer, ed. Academic Press. New York, NY.
7. When diffusive modes are involved, the quantity E_N is a function of the positional and orientational coordinates of all N molecules of the system. To simplify the notation in this article we consider the dependence on the configurational coordinates implicit in the notation "E_N" rather than mentioning it explicitly in each equation as in our previous articles.
8. METROPOLIS, N., A. W. ROSENBLUTH, M. N. ROSENBLUTH, A. H. TELLER & E. J. TELLER. 1953. J. Chem. Phys. **21:** 1987.
9. MEZEI, M., S. SWAMINATHAN & D. L. BEVERIDGE. 1979. J. Chem. Phys. **71:** 3366.
10. MEHROTRA, P. K., M. MEZEI & D. L. BEVERIDGE. 1983. J. Chem. Phys. **78:** 3156.
11. BEVERIDGE, D. L., M. MEZEI, P. K. MEHROTRA, F. T. MARCHESE, G. RAVISHANKER & S. SWAMINATHAN. 1983. Advances in Chemistry, Vol. **204:** Molecular-Based Study of Fluids. J. M. Haile & G. A. Mansoori, Eds. American Chemical Society. Washington, DC.
12. RAO, M., C. S. PANGALI & B. J. BERNE. 1979. Mol. Phys. **37:** 1773.
13. PATEY, G. N. & J. P. VALLEAU. 1975. J. Chem. Phys. **63:** 2334.
14. TORRIE, G. M. & J. P. VALLEAU. 1977. J. Comp. Phys. **23:** 187.
15. PAINE, G. H. & H. A. SCHERAGA. 1985. Biopolymers **24:** 1391.
16. MEZEI, M. J. Comp. Phys. In press.
17. KIRKWOOD, J. G. 1968. *In* Theory of Liquids. B. J. Alder, Ed. Gordon and Breach. New York, NY.
18. DE DONDER, Th. 1927. L'affinite, Gauther-Villars. Paris; Th. DE DONDER and P. VAN RYSSELBERCHE. 1936. Affinity. Stanford University Press, Stanford, CA. We thank Professor D. A. McQuarrie for guiding us to these references.
19. MEZEI, M., P. K. MEHROTRA & D. L. BEVERIDGE. 1985. J. Am. Chem. Soc. **107:** 2239.
20. WARSHEL, A. 1982. J. Phys. Chem. **86:** 2218.
21. WARSHEL, A. 1984. Proceedings of the Working Group on Specificity in Biological Interactions. Pontificiae Academiae Scientiarium Scripta Varia.
21. WARSHEL, A. 1984. Proc. Natl. Acad. Sci. USA **81:** 444.
23. CHANDRASEKHAR, J., F. S. SMITH & W. L. JORGENSEN. 1984. J. Am. Chem. Soc. **106:** 3049.
24. CHANDRASEKHAR, J. & W. L. JORGENSEN. 1985. J. Am. Chem. Soc. **107:** 2974.
25. GLASSTONE, S., K. J. LAIDLER & H. EYRING. 1941. The Theory of Rate Process. McGraw-Hill. New York, NY.
26. WILSON, K. R. Private communication.
27. JORGENSEN, W. L. & C. RAVIMOHAN. 1985. J. Chem. Phys. **83:** 3050.
28. TEMBE, B. L. & J. A. McCAMMON. 1984. Computers in Chemistry **8:** 281.
29. LYBRAND, T. P., J. A. McCAMMON & G. WIPF. 1986. Proc. Natl. Acad. Sci. USA **83:** 833.
30. MEZEI, M. 1982. Mol. Phys. **47:** 1307.
31. MRUZIK, M. R., F. F. ABRAHAM, D. E. SCHREIBER & G. M. POUND. 1976. J. Chem. Phys. **64:** 481.
32. ANDERSON, H. C. Private communication.
33. WILSON, K. R. Private communication.
34. MEZEI, M., S. SWAMINATHAN & D. L. BEVERIDGE. 1978. J. Am. Chem. Soc. **100:** 3255.
35. V. H. FRISCHLEDER. 1978. Wiss. Z. Karl-Marx-Univ. Leipzig, R., 28. Jg., H.6.
36. MEZEI, M. Unpublished results.
37. SWOPE, W. C., H. C. ANDERSEN, P. H. BERENS & K. R. WILSON. 1982. J. Chem. Phys. **76:** 637.

38. BERENS, P. H., D. H. J. MACKAY, G. M. WHITE & K. R. WILSON. 1983. J. Chem. Phys. **79:** 3375.
39. POSTMA, J. P. M. 1985. Ph.D. thesis, University of Groningen.
40. CROSS, A. 1985. A comment on Hamiltonian parameterization in Kirkwood free energy calculations. This volume.
41. BENNET, C. H. 1976. J. Comp. Phys. **22:** 245.
42. SCOTT, H. L. & C. Y. LEE. 1980. J. Chem. Phys. **73:** 4591.
43. SARKISOV, G. N., V. G. DASHEVSKY & G. G. MALENKOV. 1974. Mol. Phys. **27:** 1249.
44. OWICKI, J. & H. A. SCHERAGA. 1977. J. Am. Chem. Soc. **99:** 8382.
45. MIYAZAKI, J., J. A. BARKER & G. M. POUND. 1976. J. Chem. Phys. **64:** 3364.
46. OWICKI, J. & H. A. SCHERAGA. 1978. J. Phys. Chem. **82:** 1257.
47. OKAZAKI, S., K. NAKANISHI, H. TOUHARA & Y. ADACHI. 1979. J. Chem. Phys. **71:** 2421.
48. POSTMA, J. P. M., H. J. C. BERENDSEN & J. R. HAAK. 1982. Faraday Symp. Chem. Soc. **82:** 55.
49. SUSSMAN, F., J. GOODFELLOW, P. BARNES & J. L. FINNEY. 1985. Chem. Phys. Lett. **113:** 372.
50. LYBRAND, T. P., I. GOSH & J. A. MCCAMMON. 1986. J. Am. Chem. Soc. **107:** 7793.
51. BEN-NAIM, A. 1974. Water and Aqueous Solutions. Plenum Press. New York, NY.
52. PANGALI, C. S., M. RAO & B. J. BERNE. J. Chem. Phys. 1979. **71:** 2975.
53. BERKOVITZ, M., J. A. KARIN, A. MCCAMMON & P. J. ROSSKY. 1984. Chem. Phys. Lett. **105:** 577.
54. RAVISHANKER, G., M. MEZEI & D. L. BEVERIDGE. 1982. Faraday Symp. Chem. Soc. **17:** 79.
55. RAVISHANKER, G. & D. L. BEVERIDGE. 1985. J. Am. Chem. Soc. **107:** 2565.
56. REBERTUS, D. W., B. J. BERNE & D. CHANDLER. 1979. J. Chem. Phys. **75:** 3395.
57. ROSENBERG, R. O., R. MIKKILINENI & B. J. BERNE. 1982. J. Am. Chem. Soc. **104:** 7647.
58. JORGENSEN, W. L. 1982. J. Chem. Phys. **77:** 5757.
59. JORGENSEN, W. L. 1983. J. Phys. Chem. **87:** 5304.
60. CASE, D. A. & J. A. MCCAMMON. Dynamical simulation of oxygen binding to myoglobin. This volume.
61. JAYARAM, B., M. MEZEI & D. L. BEVERIDGE. In preparation.
62. MEZEI, M., S. W. HARRISON, G. RAVISHANKER & D. L. BEVERIDGE. In preparation.
63. HAGLER, A. T., P. S. STERN, R. SHARON, J. M. BECKER & F. J. NAIDLER. 1979. J. Am. Chem. Soc. **101:** 6842.
64. KARPLUS, M. & J. N. KUSCHIK. 1981. Macromolecules **14:** 325.
65. LEVY, R. M., M. KARPLUS, J. KUSCHIK & D. PERAHIA. 1984. Macromolecules **17:** 1370.
66. KARPLUS, M. & BRADY. Preprint.
67. RAVISHANKER, G., M. MEZEI & D. L. BEVERIDGE. 1986. J. Comp. Chem. **7:** 345.
68. LEVY, R. M., O. ROJAS & R. A. FREISNER. 1984. J. Phys. Chem. **88:** 4233.
69. BROOKS, B., R. E. BRUCCOLERI, B. D. OLAFSON, D. J. STATES, S. SWAMINATHAN & M. KARPLUS. 1984. J. Comp. Chem.: 4233.
70. MADISON, V. & K. D. KOPPLE. 1980. J. Am. Chem. Soc. **102:** 4855.
71. BLACKMAN R. B. & J. W. TUCKEY. 1958. The Measurement of Power Spectra. Dover. New York, NY.
72. WOOD, W. W. 1968. *In* Physics of Simple Liquids. H. N. V. Temperley, J. S. Rowlinson & G. S. Rushbrooke, Eds. North Holland. Amsterdam.
73. WIDOM, B. 1963. J. Chem. Phys. **39:** 2808.
74. ROMANO, S. & K. SINGER. 1979. Mol. Phys. **37:** 1765.
75. POWLES, J. G., W. A. B. EVANS & N. QUIRKE. 1982. Mol. Phys. **46:** 1347.
76. SHING, K. S. & K. E. GUBBINS. 1981. Mol. Phys. **43:** 717.
77. SHING, K. S. & K. E. GUBBINS. 1982. Mol. Phys. **46:** 1109.
78. NORMAN, G. E. & V. S. FILINOV. 1969. High Temp. USSR **7:** 216.
79. ADAMS, D. J. 1975. Mol. Phys. **29:** 311.
80. MEZEI, M. 1980. Mol. Phys. **40:** 901.
81. JACUCCI, G. & N. QUIRKE. 1982. Mol. Phys. **40:** 1005.
82. QUIRKE, N. & JACUCCI, G. 1982. Mol. Phys. **45:** 823.
83. VOTER, A. F. 1985. J. Chem. Phys. **82:** 1890.
84. BASH, P., U. C. SINGH, R. LANGRIDGE & P. KOLLMAN. In preparation.

Computer Simulations of Macromolecular Dynamics: Models for Vibrational Spectroscopy and X-Ray Refinement[a]

RONALD M. LEVY

Department of Chemistry
Rutgers University
New Brunswick, New Jersey 08903

INTRODUCTION

The use of computer simulations to study the internal dynamics of proteins and nucleic acids has attracted a great deal of attention in recent years. These simulations constitute the most detailed theoretical approach available for studying internal motions and structural flexibility of proteins.[1,2] It is essential that there be a continuing effort to develop procedures for comparing the results of simulations with a wide variety of experimental measurements. This has been a central focus of our work during the past few years. Such studies are necessary if the methodology is to be reliably used to study properties that are only indirectly accessible to experiment. Equally important, these studies lead to deeper insights into the relationship between experimental measurements and underlying molecular processes. Computer simulations, therefore, provide a fundamental connection between experimental probes of dynamics and analytical theories that are used to interpret experiments. In this paper we review recent work concerned with the use of computer simulations for the interpretation of experimental probes of macromolecular structure and dynamics.

Molecular dynamics simulations using simple potential functions have been demonstrated to provide an accurate description of structural and dynamic properties of both unassociated and complex liquids. Packing considerations play a major role in determining the molecular properties of both liquids and proteins; because packing can be described by simple and reliable potential functions there is a strong foundation for simulation studies of protein dynamics. However, many factors distinguish molecular dynamics simulations of proteins from liquid-state simulations so that it is difficult to use experience gained from molecular dynamics simulations of liquids to estimate the precision inherent in the protein simulations. For liquid simulations the basic system contains at least 100 identical molecules so that it it possible to take advantage of considerable statistical averaging in the calculation of quantities for comparison with experiment. For protein molecular dynamics simulations in contrast, the computational effort required to evaluate the large number of interatomic interactions within a single protein molecule limits the simulated system to one or at most a very small number of (macro) molecules. The highly anisotropic nature of the protein interior and

[a]This work was supported in part by the National Institutes of Health, the Petroleum Research Fund of the American Chemical Society, and the Alfred P. Sloan Foundation.

24

intrinsic interest in extracting site-specific information further complicates the computational problem.

Additional features of macromolecular simulations that make them different from and more difficult than simulations of liquids and solids include the difficulty in obtaining exact results for comparison with trajectory averages, the slow convergence of the macromolecular simulations, and the problem of treating quantum effects for large systems which lack a high degree of symmetry. Despite these difficulties, computer simulations of biological macromolecules are playing an increasingly important role in biophysical chemistry. One important use involves the modeling of biochemical processes via computer simulation, e.g., the binding of a substrate to an enzyme.[3,4] Another major area of research concerns the relationship between computer simulations and experiment.[5] There are four aspects of this area of research that have been stressed: (1) the use of experimental results to refine empirical potentials; (2) the use of simulations as a testing ground for models used to interpret experiment; (3) the use of simulations to better understand the molecular information contained in experiment; and (4) the use of simulations to suggest new experiments. The development of methods for analyzing NMR relaxation and fluorescence depolarization experiments on proteins and nucleic acids using the results of molecular dynamics computer simulations have been reviewed recently.[5] In this paper, we discuss the development of new methods for simulating vibrational spectra using detailed molecular simulations. In connection with this work we have developed new and more powerful algorithms for performing quantum Monte Carlo simulations,[6,7] and these methods are reviewed. In the final section of this paper we review recent work concerned with the use of molecular dynamics simulations of proteins and nucleic acids to analyze restrained parameters least-squares X-ray refinement models for macromolecules.[8,9]

VIBRATIONAL SPECTROSCOPY

Vibrational spectroscopy has played a very important role in the development of potential functions for molecular-mechanics studies of proteins. Bond length, bond angle, and torsional force constants which appear in the energy expressions are heavily "parameterized" from infrared and Raman studies of small model compounds. Considerable information concerning molecular structure, and potential surfaces, is contained in high-resolution FTIR and resonance Raman studies of polypeptides and proteins. One approach to the interpretation of vibrational spectra for biopolymers has been a harmonic analysis whereby spectra are fit by geometry and/or force constant changes. There are a number of reasons for developing other approaches. The consistent force field (CFF) type potentials used in computer simulations are meant to model the motions of the atoms over a large range of conformations and, implicitly, temperatures, without reparameterization.[10] It is also desirable to develop a formalism for interpreting vibrational spectra which takes into account the variation in the conformations of the chromophore and surroundings which occur due to thermal motions. Much of the interesting structural information that can be extracted from experiments involves understanding the origin of spectral shifts as a function of some experimentally adjustable parameter, often the temperature. If computer simulations on realistic potential surfaces are to be useful for interpreting these experiments, it is important that methods be developed that incorporate these anharmonic effects on the spectra. Semiclassical trajectories provide one possible approach to this problem.[11,12] Here we present alternative approaches.

Quasiharmonic Method for Calculating Vibrational Spectra from Classical Simulations on Anharmonic Potentials

We have introduced a new method for calculating vibrational spectra from classical molecular dynamics or Monte Carlo simulations.[13] The method involves a quasiharmonic oscillator approximation in which a temperature-dependent quadratic Hamiltonian is parameterized from the results of a simulation on the complete (anharmonic) potential. The parameterization is accomplished by fitting the first and second moments of the coordinate and momentum distributions obtained from a simulation on the exact surface to a harmonic model. The model provides a method for partially incorporating anharmonicity in the evaluation of optical and thermodynamic properties and estimating quantum corrections to the classical simulations. Furthermore, the approximation can be systematically improved and is extremely useful for the development of new quantum computer simulation methods.

As an illustration of the method, we recently reported the results of a vibrational analysis of a small molecule (butane) with six internal degrees of freedom using the quasiharmonic oscillator model.[13] The empirical potential contained all the terms present in the potential for macromolecules, namely, bond stretching, bending, and torsional terms as well as nonbonded interactions. A novel aspect of the simulation procedure was the use of normal-mode eigenvectors as the independent coordinates for Monte Carlo sampling, which was demonstrated to substantially increase the convergence rate of the simulation. From a conventional normal mode analysis we extracted the frequencies of the model which ranged from 119 cm^{-1} for a pure torsional vibration to 1044 cm^{-1} for a mixed bond stretch-angle bend vibration. Classical simulations were performed on the complete surface at a series of temperatures between 5 K and 300 K. We demonstrated how anharmonic effects at higher temperatures can rotate the

TABLE 1. Quasiharmonic Frequencies (in cm^{-1}) Calculated from Monte Carlo Trajectories[a] on the Exact Potential Surface for *Trans* and *Gauche* Butane

Trans Butane			Gauche Butane
100 K	200 K	300 K	300 K
113 (5)[b]	102 (17)	91 (25)	99 (33)
407	406	407	419
435	437	436	602
899	911	892	857
1008	1014	1002	966
1045	1055	1043	1034

[a]Monte Carlo trajectories constructed with Q_k (mass-weighted Cartesian) as independent coordinates.

[b]Numbers in parentheses indicate percent deviation from harmonic normal-mode eigenvalues; only deviations greater than 1% indicated.

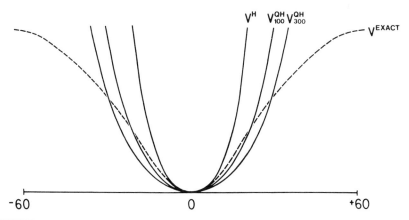

FIGURE 1. Schematic illustration of the exact potential V^{exact}, the harmonic approximation V^H, and quasiharmonic approximations at 100 K, V^{QH}, and 300 K, V^{QH}, for the torsional coordinate Q of *trans* butane. The anharmonicity of the exact potential results in the decreasing curvature of the quasiharmonic potentials with increasing temperature. (See Ref. 13.)

normal coordinates and shift the frequencies with respect to the harmonic values. For the lowest frequency mode (a torsion) increasing the temperature lowered the effective frequency and this was rationalized in terms of the shape of the quasiharmonic torsional potential. The quasiharmonic frequencies calculated from Monte Carlo trajectories on the anharmonic potential surface for *trans* and *gauche* butane are listed in TABLE 1. The effective frequency of the torsional mode is lowered by 25 cm^{-1} to 91 cm^{-1} at 300 K. The anharmonicity of the exact potential results in the decreasing curvature of the quasiharmonic potential and the lowering of the effective torsional frequency with temperature. FIGURE 1 shows a schematic illustration of the exact potential V_{exact}, the harmonic approximation V^H, and quasiharmonic approximations at 100 K and 300 K for the torsional coordinate Q of *trans* butane.

We have recently initiated a study of intramolecular vibrations in liquid water using the quasiharmonic analysis method.[14] For the initial simulations we are using a 3-point model[15,16] for the intermolecular water-water potential and a central force model[17] for the intramolecular vibrations. Since these potentials, in contrast to a more recent complete water potential,[18] were not parameterized to be used together, only qualitative conclusions can be drawn from this initial study. Of interest, the addition of intramolecular vibrations to the water model has a negligible effect on the calculated intermolecular pair correlation functions.[14] The important experimental observation is that on going from the gas phase to the liquid the OH stretching vibration is red-shifted by 300 cm^{-1}, while the HOH bending vibration is blue-shifted by 100 cm^{-1}. In the simplest form of quasiharmonic analysis the effective frequencies are constructed from the probability distributions for the vibrational displacements of the OH bond and HOH bond angle in the gas and liquid phases, respectively. In the gas phase these probability distributions were calculated by direct numerical integration of the partition function, while for the liquid, the distributions were constructed from a 5-ps simulation of 216 water molecules with periodic boundary conditions in the TVN ensemble. The temperature-dependent effective frequency is then given by:

$$\omega_{eff}^2 = \frac{K_B T}{m_{eff} \langle (\Delta q)^2 \rangle}$$

The effective mass of the OH stretching and HOH bending vibrations were calculated to be 0.94 au and 0.48 au, respectively. For the OH stretching vibration, the gas-phase quasiharmonic frequency is 3970 cm^{-1}, the liquid-phase vibrational frequency is calculated to be 3792 cm^{-1}, resulting in a red-shift of $\sim$200 cm^{-1}. In contrast, the HOH bending quasiharmonic frequency is blue-shifted by 75 cm^{-1} in solution compared to the gas-phase calculation ($\omega_{eff} = 1274$ cm^{-1} gas phase, $\omega_{eff} = 1348$ cm^{-1} liquid). It is very encouraging that the simplest quasiharmonic approximation qualitatively reproduces the directions of the water vibrational frequency shifts on going from the gas to the liquid state. Furthermore, it should be possible to use this simple approximation to estimate from classical water simulations the inhomogeneous contributions to the vibrational linewidths by constructing a distribution of quasiharmonic frequencies from a quasiharmonic analysis of individual water molecules in the simulation.

Path Integral Quantum Monte Carlo Simulations and Vibrational Spectra

The approach to the evaluation of vibrational spectra described above is based on classical simulations, for which quantum corrections are possible. The incorporation of quantum effects directly in simulations of large molecular systems is one of the most challenging and actively pursued research areas in theoretical chemistry today. The development of quantum simulation methods is particularly important in the area of molecular spectroscopy for which quantum effects are often important and where the goal is to use simulations to help understand the structural and dynamical origins of spectral lineshapes and changes in lineshapes with environmental variables, e.g., the temperature. Although it has long been known that path integrals provide one method for calculating quantum statistical-mechanical properties of polyatomic systems, only recently has attention been turned to the development of computationally tractable methods for evaluating path integrals.[19] In this section we introduce some new methods that we are developing for the rapid and accurate evaluation of path integrals and we outline the procedure we are using to construct vibrational lineshapes from the quantum simulations. We have developed and initially tested the methodology on one-dimensional systems, e.g., the quartic and morse oscillators and double-well potential. These methods are being developed because their generalization to large polyatomic systems is particularly promising, and this work is now in progress.

Moment Method for Evaluating Vibrational Lineshapes from Quantum Monte Carlo Path Integral Simulations

The direct evaluation of quantum time-correlation functions for anharmonic systems is extremely difficult. Both wave packet[20] and path integral methods[21] have been used to evaluate quantum time-correlation functions directly, but because of numerical difficulties associated with these techniques it seems unlikely that these methods will be successfully applied to large molecules. Our approach to the evaluation of finite temperature anharmonic effects on vibrational lineshapes is derived from the fact that the moments of the vibrational lineshape spectrum can be expressed as functions of expectation values of positional x_i, momentum p_j and mixed $x_i p_j$ operators. The expectation values can be evaluated using extremely efficient techniques that we have developed to evaluate equilibrium discretized path integrals. The main points are summarized here. The infrared vibrational lineshape is given as the Fourier transform of the dipole moment correlation function:

$$I(\omega) = \frac{1}{2\pi} \int_{-\infty}^{\infty} \langle u(0)u(t) \rangle \, e^{i\omega t} \, dt \tag{1}$$

By inverse Fourier transformation of Equation 1 and expansion of both sides in a Taylor series we obtain[22]

$$\sum_{n=0}^{\infty} \frac{t^n}{n!} \left[\frac{d^n}{dt^n} \langle u(0)u(t) \rangle \right] = \sum_{n=0}^{\infty} \frac{(it)^n}{n!} \int \omega^n I(\omega) \, d\omega \tag{2}$$

Equating coefficients of powers of t,

$$\frac{d^n}{dt^n} \langle u(0)u(t) \rangle_{t=0} = (i)^n \int \omega^n I(\omega) d\omega \tag{3}$$

Thus, the nth vibrational spectral moment is equal to an equilibrium correlation function, the nth derivative of the dipole moment autocorrelation function evaluated at $t = 0$.[22] By using the repeated application of the Heisenberg equation of motion

$$\frac{du}{dt} = \frac{i}{\hbar} [H, u] \tag{4}$$

and substitution in the l.h.s. of Equation 3, we can express the nth vibrational spectral moment as an expectation value of nested commutators of H with the dipole moment operator:

$$\int \omega^n I(\omega) \, d\omega = \hbar^{-n} \langle u(0) \cdot [H, [H, \ldots [H, u]]] \rangle \tag{5}$$

The expectation values on the right-hand side of this equation depend only on the ensemble averages of position and momentum operators, which can be evaluated using a new quantum Monte Carlo method described below:

Path integral reference systems. Our approach to the evaluation of finite temperature anharmonic effects on vibrational lineshapes requires the evaluation via computer simulation of expectation values of the moments of positional x_i, momentum p_j, and mixed $x_i p_j$ operators. These expectation values are evaluated using path integral techniques. The quantum partition function Z in the coordinate representation is:

$$Z = \int dx_1 \langle x_1 | e^{-\beta H} | x_1 \rangle \tag{6}$$

To obtain the discretized path integral representation for Z we use the identity:

$$e^{-\beta H} = (e^{-\beta H/P})^P \tag{7}$$

and insert complete sets of states $I = \int |x_i\rangle \langle x_1| dx_i$ p times:

$$Z = \int dx_1 \dots dx_p \langle x_1|e^{-\beta H/P}|x_2\rangle \langle x_2|e^{-\beta H/P}|x_3\rangle \dots \langle x_p|e^{-\beta H/P}|x_1\rangle \quad \textbf{(10)}$$

a discretized path integral with p points. It is not possible in general to evaluate the matrix elements in the path integral for arbitrary H. We define the exact Hamiltonian H and a reference Hamiltonian H_0 by:

$$H = \frac{P^2}{2m} + V \quad \textbf{(11a)}$$

$$H_0 = \frac{P^2}{2m} + V_0 \quad \textbf{(11b)}$$

where V is the exact anharmonic potential and V_0 is a quadratic "reference" potential discussed below. Clearly:

$$H = H_0 + (V - V_0) = H_0 + V' \quad \textbf{(12)}$$

We wish to separate the matrix elements involving H in Equation 10 into matrix elements of H_0 and V. For p "large enough" we have:

$$\langle x_i|e^{-\beta(H_0+V')}|x_{i+1}\rangle \approx e^{-\beta V1(x_1)/2P} \langle x_1|e^{-\beta H_0/P}|x_{i+1}\rangle e^{-\beta V'(x_{i+1})/2P} \quad \textbf{(13)}$$

A lot has been accomplished in the separation, Eq. 13, because for quadratic Hamiltonians the matrix elements can be evaluated analytically. As $p \to \infty$, Equation 13 becomes exact. For finite, and in particular small values of p, the error depends on the magnitude of the commutator of H_0 and V'. In a p point discretized path integral Monte Carlo simulation each quantal degree of freedom is simulated by p "classical" degrees of freedom. If we wish to apply discretized path integral methods to simulate polyatomic systems, it is essential that Equation 6 hold for small values of p. The approximation of the path integral by p discretized points for small p depends on the construction of an appropriate reference Hamiltonian H_0. We discuss below three choices for H_0:

(i) The free particle reference system

$$H_0 = \frac{P^2}{2m} \quad \textbf{(14a)}$$

$$V' = V \quad \textbf{(14b)}$$

This is the standard reference system for the evaluation of path integrals. The partition function in this reference system is

$$Z = \int dx_1 \dots dx_P \prod_{i-1}^{P} e^{-(mP/2\hbar^2\beta)(x_{i+1} - x_i)^2} e^{-\beta V(x_i)/P} \quad \textbf{(15)}$$

As has been pointed out, this is equivalent to a classical configurational partition function for a polymer with p beads. Adjacent beads are connected to each other by "springs," with spring constant $K = \{mp/\hbar^2\beta\}$ and each bead interacts with an external potential $V(x_i)$.[19]

(ii) Quasiharmonic path integral reference system

We have recently proposed the use of a temperature-dependent harmonic (quasiharmonic) reference system for the evaluation of discretized path integrals of anharmonic systems.[6] For the quasiharmonic reference system:

$$H_0 = \frac{P^2}{2m} + \omega_{eff}^2 x^2 \tag{16a}$$

$$V'(x) = V(x) - \omega_{eff}^2 x^2 \tag{16b}$$

The effective frequencies and the normal modes are obtained from the classical quasiharmonic normal mode analysis described in the previous section. The quasiharmonic potential is defined so that $V'(x)$ is small in the important regions of x space sampled at temperature T. This implies that Eq. 13 will be valid for small p and that for intramolecular vibrations, the quasiharmonic reference system is superior to the free-particle reference. The harmonic propagator is given by:

$$G_0(x, x', \beta/p) = \left\{\frac{m\omega}{2\pi\hbar\sinh(\beta\hbar\omega/p)}\right\}$$

$$\cdot \exp\left\{\left[\frac{-m\omega}{2\sinh(\beta\hbar\omega/p)}\right]\left[(x^2 + x'^2)\cosh\left[\frac{\beta\hbar\omega}{p}\right] - 2xx'\right]\right\} \tag{17}$$

The p point discretized path integral expression for the partition function is:

$$Z = \int dx_1 \dots dx_P \prod_{i=1}^{P} G_0(x_i, x_{i+1}, \beta/P) \, e^{-\beta V(x_i)/P} \tag{18}$$

Defining the integrand of Eq. 10 to be $K(X_1, \dots X_p)$, the expectation value of an operator A in the quasiharmonic discretized path integral (QHDPI) formulation is given by:

$$\langle A \rangle = \frac{\int dx_1 \dots dx_P \, A(x_i) \, K(x_1, \dots x_P)}{Z} \tag{19}$$

The kernal $K(X_1, \dots X_p)$ plays the role of the Monte Carlo weight function that $\exp(-\beta H)$ plays in classical simulations.

(iii) Variable-frequency quasiharmonic reference systems

The quasiharmonic reference system described above represents an optimized fixed-frequency reference for the evaluation of discretized path integrals. It is reasonable to attempt to construct a variable frequency harmonic reference system which (1) has an analytic form for the propagator from x_i to $_{i+1}$, and (2) minimizes the value of $V'(X)$ along the most important paths from X_i to x_{i+1}. In a forthcoming publication we discuss methods for constructing variable-frequency quadratic reference systems.[7] The reference Hamiltonian is given by:

$$H_0(x_i, x_{i+1})(x) = \frac{P^2}{2m} + \omega^2(x_i, x_{i+1}) x^2 + bx + c \tag{20}$$

The notation is meant to suggest that the frequency is variable and depends on the propagator matrix elements. We have shown that $\omega^2(x_i, x_{i+1})$ can be defined by the equation:

$$\int_{x_i(0)}^{x_{i+1}(\beta\hbar/P)} \{V_0(x(t)) - V(x(t))\} \, dt = 0 \tag{21}$$

where the integral is along the classical path for the reference system between x_i at time o and x_{i+1} at time $t = \beta\hbar/p$. Notice that the path depends on the temperature. We now compare the results of path integral simulations for a model quartic oscillator Hamiltonian using the free particle, quasiharmonic, and variable harmonic reference systems.

TABLE 2a.

Reference System	Exact Quantum Result	Discretized Path Integral Quadrature Points		
		1	4	8
	Evaluation of $\langle X^2 \rangle$			
A. $\beta\hbar\omega = 10$				
Anharmonicity b = 0.05				
Basis set	0.445			
Free particle		0.093	0.262	0.358
Quasiharmonic		0.320	0.397	0.431
Variable harmonic		0.446		
B. $\beta\hbar\omega = 2$				
Anharmonicity b = 5				
Basis set	0.161			
Free particle		0.094	0.143	0.159
Quasiharmonic		0.140	0.155	0.159
Variable harmonic		0.165		
	Evaluation of $\langle X^4 \rangle$			
A. $\beta\hbar\omega = 10$				
b = 5				
Basis set	0.071			
Free particle		0.003	0.017	0.029
Quasiharmonic		0.010	0.023	0.034
Variable harmonic		0.068		

Path integral simulations of the quartic oscillator: results for position, momentum, and spectral moments. In TABLE 2a we compare the results for the evaluation of the expectation value of $\langle x^2 \rangle$ using the different reference systems. Considering first the result for $\beta\hbar\omega = 10$ and anharmonicity b = 0.05, the exact quantum value for the second moment is $\langle x^2 \rangle = 0.445$. The classical value (which is the result for a p = 1-point quadrature discretized path integral using the free-particle reference) is 0.093. For p = 1 the quasiharmonic path integral evaluation of $\langle x^2 \rangle$ is 0.320; when the variable harmonic reference is employed the result $\langle x^2 \rangle = 0.446$ is very close to the exact value. In contrast, even for p = 8 the free particle reference is in error by 20% ($\langle x^2 \rangle = 0.358$), while the quasiharmonic reference path integral evaluation yields an improved result ($\langle x^2 \rangle = 0.431$; the error is 4%). Similar results are shown in TABLE 2a for larger values of the anharmonicity and higher temperatures. Thus for the

TABLE 2b Evaluation of Higher Moments

Property	Exact Quantum	Variable Quadratic
$\langle X^8 \rangle$	1.45	1.15 (p = 1)
		1.52 (p = 2)
$\langle P^2 \rangle$	0.680	0.597 (p = 1)
		0.645 (p = 2)
$\langle P^2 X^2 \rangle$	−0.198	−0.175 (p = 1)
		−0.192 (p = 2)

evaluation of $\langle x^2 \rangle$ using the variable harmonic reference system, a 1-point quadrative is very accurate. This means that the computational effort required to calculate quantum expectation values for this model system is only slightly greater than that required to evaluate classical ensemble averages. In TABLE 2b we show the results for the path integral evaluation of higher positional moments and momentum moments. It is clear that for p = 1 or 2, the approximate values obtained using the variable harmonic reference method to evaluate the path integral are close to the exact results. Equally promising results using this new path integral method have been obtained for the double-well and morse oscillator Hamiltonians. We are presently modifying the programs to treat polyatomic model systems.

In TABLE 2c we present the first and second moments of the vibrational spectrum of the quartic oscillator calculated by the moments method. The quantum results were obtained from path integral Monte Carlo simulations using the variable quadratic reference. For comparison, the average frequency and linewidth obtained from classical Monte Carlo evaluation of the moments is also listed. Spectral features for two values of the temperature ($\beta\hbar\omega$ = 5 and 1.0) and two anharmonicities (b = 0.05, b = 1.0) are listed. As the anharmonicity or the temperature is increased, the oscillator frequency and linewidth increase for both the classical and quantum simulations. The important point is that the classical spectrum is shifted less and broadened more as the temperature and anharmonicity increase. For example, when $\beta\hbar\omega$ = 1 and b = 0.05 are chosen, the classical spectrum is almost twice as broad as the quantum spectrum. The results presented in TABLE 2c demonstrate that for realistic values of the temperature and anharmonicity, quantum effects on the vibrational spectrum are important. However, the strong coupling of broadening with frequency shifts is a limitation of one-dimensional models for which the only broadening mechanism is anharmonicity in the vibrational degree of freedom. Significant population of vibrationally excited states is required for thermal broadening. In contrast, it is the coupling

TABLE 2c. Vibrational Spectrum of Quartic Oscillator by Moments Method

	Spectral Moment[a]	Quantum	Classical
A. $\beta\hbar\omega$ = 5	$\langle \omega_1 \rangle$	1.12	1.05
b = 0.05	$\langle \omega^2 \rangle^{1/2}$	0.03	0.06
B. $\beta\hbar\omega$ = 5	$\langle \omega \rangle$	1.96	1.4
b = 1.0	$\langle \omega^2 \rangle^{1/2}$	0.3	0.4
C. $\beta\hbar\omega$ = 1	$\langle \omega \rangle$	1.19	1.12
b = 0.05	$\langle \omega^2 \rangle^{1/2}$	0.12	0.20

[a] $\omega_0 = 1$, $\hbar = 1$, m = 1.

of the vibrational degree of freedom to other modes, protein and/or liquid that is of primary interest with respect to an understanding of the structural information content of chromophore lineshapes. The coupling of a chromophore vibration to a bath results in the time and spatial modulation of the energy spacing of the first few vibrationally excited states. This broadening mechanism can give rise to line broadening without significant frequency shifting. Path integral computer simulations of models for chromophoric molecules coupled anharmonically to additional modes, including heat bath models, are presently under way.

We have described in this section methods for calculating vibrational lineshapes for anharmonic systems which incorporate quantum properties in a fundamental way. The methods have been demonstrated to be very powerful when applied to a variety of one-dimensional problems. However, for one-dimensional models alternative methods for calculating spectra are more direct. We are pursuing path integral approaches to the problem of calculating lineshapes because we believe that there numerical methods can be generalized to polyatomic systems and that they will be much more stable and accurate than alternative approaches. One particularly attractive aspect of our approach is the ability to combine quantum and classical Monte Carlo algorithms in a single simulation. For example the simulation of the vibrational spectrum of a chromophore with 10–20 degrees of freedom imbedded in a protein could be accomplished by combining quantum Monte Carlo methods for the chromophore coordinates with classical Monte Carlo trajectories for the protein atoms.

Restrained X-Ray Refinement of Nucleic Acids and Proteins

Crystallographic refinement of proteins and nucleic acids at high resolution is being used to obtain a wealth of information concerning the atomic fluctuations and the flexibility of these macromolecules. The information concerning atomic mobility is contained in the Debye-Waller temperature factors. In proteins, correlations between temperature factors and biological functions, like ligand access to the protein interior[23,24] or antigenic recognition,[25,26] have been noted. The availability of high-resolution X-ray structures of DNA oligomers has opened up a new era in the study of the structure and biological function of DNA and its interactions with proteins.[27,28] The goal is to determine the roles of DNA structure and flexibility in the specificity of DNA interactions with ligands. The correct interpretation of high-resolution crystallographic data depends upon a thorough understanding of the effects of the refinement model upon derived quantities such as temperature factors. Temperature-dependent crystallographic studies can, in principle, provide detailed information concerning the potential surface on which the atoms move.[23,29,30] For these studies, it is necessary to determine the accuracy of the refinement model over a wide range of temperatures.

Molecular dynamics simulations of proteins and nucleic acids provide a very powerful method for testing crystallographic refinement models. The simulations constitute the most detailed theoretical approach available for studying the internal motions of these macromolecules.[31] From the time evolution of the atomic positions, time-averaged X-ray intensities can be calculated and treated as data for crystallographic refinement. The final structure and temperature factors obtained from the refinement can then be compared with the "exact results" obtained directly from the trajectory. Kuriyan et al.[8] have recently carried out a detailed analysis of a protein refinement model at a single temperature using a molecular dynamics simulation of myoglobin[32] to generate the X-ray data. In this review, we discuss the results of an analysis[9] of the temperature-dependent molecular dynamics and X-ray refinement of a Z-DNA hexamer 5BrdC-dG-5BrdC-dG-5BrdC-dG for which the experimental X-ray data are available and whose crystal structure has been refined to high resolution.[33]

Methods for Simulating Restrained X-Ray Refinement Data from Molecular Dynamics Trajectories

Molecular dynamics simulations were carried out on the 248 atom Z-DNA hexamer, using the AMBER nucleic acid force field[34] with a distance-dependent dielectric and excluding counterions. Although the model treats electrostatic effects only in a qualitative way, recent molecular dynamics simulations for both peptides[35] and nucleic acids[36,37] have demonstrated that for localized conformations sampled during short molecular dynamics simulations, average properties are not very sensitive to the electrostatic model; it is the packing and hydrogen-bonding terms that together with the vibrational potential (bond, bond angle and torsional stretching) dominate the calculated equilibrium and dynamical properties. The crystal structure of the Z-DNA hexamer was first energy-minimized with 200 conjugate gradient steps to relieve any initial strain in the structure before the molecular dynamics simulations were begun. The rms displacement between the crystal and energy-minimized coordinates is less than 0.1 Å. Simulations were performed at a series of temperatures, defined by the mean kinetic energy of the system, between 100 K and 300 K. For each temperature, 10 trajectories, each 2 psec in length, were calculated by solving simultaneously the classical equations of motion for the atoms of the helix. The use of multiple short trajectories instead of a single long trajectory has been found to be a more efficient method for sampling conformations for macromolecular systems containing many harmonic degrees of freedom.[38,39]

Crystallographic refinement is a procedure that iteratively improves the agreement between structure factors derived from X-ray intensities and those derived from a model structure. For macromolecular refinement, the limited diffraction data have to be complemented by additional information in order to improve the parameter-to-observation ratio. This additional information consists of restraints on bond lengths, bond angles, aromatic planes, chiralities, and temperature factors.

In the restrained refinement procedure[40] a function of the form:

$$\Phi = \sum_Q W_Q \big| |F_0(Q)| - |F_c(Q)| \big|^2 + \sum_i W_i \Delta_i^2 \qquad (22)$$

is minimized. W_Q is the weight assigned to the structure factors and it varies linearly with Q with coefficients adjusted so that low-resolution structures are weighted more strongly than high-resolution ones. $F_0(Q)$ and $F_c(Q)$ are, respectively, the observed and calculated structure factors. The second term in Equation 1 contains the stereochemical restraint information. Δ is the deviation of a restrained parameter (bond lengths, bond angles, volumes, nonbonded contacts, and temperature factors) from its ideal value and W_i is the weight assigned to the restraint. The form of Equation 1 is such that the weights W_i correspond to the inverse of the variance Δ^2 for each set of observations. The weights on the various classes of restraints in the simulated refinements described below were between 0.02 and 0.04Å for distances, 0.01Å for planarity, 0.04Å for chirality, and 0.25Å for nonbonded contacts. The refinements were carried out with no and tight (less than $0.1Å^2$ at 165 K or less than $1.0Å^2$ at 300 K) restraints on the temperature factors.[9]

The structure factor F(Q) in X-ray crystallography is the Fourier transform of the electron density for the molecule:

$$F(Q) = \int dr \, \rho(r) e^{iQ \cdot r} \tag{23}$$

where $\rho(r)$ is the electron density at r. In a crystallography experiment the electron density varies with time due to thermal motion and the observed structure factor amplitude is the time average of Equation 2:

$$F_0(Q) = \langle F(Q) \rangle = \int dr \, \langle \rho(r) \rangle e^{iQ \cdot r} \tag{24}$$

In order to generate a set of calculated structure factors $F_c(Q)$ from a set of coordinates it is necessary to introduce a model for the time variation of the electron density. The usual assumptions in macromolecular crystallography include harmonic isotropic motion of the atoms, and, in addition, the molecular scattering factor is expressed as a superposition of atomic scattering factors. With these assumptions the calculated structure factor (Equation 1) is given by[42]:

$$F_c(Q) = \sum_{j=1}^{N} e^{iQ \cdot r} e^{W_j(Q)} \tag{25}$$

where $F_j(Q)$ is the atomic scattering factor for atom j and r_j is the position of atom j in the model structure. The thermal averaging of atomic motion is contained in the atomic Debye-Waller factor exp $(W_j(Q))$. W_j is given by:

$$W_j(Q) = -B_j |Q|^2 \tag{26}$$

where B_j is the atomic temperature factor. The mean square atomic fluctuation $(\Delta r_j)^2$ for atom j is obtained from the refined temperature factors through the relation[42]:

$$\langle (\Delta r_j)^2 \rangle = \frac{3}{8\pi^2} B_j \tag{27}$$

There are therefore four adjustable parameters per atom in the refinement (x_j, y_j, z_j, B_j). In the computer experiments we have carried out to test the assumptions of the nucleic acid refinement model we have generated sets of "observed" structure factors $F_0(Q)$, from the Z-DNA molecular dynamics trajectories. The thermal averaging implicit in Equation 26 is accomplished by averaging the atomic structure factors obtained from coordinate sets sampled along the molecular dynamics trajectories at each temperature:

$$F_0(Q) = \langle F(Q) \rangle = \frac{1}{m} \sum_{k=1}^{M} \sum_{j=1}^{N} F_j(Q) e^{iQ \cdot r_j^k} \tag{28}$$

where r_j^k is the position of the jth atom in the kth coordinate set along the trajectory and M is the number of coordinate sets sampled. In the present study structure factors corresponding to 3195 reflections between 10Å and 1.7Å were calculated for each of 50 coordinate sets at each temperature. Only the 246 heavy atoms of the hexamer were included in the structure factor calculations; hydrogen atoms were not included in the refinement.

Restrained X-Ray Refinement of the Z-DNA Molecular Dynamics Trajectories

Refinement of the MD average structures against X-ray intensities calculated from the trajectories. Refinement of molecular dynamics average structures against simulated X-ray diffraction intensities was carried out at four temperatures between 165 K

TABLE 3. Parameters for Refinement of Simulated X-Ray Intensities of Z-DNA Hexamer

Parameter	165 K	300 K
Number of bad distances[a]		
Before starting refinement	1	50
After refinement with strong B restraints	0	76
After refinement without B restraints	0	50
R-factor		
Before starting refinement	1.6	22.0
After refinement with strong B restraints	1.6	7.0
After refinement without B restraints	0.9	6.2
Average temperature factor (A^2)		
Before starting refinement	0.5	5.9
After refinement with strong B restraints	0.7	5.1
After refinement without B restraints	0.5	5.5
RMS of delta B[b]		
Before starting refinement	0.3, 0.3, 0.4, 0.3	2.4, 3.5, 8.2, 5.6
After refinement with strong B restraints	.06, .08, .10, .13	0.6, 0.7, 0.2, 0.6
After refinement without B restraints	0.3, 0.3, 0.3, 0.3	3.0, 3.5, 3.8, 4.1

[a]Distances that deviate from ideality by more than two standard deviations.
[b]The four values correspond to the differences in temperature factors for atoms connected by a bond length, for atoms connected by a bond angle, for P—O bond lengths, and for phosphate atoms connected by a bond angle or for atoms involved in hydrogen bonding.

and 300 K with no restraints and with strong restraints on temperature factors. For the initial temperature factors used to start the refinement we tried both individual atomic temperature factors calculated from the dynamics and a single average temperature factor obtained by averaging the molecular dynamics results over all atoms at each temperature. Similar results were obtained for both choices of initial temperature factors. The results reported below correspond to refinements started with individual atomic temperature factors taken from the molecular dynamics results.

The R factors, average temperature factors, rms deviations of the temperature factors, and the number of "bad" distances obtained for the refinement of the molecular dynamics average structures against the simulated X-ray intensities at 165 K and 300 K are listed in TABLE 3. At 165 K the initial R factor and number of bad distances before refinement are both very small and are not changed significantly after refinement. At 300 K the initial R factor and number of bad distances are 22% and 50, respectively. With tight restraints on temperature factors, the R factor decreased to 7% after three refinement cycles and the number of bad distances increased to 76. With no restraints on B, the R factor decreased to 6.3% after three refinements cycles with 50 bad distances.

At the low temperature (165 K) the effect of refinement with strong temperature factor restraints is to increase the average temperature factor from 0.5 Å^2 (exact result) to 0.7 Å^2 and to decrease the variances in the temperature factors for the different classes of stereochemical constraints. With strong restraints on B, the refinement resulted in sharp differences between the cytosines and the guanines both for the sugars and bases which were not present in the temperature factors calculated directly from the 165 K trajectories. For example, the ratio of the temperature factors for guanine bases to cytosine bases, which is 1.1 calculated directly from the molecular dynamics simulation, increases to 2.6 after refinement with strong B restraints. In contrast, when refinement is done without temperature factor restraints at 165 K, the average temperature factors and the variances in the temperature factors are very close to the exact molecular dynamics values.

At 300 K the effect of the refinement both with and without strong restraints on temperature factors is to decrease the average thermal factor compared to the exact values. The temperature factor averaged over all atoms calculated directly from the room temperature trajectories is 5.9 Å^2 and is reduced by 15% to 5.1 A after refinement with strong B restraints. The average temperature factor (5.5 Å^2) obtained from the 300 K refinement without temperature factor restraints is closer to the exact value. The errors in the temperature factors introduced by the refinement at 300 K is largest for the atoms with the largest thermal fluctuations, the phosphates. This result is clearly demonstrated in FIGURE 2, which compares the phosphate temperature factors calculated directly from 300 K trajectories with the results of the two refinements, with and without temperature factor restraints. For example, the two atoms with the largest thermal fluctuations are the C5 and C11 phosphates. The exact B values for these atoms are 15.1 Å^2 and 17.0 Å^2, respectively; after the refinement without temperature factor restraints the B values are reduced to 10.8 Å^2 and 10.9 Å^2, and they are reduced even further to 8.8 Å^2 and 8.3 Å^2, respectively, after the refinement with temperature factor restraints. The errors introduced by the refinement are also apparent in the effect on the temperature factor variances. At room temperature the actual variances in temperature factors computed for each stereo-chemical class are considerably greater than 1 Å^2 (they range from 2.4 Å^2 for atoms connected by bonds not involving a phosphorous to 8.2 Å^2 for P and O atoms connected by a P—O bond). The refinement of the simulated X-ray intensities at 300 K with temperature factor restraints greatly reduced the variances in the B values for each of the stereochemical classes so that the final variances are less than 1 Å^2 (TABLE 3). These results are in accord with a recent analysis of the use of restraints in temperature factor refinements for proteins[8] for which it was shown that the weights used to restrain neighboring atom temperature factors are much more restrictive than the variation in neighboring temperature factor values obtained from protein molecular dynamics simulations. Restraints on the differences in temperature factors between bonded atom pairs have been shown to be uncorrelated with variances in the corresponding bond

Phosphate Temperature Factors

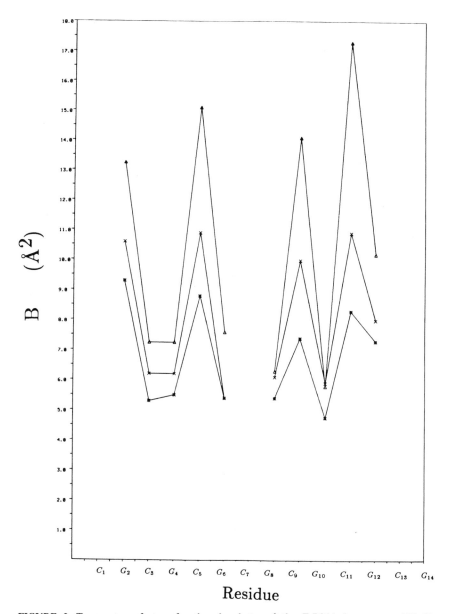

FIGURE 2. Temperature factors for the phosphates of the Z-DNA hexamer at 300 K. Temperature factors were evaluated using the mean square atomic fluctuations calculated directly from the molecular dynamics trajectories; x = temperature factors calculated from the refinement without B-factor restraints; * = temperature factors calculated from the refinement with B-factor restraints.

length distributions.[43] From the present results concerning errors in predicted temperature factor restraints we conclude that the commonly used value of 1 Å^2 or less between temperature factors on adjacent atoms[9,44] is too restrictive.

Although, as discussed earlier, there are quantitative errors in temperature factors introduced by the refinement procedure, the temperature-dependence of the atomic mobilities as estimated by the refined temperature factors provides a reasonably accurate description of the true temperature-dependence of the system. In FIGURES 3b and c the mean square atomic fluctuations extracted from the refinement simulations at each temperature and averaged by group are plotted as a function of temperature for comparison with the exact results shown in FIGURE 3a. As to the refinement without temperature factor restraints (Fig. 3b), except for the phosphates at the highest temperature, the extent of anharmonicity (curvature) is in good agreement with the exact result despite the fact that the refinement model assumes isotropic, harmonic motion. The ordering of the atomic fluctuations by groups (bases < deoxyriboses < CpG phosphates < GpC phosphates) at the two higher temperatures (275 K and 300 K) agrees with the exact results, although the agreement is not as good as that at 165 K and 225 K, even though the harmonic model would be expected to be more accurate at low temperature. It is clear from FIGURE 3c that when strong temperature factor restraints are introduced in the refinement, differences in the temperature-dependence of the atomic fluctuations among the groups are suppressed, although the shapes of the curves agree qualitatively with the results calculated directly from the simulations (Fig. 3a). The present results provide a theoretical foundation for the use of Debye-Waller factors obtained from refinements at several temperatures to extract information concerning the anharmonicity of the atomic displacements and underlying potential surface.

SUMMARY

The use of computer simulations to study the internal dynamics of globular proteins and nucleic acids has attracted a great deal of attention in recent years. These simulations constitute the most detailed theoretical approach available for studying internal motions and structural flexibility of biological macromolecules. In this paper we review recent work concerned with the use of computer simulations for the interpretation of, and comparison with, experimental probes of molecular dynamics. New methods for calculating vibrational spectroscopic lineshapes from computer simulations are discussed. A quasiharmonic approximation is described by which classical computer simulations on multidimensional potential surfaces can be used to estimate the effects of anharmonicity on vibrational spectra. A novel aspect of the method is the use of normal-mode eigenvectors as the independent coordinates for Monte Carlo sampling. Results for isolated small molecules and liquid water are reviewed. The construction of vibrational lineshapes from quantum computer simulations using path integral methods with new (quasiharmonic and variable quadratic) reference systems are discussed. Results for small model systems are presented and extension of the methods to large molecules is discussed. The recent use of molecular dynamics simulations to analyze X-ray refinement models for proteins and nucleic acids is also reviewed.

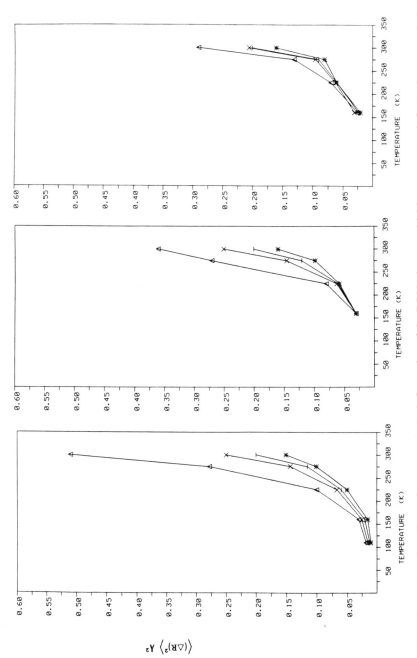

FIGURE 3. Temperature-dependence of mean square atomic fluctuations of the Z-DNA hexamer. (**a**) Mean square atomic fluctuations were calculated directly from the molecular dynamics trajectories. (**b** and **c**) Mean square atomic fluctuations were calculated using $(\Delta r)^2 = (3/8\pi^2)B$ with the Debye-Waller temperature factors obtained from the X-ray refinement of the molecular dynamics trajectories with no restraints (**b**) and with tight restraints (**c**). Symbols: * = bases; — = sugars, x = CpG phosphates; Δ = GpC phosphates.

ACKNOWLEDGMENTS

I am grateful to Richard Friesner and Attila Szabo for collaboration and discussion concerning the quasiharmonic method and to Eric Westhof, John Kuriyan, Martin Karplus, and Greg Petsko for collaboration and discussion concerning X-ray refinement simulations.

REFERENCES

1. KARPLUS, M. & J. A. McCAMMON. 1981. CRC Crit. Rev. Biochem. **9:** 293.
2. McCAMMON, J. A. 1984. Rep. Prog. Phys. **47:** 1.
3. ALLISON, S. A. & J. A. McCAMMON. 1985. J. Phys. Chem. **89:** 1072.
4. BRUNGER, A., C. L. BROOKS & M. KARPLUS. 1985. Proc. Natl. Acad. Sci. USA **82:** 8458.
5. LEVY, R. M. & J. KEEPERS. 1986. Comments on Molecular and Cellular Biophys. **3:** 273.
6. FRIESNER, R. A. & R. M. LEVY. 1984. J. Chem. Phys. **80:** 4488.
7. ZHANG, P., R. M. LEVY & R. A. FRIESNER. Submitted for publication.
8. KURIYAN, J., G. A. PETSKO, R. M. LEVY & M. KARPLUS. 1986. J. Mol. Biol. **90:** 227.
9. WESTHOF, E., S. GALLION, P. WEINER & R. M. LEVY. 1976. 1986. J. Mol. Biol. **91:** in press.
10. WARSHEL, A. J. Chem. Phys. 1971. **55:** 3327.
11. WARSHEL, A. & M. KARPLUS. 1972. Chem. Phys. Lett. **17:** 7.
12. MYERS, A., R. MATHIES, S. TANNOR & E. J. HELLER. 1982. J. Chem. Phys. **77:** 3857.
13. LEVY, R. M., O. ROJAS & R. A. FRIESNER. 1984. J. Phys. Chem. **88:** 4233.
14. NAKAI, S., F. HIRATA & R. M. LEVY. To be published.
15. JORGENSEN, W. 1981. J. Am. Chem. Soc. **103:** 335.
16. BERENDSEN, H., J. P. M. POSTMA, W. F. VAN GUNSTEREN & J. HERMANS. 1981. *In* Intermolecular Forces. B. Pullman, Ed.: 331. Reidel. Dordrecht, Holland.
17. RAHMAN, A. & F. STILLINGER. 1975. J. Chem. Phys. **12:** 5223.
18. REIMERS, J. R. & A. O. WATTS, 1984. Chem. Phys. **91:** 201.
19. CHANDLER, D. & P. G. WOLYNES. 1981. J. Chem. Phys. **74:** 4078.
20. HELLER, E. 1981. Accts. Chem. Res. **14:** 368.
21. THIRUMOLAI, D. & B. BERNE. 1983. J. Chem. Phys. **79:** 5029.
22. GORDON, R. G. 1965. J. Chem. Phys. **43:** 1307.
23. FRAUENFELDER, H., G. PETSKO & D. TSERNOGLU. 1979. Nature **280:** 558.
24. ARTYMIUK, P. J., C. BLAKE, D. GRAU, S. OATLEY, D. C. PHILLIPS & M. STERNBERG. 1979. Nature **280:** 563.
25. WESTHOF, E., D. ALTSCHUH, A. C. BLOOMER, A. MONDRAGON, A. KLUG & M. H. VAN REGENMORTEL. 1984. Nature **311:** 123.
26. TAINER, J., E. D. GETZOFF, H. ALEXANDER, R. A. HOUGHTEN, A. J. OLSON, A. J. LERNER & W. A. HENDRICKSON. 1984. Nature **312:** 127.
27. A. H.-J., G. J. QUIGLEY, F. J. KOLPAK, J. L. CRAWFORD, J. H. VAN BOOM, G. VAN DER MAREL & A. RICH. 1979. Nature **282:** 680.
28. WING, R. M., H. R. DREW, T. TAKANO, C. BROKA, S. TANAKA, K. ITAKURA & R. E. DICKERSON. 1980. Nature **287:** 755.
29. HARTMANN, H., F. PARAK, W. STEIGMAN, G. A. PETSKO, P. D. RINGE & H. FRAUENFELDER. 1982. Proc. Natl. Acad. Sci. USA **79:** 4967.
30. KNAPP, E. W., S. F. FISCHER & F. PARAK. 1982. J. Phys. Chem. **86:** 5042.
31. KARPLUS, M. & J. A. McCAMMON. 1981. CRC Crit. Rev. Biochem. **9:** 293.
32. LEVY, R. M., R. P. SHERIDAN, J. W. KEEPERS, G. S. DUBEY, S. SWAMINATHAN & M. KARPLUS. Biophys. J. 1985. **48:** 509.
33. WESTHOF, E., T. PRANGE, B. CHEVRIER & D. MORAS. 1985. Biochimie. **67:** 811.
34. WEINER, S., P. KOLLMAN, D. CASE, U. SINGH, G. GHIO, G. ALAGON, S. PROFETA & P. WEINER. 1984. J. Am. Chem. Soc. **106:** 765.
35. PETTIT, J. & M. KARPLUS. 1985. J. Am. Chem. Soc. **107:**1166.
36. LEVITT, M. 1982. Cold Spring Harbor Symp. Quant. Biol. **47:** 251.

37. SINGH, U. C., S. J. WEINER & P. KOLLMAN. 1985. Proc. Natl. Acad. Sci. USA **82:** 755.
38. LEVY, R. M., D. PERAHIA & M. KARPLUS. 1982. Proc. Natl. Acad. Sci. USA **79:** 1346.
39. KEEPERS, J. & R. M. LEVY. To be published.
40. KONNERT, J. H. & W. A. HENDRICKSON. 1980. Acta Crystallog. **A36:** 344.
41. WARSHEL, A. 1977. Ann. Rev. Biophys. Eng., **6:** 273.
42. WILLIS, B. T. & W. PRYOR. 1975. Thermal Vibrations in Crystallography. University Press. London.
43. YU, H., M. KARPLUS & W. A. HENDRICKSON. 1985. Acta. Crystallog. **B41:** 191.
44. WESTHOF, E., P. DUMAS & D. MORAS. 1985. J. Mol. Biol. **184:** 119.

Nuclear Magnetic Resonance Relaxation and the Dynamics of Proteins and Membranes: Theory and Experiment

ATTILA SZABO

Laboratory of Chemical Physics
National Institute of Diabetes and Digestive
and Kidney Diseases
National Institutes of Health
Bethesda, Maryland 20892

INTRODUCTION

Nuclear magnetic relaxation studies of proteins, nucleic acids, and membranes can provide reliable information concerning the nature of the internal dynamics in these systems. This article reviews some developments[1,2] in the theory required to analyze such experiments that has allowed meaningful testing[3] of the predictions of a molecular dynamics simulation[4] on pancreatic trypsin inhibitor (PTI). In addition, we discuss some ambiguities that arise in connection with the analysis of relaxation studies[5,6] of acyl chain dynamics in lipid bilayers. These difficulties highlight the need for computer simulations to help one choose among different possible interpretations of the data.

To provide the necessary background, let us consider a ^{13}C NMR relaxation study of a methyl group in a protein in solution. The relaxation of the carbon nucleus is determined by the fluctuation of the carbon-hydrogen dipole-dipole interaction. This interaction is a random function of time because the orientation of a ^{13}C-H vector fluctuates with respect to the external magnetic field. These vectors reorient not only because the protein undergoes overall rotational diffusion, but also because the methyl group rotates about its symmetry axis, which in turn can "wobble" with respect to the protein. The observables (e.g., the inverse of the spin-lattice relaxation time, $1/T_1$) are determined by the Fourier transform [the spectral density, $J(\omega)$] of a time-correlation function [$C(t)$] evaluated at various combinations of the Larmor frequencies of the two nuclei. Since these frequencies depend on the magnetic field strength, it is important to perform multifield NMR studies. For a spherical protein with rotational correlation time τ_M, $C(t)$ is given by[1]

$$C(t) = e^{-t/\tau_M} \langle P_2[\hat{\mu}(0) \cdot \hat{\mu}(t)] \rangle \tag{1}$$

where $P_2(x)$ is the second Legendre polynomial and $\hat{\mu}$ is a unit vector along the ^{13}C-H bond as viewed from a coordinate system rigidly attached to the protein.

The usual approach to analyzing relaxation data is based on some model described by certain adjustable parameters (e.g., the rates and angular amplitudes of the internal motions). The correlation function $C(t)$ is evaluated within the framework of the model; it is Fourier-transformed to obtain the spectral density; and the parameters are

44

adjusted to maximize the agreement between the calculated and observed relaxation data (e.g., the T_1s). Although such analyses can be useful, there is the danger of overinterpretation. Physically realistic models have algebraically complicated spectral densities[7] that contain many adjustable parameters, so it is always possible that many different combinations of these parameters fit the data equally well. Moreover, the same data can often be described by equally reasonable but physically distinct models. Finally, it is sometimes possible to fit experimental data extremely well with a model that is clearly unphysical. For example, in a classic model,[8] a ^{13}C-H vector is assumed to rotationally diffuse about an axis which is rigidly attached to an isotropically reorienting protein. One would not expect such a model to adequately describe the relaxation of a methyl group whose symmetry axis is connected to the relatively rigid α-carbon backbone by several single bonds. This is generally the case when one adopts the standard tetrahedral geometry. However, by "fudging" this angle (i.e., using 120° instead of 109°), this model can work like a charm. Such considerations naturally raise a number of interesting questions. What are the essential common features of physically plausible but distinct models that allow them to reproduce experimental data equally well? How can a physically unrealistic model work so well? It must be doing something right. What is the unique information content of a set of relaxation data?

THE MODEL-FREE APPROACH TO THE INTERPRETATION OF NMR RELAXATION

When the internal motions are sufficiently fast, the answers to the above questions are remarkably simple. It can be shown[1] that in this limit the information content of multifield NMR relaxation data is completely specified by a generalized order parameter, $\mathcal{S}$, and an effective correlation time τ_e, which are measures of the spatial restriction and the time scale of motions, respectively. These quantities can be defined in a model-independent way and can be extracted from experimental data quite simply (see below). If a physical picture of the motion is desired, one can readily interpret the numerical values of these quantities in a variety of models. The common feature of physically distinct models that work equally well is that they are able to reproduce the same values of $\mathcal{S}$ and τ_e, albeit using different combinations of model parameters. This is what happened in the example discussed above involving the methyl group with the unrealistic geometry (i.e., the model had sufficient mathematical flexibility to do the job, but the model parameter turned out to be unphysical).

The square of the generalized order parameter is equal to the long time limit of the correlation function describing the internal motions[1]

$$\mathcal{S}^2 = \lim_{t \to \infty} \langle P_2[\hat{\mu}(0) \cdot \hat{\mu}(t)] \rangle$$
$$= \langle P_2(\cos \theta) \rangle^2 + 2|\langle C_{21}(\Omega) \rangle|^2 + 2|\langle C_{22}(\Omega) \rangle|^2 \qquad (2)$$

where Ω describes the orientation of $\hat{\mu}$ in a macromolecule-fixed frame, $C_{2m}(=(5/4\pi)^{1/2} \cdot Y_{2m})$ are modified spherical harmonics,[9] and the angular brackets denote average overall possible orientations accessible on a timescale shorter than that of the overall reorientation of the macromolecule. When the internal motion is isotropic, $\mathcal{S} = 0$, whereas when it is completely restricted, $\mathscr{S} = 1$. When the motion is azimuthally symmetric about an axis, Equation 2 simplifies to

$$\mathscr{S} = \langle P_2(\cos \theta) \rangle \equiv S \tag{3}$$

where S is the familiar order parameter that plays an important role in the analysis of solid-state NMR lineshapes. For example, the quadrupole splitting of a deuterated CD_2 group of a phospholipid in the liquid crystalline phase of a membrane is proportional to S if the lipid is undergoing axial rotation (θ in Equation 3 is the angle between the C-D bond and the normal to the membrane). The effective correlation time is defined as[1]

$$\tau_e(1 - \mathscr{S}^2) = \int_0^\infty [\langle P_2[\hat{\mu}(0) \cdot \hat{\mu}(t)] \rangle - \mathscr{S}^2] dt \tag{4}$$

i.e., it is proportional to the area under the correlation function.

How does one extract $\mathscr{S}$ and τ_e from experimental data? For an isotropically reorienting protein, the following simple expression for the spectral density,

$$J(\omega) = \frac{\mathscr{S}^2 \tau_M}{1 + (\omega \tau_M)^2} + \frac{(1 - \mathscr{S}^2) \tau}{1 + (\omega \tau)^2} \tag{5}$$

where $\tau^{-1} = \tau_M^{-1} + \tau_e^{-1}$ and τ_M is the overall correlation time, is used to calculate the observed quantities and the numerical values of $\mathscr{S}$ and τ_e are optimized. When the internal motions are sufficiently fast so as to be in the extreme narrowing limit, Equation 5 reduces to

$$J(\omega) = \frac{\mathscr{S}^2 \tau_M}{1 + (\omega \tau_M)^2} + (1 - \mathscr{S}^2) \tau_e \tag{6}$$

which is *exact* under the above conditions.[1] The extreme narrowing limit is reached when the internal motions are faster than the reciprocal of the Larmor frequencies (faster than 200 psec at currently available field strengths). When this condition is not fulfilled, Equation 5 serves as a useful approximation. The range of validity of this equation has been established by using it to analyze simulated data generated from sophisticated models of the internal dynamics.[1] It has been shown[2] that a wide variety of experimental data can be uniquely interpreted using the model-free approach.

In summary, the information content of *solution* NMR relaxation experiments concerning fast internal motions is limited to two quantities characterizing the timescale and spatial restriction of the motion. NMR studies in the solid state contain more information.[10,11] For example, in solution there is no simple way of establishing whether the reorientation of a methyl group is better described by unrestricted axial rotational diffusion or by discrete three-site jumps. In the solid state, on the other hand, even if the motion is in the extreme narrowing limit, the two models predict very different relaxation behavior of the lineshape.[10]

COMPARISON OF SIMULATIONS WITH EXPERIMENT

Pancreatic trypsin inhibitor is a small protein that has been studied extensively using both NMR[12] and molecular dynamics[3] techniques. Wüthrich and coworkers[12] assigned the resonances of twelve methyl groups and determined their relaxation parameters at two magnetic fields. Karplus and McCammon[4] in their pioneering work performed a 96-psec molecular dynamics simulation of this protein in vacuum. Since they incorporated the hydrogens into the heavy atoms, it is not possible to evaluate the correlation function in Equation 1 for methyl groups from their trajectory. This obstacle can be circumvented as follows.[3] Using the model-free approach, values of $\mathscr{S}$ and τ_e are extracted from the relaxation data.[2] These values reflect motion of and about the symmetry axis of the methyl groups. The resulting τ_e values (19–70 psec) are clearly in the extreme narrowing limit and hence the model-free spectral density is essentially exact for this system. If it is assumed that the motions of and about the methyl symmetry axis are uncoupled, one can calculate the generalized order parameter for the motion of this axis ($\mathscr{S}_{axis}$) from $\mathscr{S}$ using the geometry of the methyl group. Since it involves a bond between heavy atoms, $\mathscr{S}_{axis}$ can then be calculated from the trajectory via Equation 2.

A comparison[3] of the theoretical and experimental order parameters shows that the relative flexibility of the various residues is quite well described by the simulation. However, with two exceptions, the experimental order parameters are smaller than the theoretical ones, indicating that the angular amplitudes are *under*estimated in the simulation. There are several possible reasons for this, including the quality of the potential functions, the neglect of solvent and the finite length of the trajectory (i.e., the simulation may have explored only a few of the conformational minima accessible to the protein). To get a feeling for the plausibility of the last explanation, consider an idealized situation in which a bond vector can interconvert between two discrete, but energetically equivalent, orientations. If the internal correlation function decays with a time constant of 25 psec, it can be shown that the conditional probability of starting and remaining in one of the sites is $\exp(-t/50 \text{ psec})$, so that there is a 14% chance that a 100-psec trajectory never even explored the other site. To obtain the correct order parameter, the trajectory would have to spend the same time in both sites.

It is interesting to note that if the analogous comparison is made between the methyl carbon temperature factors calculated from the simulation and those obtained from refinement of the X-ray data, the situation is reversed (i.e., the simulation appears to predict that the protein is too flexible). A possible resolution of this apparent paradox is that refinement procedures may underestimate the root-mean-square amplitudes, as recently discussed by Kuryan et al.[13] However, it must be kept in mind that order parameters and temperature factors do not reflect motional flexibility in identical ways and that motional amplitudes may differ in solution and solid phases.

DYNAMICS OF LIPID BILAYERS

The most extensive multifield NMR relaxation study in the literature is that of Brown and coworkers[5,6] on phospholipid methylenes in bilayer vesicles. These authors determined the ^{13}C spin-lattice relaxation times (T_1) as a function of acyl chain segment position at *seven* magnetic field strengths corresponding to (proton) spectrometer frequencies ranging from 60 to 500 MHz. The T_1s reflect exclusively internal motions of the phospholipid including *trans-gauche* isomerations, axial rotation of the entire molecule, and so-called director fluctuations (i.e., deviations of the instantaneous long axis of the lipid from the macroscopic bilayer normal). The overall rotational motion of the vesicle is much too slow to contribute significantly. The T_1s were found to depend on the spectrometer frequency, indicating that at least some of the internal motions are slow on the NMR time scale (i.e., *not* in the extreme narrowing limit). As shown in FIGURE 1, the reciprocals of the T_1s are proportional to the inverse square root of the spectrometer frequency within experimental error. Such behavior is predicted by a model that involves not only fast local motions but also long-range *collective* director fluctuations resulting from elastic deformations of the membrane. In this model, a methylene segment ($[CH_2]_n$) is described by non-

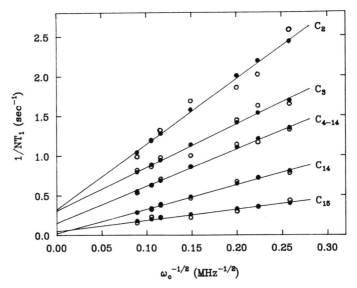

FIGURE 1. The reciprocal of the spin-lattice relaxation times of $(CH_2)_n$ segments of 1,2-dipalmitoylphosphatidylcholine at 50°C as a function of the inverse square root of the carbon Larmor frequencies. The *open circles* are the experimental result; the *solid line* is the linear least-squares fit of these. The *solid circles* are calculated by fitting the experimental results using the Lorentzian-like spectral density in Equation 8.

Lorentzian spectral density of the form

$$J_n(\omega) = B_n\omega^{-1/2} + \tau_n \qquad (7)$$

where τ_n is the correlation time for fast local motions. When this equation is used to fit the data, it is found[5,6] that B_n is proportional to the square of the corresponding order parameter obtained from deuterium NMR lineshapes (i.e., S_n^2).

The model-free approach was developed to analyze relaxation behavior of macromolecules having relatively slow, unrestricted overall motion and fast restricted internal motions. Thus, one cannot expect *a priori* that it is applicable to situations where relaxation is due exclusively to slow and fast restricted internal motions. Nevertheless, let us see what a spectral density of the model-free form can do. We assume that each $(CH_2)_n$ segment can be described by

$$J_n(\omega) = \frac{A_n^2 \tau_s}{1 + (\omega\tau_s)^2} + (1 - A_n^2) \tau_n \tag{8}$$

(c.f. Equation 6). Note that the slow-motion correlation time (τ_s) is taken to be the same for all segments. The resulting fit is shown in FIGURE 1 (solid circles). It is remarkable that the calculated values, which are of course obtained from the Lorentzian-like spectral density given in Equation 8, are almost perfectly described over the range of spectrometer frequencies by a spectral density with $\omega^{-1/2}$ dependence. Thus, even if one exploits the full range of commercially available magnetic field strengths, one cannot choose between Equations 7 and 8.

The values of τ_n in Equation 8 obtained from the fit were between 7 and 42 ps; the numerical values of A_n^2 were similar to the squares of the deuterium order parameters (S_n^2). The slow correlation time was 1.8 nsec. What is the nature of this motion that is common to all segments and is responsible for the frequency dependence of the T_1s? This is a nontrival problem because, as mentioned previously, there is not an *a priori* justification for using a spectral density of the model-free form. A satisfactory answer to the above question must account for the empirical result that A_n^2 is similar to S_n^2. An attractive possibility is that the slow motion is simply the axial rotation of the phospholipid.

Consider a spherical vesicle with rotational correlation time τ_V. Let the axial rotation diffusion coefficient of the lipid be $D_\parallel$ $(\tau_\parallel = D_\parallel^{-1})$ and suppose that local motion (isomerization, torsion) of the nth methylene group has correlation time τ_n. Assuming that (1) the axial and local motions are uncoupled, (2) $\tau_V \gg \tau_\parallel \gg \tau_n$ and (3) τ_n is in the extreme narrowing limit, one can show that the corresponding spectral density of $(CH_2)_n$ is

$$J_n(\omega) = \frac{S_n^2 \tau_v}{1 + (\omega\tau_v)^2} + \frac{2|\langle C_{21}\rangle_n|^2 \tau_\parallel}{1 + (\omega\tau_\parallel)^2} + \frac{2|\langle C_{22}\rangle_n|^2 \tau_\parallel/4}{1 + (\omega\tau_\parallel/4)^2} + (1 - \mathscr{S}_n^2)\tau_n \tag{9}$$

where S_n (see Equation 3) and $\mathscr{S}_n$ (see Equation 2) are the axially symmetric and generalized order parameters describing segmental motion, respectively. Since τ_V is very slow, the first term in Equation 9 does not contribute to the T_1. At first sight, this spectral density appears inconsistent with the one that has been used to fit the data (Equation 8). Recall that empirically A_n^2 was found to be proportional to S_n^2, but S_n^2 in Equation 9 modulates the contribution from the overall rotation of the vesicle rather than the terms arising from the supposedly slow axial rotation. The only way out of this difficulty is to assume that one or both of the coefficients $2|\langle C_{21}\rangle_n|^2$ and $2|\langle C_{22}\rangle_n|^2$ happen to be similar to S_n^2. Since one has little physical intuition about these, there is no real basis for making such an assumption. This is clearly a situation where computer simulations can play an important role. In collaboration with Richard Pastor and Martin Karplus, Brownian simulations of the lipid dynamics, in the framework of the mean-field Marcelja[14] model of segmental motion, are currently in progress. Preliminary estimates of the equilibrium averages $2|\langle C_{21}\rangle_n|^2$ and $2|\langle C_{22}\rangle_n|^2$ obtained by Pastor, using the Marcelja model, show that $2|\langle C_{21}\rangle_n|^2$ is very small and that magnitude of $2|\langle C_{22}\rangle_n|^2$ is similar to that of S_n^2. The results of the Brownian dynamics simulations should prove to be invaluable in testing the above conjecture concerning the importance of axial motion. They may also suggest alternate explanations.

REFERENCES

1. LIPARI, G. & A. SZABO. 1982. J. Am. Chem. Soc. **104:** 4546.
2. LIPARI, G. & A. SZABO. 1982. J. Am. Chem. Soc. **104:** 4559.
3. LIPARI, G., A. SZABO & R. M. LEVY. 1982. Nature **300:** 197.
4. KARPLUS, M. & J. A. MCCAMMON. 1979. Nature **277:** 578.
5. BROWN, M. F., A. A. RIBEIRO, & G. D. WILLIAMS. 1983. Proc. Natl. Acad. Sci. USA **80:** 435.
6. BROWN, M. F. 1984. J. Chem. Phys. **80:** 2808.
7. WITTEBORT, R. & A. SZABO. 1978. J. Chem. Phys. **69:** 1722.
8. WOESSNER, D. E. 1962. J. Chem. Phys. **36:** 1.
9. BRINK, D. M. & G. R. SATCHLER. 1962. Angular Momentum. Oxford University Press. London.
10. TORCHIA, D. A. & A. SZABO. 1982. J. Magn. Reson. **49:** 107.
11. TORCHIA, D. A. & A. SZABO. 1985. J. Magn. Reson. **64:** 135.
12. RICHARZ, R. K., K. NAGAYAMA & K. WÜTHRICH. 1980. Biochemistry **19:** 5189.
13. KURYAN, J., G. PETSKO, R. M. LEVY & M. KARPLUS. 1986. J. Mol. Biol. In press.
14. MARCELJA, S. 1974. Biochem. Biophys. Acta **367:** 165.

Molecular Dynamics Simulation Study of Polypeptide Conformational Equilibria: A Progress Report[a]

AMIL ANDERSON, MIKE CARSON,[b]
AND JAN HERMANS

Department of Biochemistry
School of Medicine
University of North Carolina
Chapel Hill, North Carolina 27514

INTRODUCTION

The recent development of both more reliable interatomic forcefields and of new forcing-potential methods has extended the scope of molecular dynamics simulations to a variety of interesting new problems of relevance to biological structure and function. Such problems include the affinity of small molecules as solutes for water and as substrates for protein molecules, conformational equilibria, and the transitions between states. Each of these plays its role in more complex processes, such as the approach of substrate to enzyme. In these problems an understanding of equilibria, i.e., *free* energy differences, is critical.

We are using molecular dynamics simulations to study the equilibria of the polypeptide backbone. We here report a study of the transition between two equilibrium conformations of the alanine dipeptide, acetyl-alanine methylamide, in aqueous solution, which uses the recently developed simple point charge model for water-protein interactions.[1,2] The alanine dipeptide has been the subject of a great many analyses by molecular mechanics and quantum mechanics. The two main degrees of conformational freedom of the molecule are the internal rotations about the two "backbone" bonds, i.e., the $N-C_\alpha$ and $C_\alpha-C$ bonds, for which the dihedral angles are designated as ϕ and ψ. Roughly speaking, conformation space divides into three areas. The largest, centered at $\phi = -120°$ and $\psi = 120°$, is the backbone conformation of polypeptides in the extended or β conformation. In this conformation the dipole moments of successive peptide groups are antiparallel. The other two conformations are the right- and left-handed α conformations, α_R and α_L, in which these dipole moments are parallel. For the α_R conformation the angles ϕ and ψ are both near $-70°$, whereas for the α_L conformation they are near $+70°$. In the alanine dipeptide, the

[a]This work was supported by a research grant from the National Science Foundation (NSF) (DMB-850-1037). Computations were supported by National Institutes of Health (NIH) Research Resource Grant RR-00898 to the UNC Molecular Graphics Laboratory, a grant of supercomputer time from the NSF, and recent equipment grants from the NSF and NIH for computer and computer graphics equipment.
[b]Present address: University of Alabama, Birmingham, Alabama.

former is found to be the most stable of these two, but less stable than the β conformation. The work of Flory and coworkers has shown that end-to-end distances of randomly coiled polypeptides are quite sensitive to the relative frequency of the α conformation.[3,4] Measured values suggest that the α_R conformation is about 10 times less prevalent than the β conformation. Analysis of experiments that report on the conformation of the alanine dipeptide in water[5] indicate a mixture of β and α conformations.

A molecular dynamics simulation of the dipeptide in water has been done by Rossky and Karplus.[6] In this study only the β conformation was simulated. Beveridge and coworkers[7] have studied the same system with objectives similar to ours, albeit with different methods. Their results are discussed at the end of this paper.

METHODS

The simulations were performed on a molecule of alanine dipeptide in a cubic volume 13.3 Å on a side containing in addition 73 water molecules. Nonbonded interactions were computed within a cutoff of 6 Å, always for pairs of electrically neutral groups (water molecules, and NH, CO, CH and CH_3 groups of the dipeptide). The CH and CH_3 groups were treated as united atoms, with use of less repulsive parameters for 1–4 interactions within the dipeptide molecule.[2] The nonbonded pairlist was updated every 20 steps in some and every 50 steps in other calculations. This choice did not appear to have a significant effect.

All bond lengths and the water bond angle were constrained to ideal values with the SHAKE algorithm.[8,10,11] Bonded geometry of the dipeptide and force constants for geometric deformation were as used before.[12,13] The volume of the system was constant during the simulations, while the temperature was constantly relaxed to a target value of 300 K as described by Berendsen et al.[8,14]; this implies use of the leapfrog algorithm. Routines for the SHAKE and molecular dynamics algorithms were kindly provided by H.J.C. Berendsen and W. van Gunsteren and incorporated by us into the CEDAR program.[15] In the simulations reported here the time step was 1 fsec, and the relaxation time for application of the constant-temperature algorithm was 50 fsec.

A cosine forcing potential was applied to change the dihedral angle ψ from around $110°$ to $-50°$, and back. The potential obeyed the equation

$$U_f = K \left[1 - \cos (\psi - \psi_0) \right] \approx (K/2)(\psi - \psi_0)^2 \tag{1}$$

The minimum of the potential is located at ψ_0, and is changed between two limits, ψ_1 and ψ_2, according to

$$\psi_0 = \psi_1 + \lambda(\psi_2 - \psi_1) \tag{2}$$

where λ gradually changes from 0 to 1, or from 1 to 0 in the reverse simulation. The free energy change for the transition follows from[16–18]

$$\Delta A = \int (\delta U_f/\delta \lambda) \, d\lambda \tag{3}$$

and is therefore equal to the mean value of the "force" due to the applied potential, as λ varies.

Corrections for the effect of applying the forcing potential were made by calculating the free energy change for imposing the potential in two different ways. The first consisted of performing a simulation in which the potential was changed according to

$$U_f = \lambda(K/2)(\psi - \psi_0)^2 \qquad (4)$$

with λ varying gradually from 0 to 1, or from 1 to 0, and again calculating the change of the free energy as the average of the "force" $\delta U_f/\delta\lambda$. In the other method we used results of an unrestrained simulation to estimate ΔA from the mean value of the Boltzmann factor for U_f over the sample of conformations in the unperturbed trajectory. This follows from

$$\Delta A = -kT \ln z_f \qquad (5)$$

where

$$z_f = \int \exp(-U_f/kT) \exp(-U/kT) \, dx / \int \exp(-U/kT) \, dx \qquad (6)$$

and the fact that the conformations are represented in the trajectory in proportion to $\exp(-U/kT)$, with U the unperturbed conformational energy. These methods gave similar answers.

In addition to forcing a change of ψ, we restrained ϕ to be near $-70°$ during the simulations, and applied corrections corresponding to this restraint to the values of ΔA. The purpose of this restraint on ϕ was to restrict the trajectory to where the energy map for this potential (*in vacuo*) shows the low-energy path between the two conformations to be, and thereby perhaps to facilitate convergence to a reversible procedure. At this time there are insufficient data to determine whether this premise was correct. Since calculations for the dipeptide molecule in which the conformation was changed in small steps by rigid-body internal rotation about the N—C_a and C_a—C bonds indicated the virtual absence of an energy barrier between the α_R and β conformations, one additional potential was used to retain the dipeptide in the α conformation for *in vacuo* simulations of the dipeptide, in which the forcing potential was being imposed or removed. (This potential was flat everywhere, except in two narrow intervals, $-80° < \psi < -75°$ and $-5° < \psi < 0°$, in which it produced an energy difference of 54 kJ/mol. In retrospect, a choice of $-90°$ as the lower limit would have been preferable.)

The value of K used for the forcing potential and the restraint of ϕ was 84 kJ/mol.

Calculations were carried out on three different computers. The first part of this work was done with a VAX 11/780 computer running under the UNIX operating system, the longest simulations were done on a Cray I, and some recent work uses a Masscomp 500 series machine. The Masscomp has two Motorola 68000-based central processors, a floating point processor and a small array processor. (The small memory of the array processor, 16,000 words, is a limitation on its optimal use.) The approximate relative speed of the CEDAR program on these machines is

VAX 11/780 (VMS operating system) = 2
VAX 11/780 (UNIX operating system) = 1
Masscomp 500 without array processor = 1
Masscomp 500 with array processor = 4
Cray I after optimization of water-water force calculation = 100

RESULTS

Dipeptide in Vacuo

As a test, our methods were first applied to the dipeptide *in vacuo*. In the left part of FIGURE 1 are shown plots of $\langle\psi\rangle - \psi_0$, the difference between the actual value of ψ and the minimum of the forcing potential, as a function of ψ_0. The difference between the curves corresponds to the different lengths of the simulations required to vary λ from 0 to 1, or back. As this time was increased, reversibility of the process improved, as evidenced by a gradual achievement of equality of the "forces" exerted, respectively, on and by the system in forward and reverse simulations.

Except for constant factors, the plots of FIGURE 1 are very nearly the same as plots of $\delta U_f/\delta\lambda$ versus λ, and therefore the area under the curves corresponds to ΔA for the transition.

The right half of FIGURE 1 shows a plot of the free energy change during the

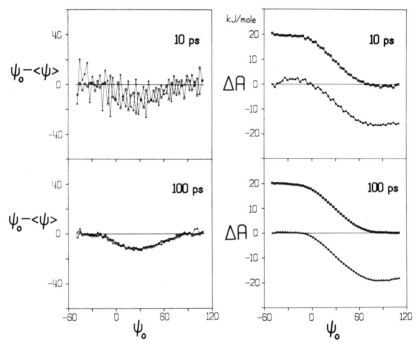

FIGURE 1. (*Left*) difference between the mean observed value of the dihedral angle ψ and the (mean) value of the center of the forcing potential over short intervals during the simulations of the alanine dipeptide *in vacuo*. Different panels correspond to different durations of the total simulated time, as indicated. *Triangles* or *squares* correspond to simulations in which the center of the forcing potential was increasing or decreasing, respectively. (*Right*) free energy of the restrained dipeptide as a result of displacement of the forcing potential, calculated by integration of the differences of FIGURE 1.

TABLE 1. Free-Energy Changes for the Imposition of the Forcing Potential, Displacement of Its Center between the Two Extremes of ψ, and Its Removal (and the Reverse Processes) for Alanine Dipeptide *in Vacuo*[a]

	Forward	Reverse
Imposition at $-50°$[b]	8.95	-9.20
Displacement from $-50°$ to $110°$	-22.03	20.90
Removal at $110°$[b]	-6.22	7.93
Net free energy difference $\alpha_R \rightarrow \beta$:	-19.3	19.6

[a]Values in kJ/mol.
[b]From perturbed trajectories, with Equations 3 and 4.

simulations in which ψ was changed. The calculated values of ΔA for this reaction, as well as those for imposition of the restraint at the endpoints, are given in TABLE 1, together with the consequent net value of ΔA for the transition of the unrestrained dipeptide from the α_R to the β conformation. One notes that the difference is very large; one will furthermore notice that the "force" has the same sign during the entire trajectory, except at the two ends, where it is zero (FIG. 1, left side). Thus, in this model, the α_R conformation is not separated from the β conformation by a significant free energy barrier (FIG. 1, right side). The free energy profile for the α_R to β transition is similar to the *energy* profile one calculates for the rigid dipeptide with use of this same potential.

We believe that these results for the dipeptide *in vacuo* actually have little relevance to physical reality, for the simple reason that the forcefield used to calculate the potential energy was specifically developed for liquid water and extended to polypeptides in water, using other results[19] appropriate for the condensed, predominantly hydrogen-bonded phase. In such a phase, dipole moments are increased; thus the dipole moment of the SPC water model is 2.3 debye,[1] higher by 0.5 debye than the experimental value for a water molecule *in vacuo*. Using this potential *in vacuo*, one obtains highly artificial estimates of the energy. The large difference in energy for the α_R and β conformations of the dipeptide is almost entirely due to the electrostatic energy for the peptide group dipoles, which are antiparallel in the former, and parallel in the latter conformation. The former conformation also allows a close approach of polar H and O atoms. In a medium of low dielectric constant, these groups would not be so strongly polarized, and the energy difference smaller. In a medium of high dielectric constant, the electric interaction would be screened, and the difference again smaller. How much smaller will be seen in the following section, where we report our results for simulation of the dipeptide in a bath of water molecules.

Simulations in Water

FIGURE 2 shows the variation of $\Delta\psi$ with ψ_0, and the corresponding free energy change obtained by integration of this difference, according to Equations 1 and 3. The free energy difference between the two conformations is seen to be much less than *in vacuo*. Furthermore, the two conformations are separated by a free energy barrier of 4.5 kJ/mol. TABLE 2 shows the free energy differences for the three component

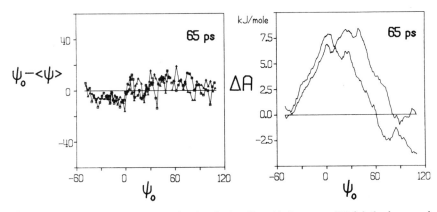

FIGURE 2. Same as FIGURE 1, but for the alanine dipeptide in water. (*Right*) the *lower* and *upper curves* correspond to increasing or decreasing ψ_0, respectively.

processes, i.e., (1) imposition of the restraint potential at one extreme value of ψ, (2) change of the potential minimum to the other extreme, and (3) removal of the restraint, as well as their sum.

It will be noted that the results for the forward and reverse transitions do not coincide quite as well as those for the same system *in vacuo*. In part, the difference is only apparent and due to the difference in scales of ΔA in FIGURES 1 and 2. However, we attribute part of the poorer reversibility to the behavior of the system in one particular interval of the simulation, namely, when the value of ψ_0 is about halfway between the extremes. Separate simulations show that relaxation of the restraining force is here exceptionally slow, taking several psec. Consequently, the restrained trajectory was obtained in three parts, in which the location of the minimum of the forcing potential was moved at different rates. For the parts nearest the starting and ending conformations, the rate was 0.2 psec/degree, whereas for the central one-half of the trajectory the rate was 0.6 psec/degree. (Given larger computer power than presently available to us, we would improve the accuracy of these results by doing the calculation at an even slower rate, and also by increasing the cutoff distance for nonbonded interactions and the number of independent solvent molecules.)

We have also computed two unrestrained trajectories for the alanine dipeptide in

TABLE 2. Free-Energy Changes for the Imposition of the Forcing Potential, Displacement of Its Center between the Two Extremes of ψ, and Its Removal (and the Reverse Processes) for Alanine Dipeptide in Water[a]

	Forward	Reverse
Imposition at $-50°$[b]	5.4	
Displacement from $-50°$ to $110°$	-3.9	0.8
Removal at $110°$[b]	-6.8	
Net free energy difference $\alpha_R \rightarrow \beta$:	-5.3	$+0.6$

[a]Values in kJ/mol.
[b]From unperturbed trajectories, with Equations 5 and 6.

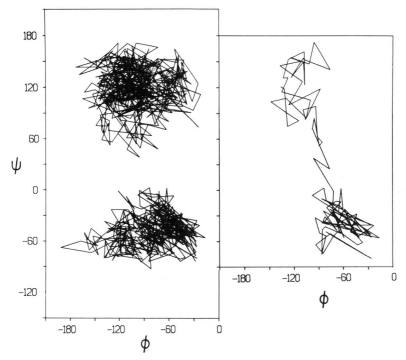

FIGURE 3. (*Left*) trajectories from two free simulations of the alanine dipeptide in water, started in different parts of conformation space. The simulation reported in the *upper portion* (β region) was more than 100 psec, and that reported in the *lower portion* (α_R region) is shown for the 63-psec interval before it crossed into the β region. (*Right*) the trajectory when it crosses the free energy barrier. Successive points are 2 psec apart.

water, starting near different minima of the free energy. Each trajectory covers 100 psec. One spontaneous crossing of the free energy barrier occurred in these trajectories, i.e., from the α_R to the β region, after 63 psec of the simulation that started in the α_R conformation. The left side of FIGURE 3 shows these trajectories (the α trajectory only up to the barrier crossing). The right side shows the trajectory in and around the crossing event.

As we have seen, the barrier is not very high and therefore a spontaneous crossing will occasionally be observed in a long trajectory. We have obtained some additional results that would indicate that the occurrence of a barrier crossing may be influenced by parameters that can affect the accuracy of the simulation, such as number of dynamics steps between recalculation of the nonbonded pairlist and the choice of dynamics time step. This problem is still under investigation.

CONCLUSION

The free energy difference for the α and β conformations of the dipeptide, calculated according to the SPC potential, is close to zero; the β conformation is more stable than the α_R conformation by a factor of roughly 3. As mentioned in the INTRODUCTION, experiments suggest that in proteins this ratio is circa 10. In our opinion the agreement between these two values is good, given the limitations of the alanine dipeptide as a model of the polypeptide backbone, and the limitations of molecular modeling, and especially of the interatomic potentials.

Beveridge et al.[7] have used Monte Carlo sampling methods, with a series of discrete umbrella potentials, to study the alanine dipeptide in water and in a vacuum, using a similar, albeit not identical, potential. With their method, they obtained an estimate of the difference of the free energy of hydration for different rigid conformations of the dipeptide molecule. These values had to be corrected for differences in the conformational energy and entropy, in order to obtain an estimate of the conformational equilibrium. In the calculation reported here, this overall free energy difference is, of course, obtained directly. (In order to compare the results of the two studies, one must recognize that the C_7, C_5 and P_{II} conformations distinguished by Beveridge et al. are in this work treated together as the β conformation. Indeed, the trajectory of FIGURE 3 does not show any features that would justify a distinction between these three conformations for the dipeptide in water.) Their values for ΔA translate into a net difference between β and α_R conformations of $+2.5$ kJ/mol. Thus, the values of the overall free energy difference for the two conformations obtained in these two studies agree well, which is in itself an interesting result, since the potential functions used in the simulations were not the same.

This study is being continued with a larger system, consisting of several alanine residues with emphasis on conformational equilibria of the terminal residue(s) and their dependence on the conformation of the remaining residues.

ACKNOWLEDGMENTS

We thank Drs. Mezei and Beveridge for communicating their unpublished results.

REFERENCES

1. BERENDSEN, H. J. C., J. P. M. POSTMA, W. F. VAN GUNSTEREN & J. HERMANS. 1981. *In* Intermolecular Forces. B. Pullman, Ed. Reidel, Dordrecht.
2. HERMANS, J., H. J. C. BERENDSEN, W. F. VAN GUNSTEREN & J. P. M. POSTMA. 1984. Biopolymers **23:** 1513–1518.
3. FLORY, P. J. 1969. Statistical Mechanics of Chain Molecules, Interscience. New York, NY.
4. BRANT, D. A. & P. J. FLORY. 1965. J. Am. Chem. Soc. **87:** 663; **87:** 2791.
5. MADISON, V. & K. D. KOPPLE. 1980. J. Am. Chem. Soc. **102:** 4855–4863.
6. ROSSKY, P. & M. KARPLUS. 1979. J. Am. Chem. Soc. **101:** 1913–1937.
7. BEVERIDGE, D. L., G. RAVISHANKER, M. MEZEI & B. GEDULIN. 1985. *In* Biomolecular Structure and Dynamics III. H. S. Ramaswami & M. H. Sarma, Eds. Adenine Press. Albany, NY.
8. VAN GUNSTEREN, W. F. & H. J. C. BERENDSEN. 1985. *In* Hermans[9]: 5–14.
9. HERMANS, J., Ed. 1985. Molecular Dynamics and Protein Structure. Polycrystal Book Service. Western Springs, IL.
10. VAN GUNSTEREN, W. F. & H. J. C. BERENDSEN. 1977. Mol. Phys. **34:** 1311–1327.

11. RYCKAERT, J. P., G. CICCOTTI & H. J. C. BERENDSEN. 1977. J. Comput. Phys. **23**: 327–341.
12. SCHERAGA, H. A. 1968. Adv. Phys. Org. Chem. **6**: 103–184.
13. FERRO, D. R., J. E. MCQUEEN, J. T. MCCOWN & J. HERMANS. 1980. J. Mol. Biol. **136**: 1–18.
14. BERENDSEN, H. J. C., J. P. M. POSTMA, W. F. VAN GUNSTEREN, A. DI NOLA & J. R. HAAK. 1984. J. Chem. Phys. **81**: 3684–3690.
15. CARSON, M. & J. HERMANS, 1985. *In* Hermans[9]: 165–166.
16. BERENDSEN, H. J. C., J. P. M. POSTMA & W. F. VAN GUNSTEREN. 1985. *In* Hermans[9]: 43–46.
17. BEVERIDGE, D. L. & M. MEZEI. 1985. *In* Hermans[9]: 53–57.
18. MEZEI, M. & D. L. BEVERIDGE. 1986. Ann. N.Y. Acad. Sci. This volume.
19. LIFSON, S., A. T. HAGLER & P. DAUBER. 1979. J. Am. Chem. Soc. **109**: 5111–5141.

Conformational Energy Calculations on Polypeptides and Proteins: Use of a Statistical Mechanical Procedure for Evaluating Structure and Properties[a]

HAROLD A. SCHERAGA AND GREGORY H. PAINE

Baker Laboratory of Chemistry
Cornell University
Ithaca, New York 14853-1301

INTRODUCTION

For many years, we have been carrying out experimental work to gain an understanding of how interatomic interactions (and intermolecular solute-solvent interactions) influence protein structure, protein folding, and higher-order structures (e.g., enzyme-substrate complexes).[1,2] About twenty years ago, we combined this experimental approach with a theoretical one[3] to achieve the same goal; the theoretical work involved the development of procedures to generate polypeptide chains and compute their (empirical) conformational energies. This research has been reviewed on several occasions, the most recent reviews appearing in References 2 and 4–10. In this article, we shall concentrate on the theoretical aspects of some of these problems.

The theoretical work has included empirical predictions schemes,[11–13] statistical-mechanical Ising-model treatments,[14–18] energy minimization,[19–21] Monte Carlo,[22–24] and molecular dynamics[25] procedures. Recently, a new statistical-mechanical procedure (to be summarized herein) has been introduced.[26] This methodology is computer-intensive and has benefited from the development of array processors[27] and supercomputers[28] based on parallelization of the computations. The large scale of the computations arises from the multiple-minima problem, i.e., the existence of many local minima in the complex, multidimensional energy surface, and the need to explore a large enough portion of the surface in order to surmount local barriers and reach the global minimum. The reader is referred to the original papers for the details.

SOME EARLIER WORK

Our early work on the role of the solvent (i.e., water) involved a statistical-mechanical treatment of a model of an aqueous solution of nonpolar solutes to treat the hydrophobic interaction.[29] Bimolecular ionic association was treated later.[30] With the development of empirical potential functions, this earlier treatment was replaced by

[a]This work was supported by research grants from the National Institutes of Health (No. GM-14312) and the National Science Foundation (No. DMB84-01811).

simulation procedures involving Monte Carlo[23] and molecular dynamics[25] methods. Problems of convergence in applying these procedures to aqueous solutions, which have not yet been completely surmounted, have been pointed out.[23,25,31] The results of these simulations, together with empirical data, were used to "parameterize" a solvent-shell model to treat the hydration of polypeptides and proteins[32-34] during the course of conformational energy calculations on these systems.

The computational method has been applied to a variety of systems to provide an understanding of the observed behavior. These include: rationalization (and prediction) of the handedness (twist) of α-helices[35-37] and β-sheets,[38-42] the helix-coil[43-45] and helix-helix[46,47] transition curves *in water* and in the solid state, the packing of α-helices with each other[48,49] and with β-sheets,[50] the packing of two β-sheets,[51] the formation of the triple-stranded form of collagen[52-55] and the packing of such triple strands,[40,56,57] the conformations of gramicidin S,[58,59] enkephalin[21,26,60] and the 20-residue membrane-bound portion of melittin,[61] the refined X-ray structures of proteins,[62] the conformations of homologous globular proteins (α-lactalbumin[63] and several neurotoxins[64]), the use of spatial geometric arrangements of loops,[65-67] distance constraints[68] and differential geometry[69] in minimizing the energies of globular proteins, and the structures of enzyme-substrate complexes.[7,9,70,71]

STATISTICAL MECHANICAL PROCEDURE WITH AN ADAPTIVE IMPORTANCE SAMPLING MONTE CARLO ALGORITHM

We recently presented a new method for determining the native, three-dimensional structure of a polypeptide.[26] This method is based on the calculation of the probable conformation of a polypeptide chain by the application of equilibrium statistical mechanics in conjunction with an adaptive, importance sampling Monte Carlo algorithm.

A polypeptide chain, with a specific amino acid sequence of N residues, is assumed to exist as an ensemble of conformations over possible energies. The partition function can be written in terms of internal variables as

$$Z(\beta) \equiv \int \ldots \int \exp\left[-\beta U\{\phi, \psi, \tilde{\chi}\}\right] d\{\phi, \psi, \tilde{\chi}\} \tag{1}$$

where $\beta = (RT)^{-1}$, R is the gas constant, T is the absolute temperature, and $U\{\phi, \psi, \tilde{\chi}\}$ is the conformational energy of a rigid-model polypeptide chain as a function of the dihedral angles $\{\phi, \psi, \tilde{\chi}\}$, where the latter is a shorthand notation for all of the backbone (ϕ, ψ) and side-chain (χ) dihedral angles. This partition function is then used to compute various properties of the system. It was first employed[26] to compute $P_i(\phi_i, \psi_i)$, the single-residue probability density of the backbone structure for the ith residue averaged over all other dihedral angles of the chain, with all side-chain dihedral angles held fixed (see below), i.e.,

$$P_i(\phi_i, \psi_i) \equiv \frac{1}{Z(\beta)} \int \ldots \int \exp\left[-\beta U\{\phi', \psi', \tilde{\chi}\}\right] d\{\phi', \psi', \tilde{\chi}\} \tag{2}$$

where the primes indicate that, during the evaluation of the conformational energy, the backbone dihedral angles of the ith residue are fixed at ϕ_i and ψ_i. A similar definition for the probability density of the side-chain dihedral angles of the ith residue can easily be formulated. For convenience of interpretation, the probability is converted into a "conditional free energy" by the relation

$$A_i(\phi_i, \psi_i) = -RT \ln [P_i(\phi_i, \psi_i)] \qquad (3)$$

The quantity $U\{\phi', \psi', \tilde{\chi}\}$ is computed by summing over the interaction energies[19,20] of all pairs of nonbonded atoms of the *whole* molecule. Equations 2 and 3 are used to compute conditional free energy ϕ–ψ contour maps, similar to standard conformational-energy ϕ–ψ contour maps, for each residue of the polypeptide. The probable conformation of the ith residue is taken as the one of lowest conditional free energy, and the probable conformation of the whole polypeptide chain is assumed to be the combination of the probable conformations of the individual residues.

Besides calculating the probable conformation, it is also possible to compute the average conformation[72] over the whole ensemble. For this purpose, we first compute the average values of ϕ_i and ψ_i for the ith residue as

$$\langle \phi_i \rangle = \frac{1}{Z(\beta)} \int \dots \int \phi_i \exp [-\beta U\{\phi, \psi, \tilde{\chi}\}] d\{\phi, \psi, \tilde{\chi}\} \qquad (4)$$

$$\langle \psi_i \rangle = \frac{1}{Z(\beta)} \int \dots \int \psi_i \exp [-\beta U\{\phi, \psi, \tilde{\chi}\}] d\{\phi, \psi, \tilde{\chi}\} \qquad (5)$$

The average conformation of the polypeptide chain is then taken as the combination of the average values of ϕ and ψ for all individual residues. The native structure is assumed to be equivalent to the conformation of highest probability, while the average conformation corresponds to the experimentally determined structure.

A statistical-mechanical approach was used previously by taking only nearest-neighbor (Ising-model type) interactions into account.[17] However, when interactions over the whole molecule are included, as is done in the procedure discussed here,[26,72] it is necessary to use a Monte Carlo procedure.

For reasons discussed elsewhere,[26] there are deficiencies in the use of the Metropolis[73] Monte Carlo procedure that has been used in the past to evaluate integrals of the type appearing in Eqs. 1, 2, 4 and 5. We have overcome these deficiencies[26,72] by evaluating these integrals with an adaptive, importance sampling technique.[74] We refer to this Monte Carlo algorithm as SMAPPS (Statistical Mechanical Algorithm for Predicting Protein Structure).

SMAPPS searches the entire conformational space, adjusting itself automatically to concentrate its sampling in regions where the magnitude of the integrand is largest ("importance sampling"). Briefly, the algorithm randomly selects M conformations $\{\xi_\alpha, \eta_\alpha, \tilde{\zeta}_\alpha\}$, of the polypeptide chain based on a probability density function $g\{\phi, \psi, \tilde{\chi}\}$ (see Eq. 27 of Reference 26). The quantities in Eqs. 1, 2, 4 and 5, after M evaluations of the integrand, are then given as

$$Z(\beta) \simeq \frac{1}{M} \sum_{\alpha=1}^{M} \frac{\exp [-\beta U\{\xi_\alpha, \eta_\alpha, \tilde{\zeta}_\alpha\}]}{g\{\xi_\alpha, \eta_\alpha, \tilde{\zeta}_\alpha\}} \qquad (6)$$

$$P_i(\phi_i, \psi_i) \simeq \frac{1}{Z(\beta)} \frac{1}{M_i} \sum_\phi{}' \sum_\psi{}' \sum_{\tilde{\chi}} \frac{\exp [-\beta U\{\xi'_\alpha, \eta'_\alpha, \tilde{\zeta}_\alpha\}]}{\overline{g}\{\xi'_\alpha, \eta'_\alpha, \tilde{\zeta}_\alpha\}} \qquad (7)$$

$$\langle \phi_i \rangle \cong \frac{1}{Z(\beta)} \frac{1}{M} \sum_{\alpha-1}^{M} \frac{\xi_\alpha \exp\left[-\beta U\{\xi_\alpha, \eta_\alpha, \tilde{\zeta}_\alpha\}\right]}{g\{\xi_\alpha, \eta_\alpha, \tilde{\zeta}_\alpha\}} \tag{8}$$

$$\langle \psi_i \rangle \cong \frac{1}{Z(\beta)} \frac{1}{M} \sum_{\alpha-1}^{M} \frac{\eta_\alpha \exp\left[-\beta U\{\xi_\alpha, \eta_\alpha, \tilde{\zeta}_\alpha\}\right]}{g\{\xi_\alpha, \eta_\alpha, \tilde{\zeta}_\alpha\}} \tag{9}$$

where $\bar{g}\{\phi, \psi, \tilde{\chi}\}$ is the probability density function associated with the estimation of $P_i(\phi_i, \psi_i)$ (see Eq. 29 of Reference 26), and M_i is the number of conformations used to estimate $P_i(\phi_i, \psi_i)$ (see Eq. 30 of Reference 26). The primed summation signs of Eq. 7 indicate that the process is a summation over all dihedral angles except for ϕ_i and ψ_i.

The initial values of $g\{\xi_\alpha, \eta_\alpha, \tilde{\zeta}_\alpha\}$ are selected randomly from the *whole*(ϕ–ψ) space of each residue, with emphasis on sampling in the (broadly defined) low-energy regions of the map (see Figure 1 of Reference 26). After the initial value of the partition function (i.e., of these integrals) is computed, the SMAPPS program utilizes information from the Monte Carlo search and adjusts $g\{\phi, \psi, \tilde{\chi}\}$ and $\bar{g}\{\phi, \psi, \tilde{\chi}\}$ so that the subsequent Monte Carlo search will emphasize the regions of high probability. The adjusted probability density function is then used to provide an improved estimation of $Z(\beta)$ (and of the quantities in Eqs. 7–9). The process of evaluation and adaptation is continued until $g\{\phi, \psi, \tilde{\chi}\}$ approximates the Boltzmann distribution function.

To test the effectiveness of the algorithm, SMAPPS was applied to the prediction of the native conformation of the backbone of the pentapeptide, Met-enkephalin, a structure whose low-energy conformation had been calculated previously[60] (by energy minimization) with essentially the same potential functions.[19,20] In the calculations, only the backbone dihedral angles (ϕ and ψ) were allowed to vary; all side-chain (χ) and peptide-bond (ω) dihedral angles were kept fixed at the values corresponding to the alleged global minimum energy previously determined[60] by direct energy minimization. With this initial application of SMAPPS, three distinct low-free energy β-bend stuctures of Met-enkephalin were found (based on Eqs. 3 and 7). In particular, one of the structures had a conformation remarkably similar to the one associated with the previously alleged global minimum energy.[60] The two additional structures of the pentapeptide had conformational energies *lower* than the previously computed[60] low-energy structure. However, the three low-energy Monte Carlo results were in agreement with an improved energy-minimization procedure.[21]

Recently, we have relaxed the constraint that the side-chain and peptide-bond dihedral angles of Met-enkephalin be kept fixed.[75] By allowing for these additional degrees of freedom, SMAPPS has led to a new structure of high probability with a conformational energy of more than 2 kcal/mol lower than any structure previously determined.[75] At present, the Monte Carlo calculations of the structure of the enkephalin molecule with full degrees of freedom have not been compared with results obtained by the improved energy-minimization procedure.[21]

These initial results on the structure of Met-enkephalin indicate that an equilibrium statistical-mechanical procedure, coupled with an adaptive Monte Carlo algorithm, can overcome many of the problems associated with the standard methods of direct energy minimization, primarily the surmounting of barriers, to reach the global minimum.

The same procedure was used to calculate the average conformation of Met-enkephalin, based on Eqs. 4, 5, 8 and 9. The calculation was repeated until a total of ten independent average conformations were established. The regions of conformational space occupied by the ten average structures compare favorably with those of low conditional free energy (highest probability). The probable, and average, structures provide suitable *initial* conformations that may be used for subsequent direct energy minimization with the ECEPP/2 (Empirical Conformational Energy Program for Peptides) algorithm.[19,20]

In assessing the significance of the probable, and average, conformations, it should be realized that solid-state and solution methods for protein-structure determination provide an *average* structure, i.e., an average over many molecules and over the long period of time of the experiment. This is because of fluctuations due to thermal excitation. Hence the protein molecule moves from one low-energy conformation to another, and probably exists as an ensemble of conformations. Consequently, it is relevant to examine two types of structures that can be distinguished, viz., the average and the probable conformation. The justification for calculating the average conformation is obviously to allow a direct comparison of theory with experiment. However, the relevance of determining the probable conformations of a polypeptide is based solely on the hypothesis that, out of the whole ensemble of conformations, the one of highest probability may be termed the native structure.

FUTURE DIRECTIONS

This procedure is being applied in several directions. First of all, it is of interest to determine the largest size polypeptide that can be treated by this method, given the computing power presently available.[27,28] For this purpose, we are applying SMAPPS to a 26-residue peptide, melittin. We had previously computed the conformation of the membrane-bound first-20-residue portion of this molecule.[61] At this early stage of the calculations, we cannot provide any preliminary results on the structure of the full melittin molecule.

Second, the method is being applied to compute the free energy of a Lennard-Jones fluid. If successful, it will then be applied to liquid water, and to aqueous solutions of polar and nonpolar solutes (keeping in mind the earlier encounter with problems of convergence when dealing with aqueous solutions[23,25,31]). The application to fluids would be of special interest because the Metropolis algorithm cannot calculate the free energy efficiently because this quantity cannot be expressed as an ensemble average.[76]

To understand the nature of this problem, we observe that the definition of the Helmholtz free energy $A(V, T)$ in the NVT-ensemble (Gibbs canonical ensemble) can be given in the following form:

$$\exp(-\beta A) = \frac{1}{\lambda^{3N}N!} Q(N, V, T), \quad \lambda = \left(\frac{\beta h^2}{2\pi m}\right)^{1/2} \tag{10}$$

where

$$Q(N, V, T) = \int_V d\mathbf{r}_1 \dots \int_V d\mathbf{r}_N \exp(-\beta E) \tag{11}$$

where m is the molecular mass, h is Planck's constant, $E(\mathbf{r}_1, \dots, \mathbf{r}_N)$ is the potential energy of the system, and $Q(N, V, T)$ is the classical configurational integral. The Metropolis algorithm cannot evaluate $Q(N, V, T)$ from Eq. 11 because $Q(N, V, T)$ is not an ensemble average quantity. Instead, in order to estimate $Q(N, V, T)$, and hence the Helmholtz free energy, by the Metropolis algorithm, the usual approach is first to calculate the average value of the function $\exp(\beta E)$, which can be expressed conveniently in terms of the configurational integral,[76] i.e.,

$$\langle \exp(\beta E) \rangle = \frac{\int_V dr_1 \ldots \int_V dr_N \exp(\beta E) \exp(-\beta E)}{Q(N, V, T)} = \frac{V^N}{Q(N, V, T)}$$

or

$$Q(N, V, T) = V^N / \langle \exp(\beta E) \rangle \tag{12}$$

A calculation of the Helmholtz free energy based on Equation 12, however, is not practical because the Metropolis Monte Carlo technique will efficiently sample the states for which $\exp(-\beta E)$ is relatively large, but these are precisely the states in which the function $\exp(\beta E)$ being averaged in Eq. 12 is relatively small.[76] Such a procedure will thus underestimate the value of the free energy.

In contrast, the procedure that we have developed[26,72] evaluates the configurational integral (Eq. 11), and hence the free energy of a physical system, directly. This is accomplished by allowing the adaptive, importance sampling Monte Carlo algorithm to search the entire configurational space of a liquid based on a probability density function constructed initially to approximate the integrand, viz., the function $\exp(-\beta E)$. After the initial Monte Carlo search, the algorithm constructs a new probability density function based on the information obtained from the previous Monte Carlo search. A subsequent Monte Carlo search is made using the new probability density function, which now emphasizes the regions where the magnitude of the integrand was largest. This process of searching and adapting is continued until the algorithm has reached the optimal probability density function, i.e., the Boltzmann factor. The optimal probability density function can then be used to evaluate the free energy of the fluid directly.

Although there have been several attempts to extend the Metropolis algorithm to calculate the free energy by such techniques as umbrella sampling,[77] integration with a coupling parameter,[78] and perturbation theory,[79] these methods suffer from the problem that a reference system is required. The reference system must be one that is a close approximation of the original system being studied; also, the value of the free energy of the reference system must be known. While this is a difficult enough problem for fluids, it is even more difficult for a polypeptide chain, for which such a reference system is not easily available. Once more, because the adaptive, importance sampling Monte Carlo algorithm is designed to search the entire configurational space and then utilize the information for a more efficient subsequent search, it has an advantage over these other methods, i.e., there is no need to introduce a reference system.

It is hoped that the application of this adaptive, importance sampling Monte Carlo algorithm will be extendible to large systems. This would enable us to surmount the large stumbling block in computing protein conformation in solution, viz., the multiple-minima problem.

SUMMARY

We are using a variety of theoretical and computational techniques to study protein structure, protein folding, and higher-order structures. Our earlier work involved treatments of liquid water and aqueous solutions of nonpolar and polar solutes, computations of the stabilities of the fundamental structures of proteins and their packing arrangements, conformations of small cyclic and open-chain peptides, structures of fibrous proteins (collagen), structures of homologous globular proteins, introduction of special procedures as constraints during energy minimization of globular proteins, and structures of enzyme-substrate complexes. Recently, we presented a new methodology for predicting polypeptide structure (described here); the method is based on the calculation of the probable and average conformations of a polypeptide chain by the application of equilibrium statistical mechanics in conjunction with an adaptive, importance sampling Monte Carlo algorithm. As a test, it was applied to Met-enkephalin.

[**Note added in proof:** We have recently completed our analysis of Met-enkephalin for which all dihedral angles of the pentapeptide were allowed to vary.[75] This analysis included an estimation of the free energy of the probable conformations and of the global minimum-energy conformation. The results indicate that the thermodynamically preferred conformation of the pentapeptide contains a γ-turn. This conformation, however, does not correspond to the structure of lowest conformational energy. Instead, the global minimum-energy conformation, recently determined by a new optimization technique[80] developed in this laboratory, contains a type II' β-bend. A similar minimum-energy conformation is found by the SMAPPS procedure. The thermodynamically preferred γ-turn structure has a conformational energy of 4.93 kcal/mol higher than that of the β-bend structure of lowest energy but, because of the inclusion of entropy in the SMAPPS procedure, is estimated to be $\cong 9$ kcal/mol *lower* in free energy.]

REFERENCES

1. SCHERAGA, H. A. 1967. Fed. Proc. **26:** 1380.
2. SCHERAGA, H. A. 1984. Carlsberg Research Commun. **49:** 1.
3. NÉMETHY, G. & H. A. SCHERAGA. 1965. Biopolymers **3:** 155.
4. PATERSON, Y., G. NÉMETHY & H. A. SCHERAGA. 1981. Ann. N.Y. Acad. Sci. **367:** 132.
5. NÉMETHY, G., W. J. PEER & H. A. SCHERAGA. 1981. Ann. Rev. Biophys. Bioeng. **10:** 459.
6. SCHERAGA, H. A. 1982. Pure Appl. Chem. **54:** 1495.
7. SCHERAGA, H. A., M. R. PINCUS & K. E. BURKE. 1982. *In* Structure of Complexes between Biopolymers and Low Molecular Weight Molecules. W. Bartmann & G. Snatzke, Eds.: **53.** John Wiley. Chichester, England.
8. SCHERAGA, H. A. 1983. Biopolymers **22:** 1.
9. SCHERAGA, H. A. 1984. Pont. Acad. Sci. Scr. Var. **55:** 21.
10. SCHERAGA, H. A. 1985. Ann. N.Y. Acad. Sci. **439:** 170.
11. MAXFIELD, F. R. & H. A. SCHERAGA. 1979. Biochemistry **18:** 697.
12. KIDERA, A., Y. KONISHI, M. OKA, T. OOI & H. A. SCHERAGA. 1985. J. Protein Chem. **4:** 23.
13. KIDERA, A., Y. KONISHI, T. OOI & H. A. SCHERAGA. 1985. J. Protein Chem. **4:** 265.
14. LEWIS, P. N., N. GŌ, M. GŌ, D. KOTELCHUCK & H. A. SCHERAGA. 1970. Proc. Natl. Acad. Sci. USA **65:** 810.
15. GŌ, N., P. N. LEWIS, M. GŌ & H. A. SCHERAGA. 1971. Macromolecules **4:** 692.

16. TANAKA, S. & H. A. SCHERAGA. 1977. Macromolecules **10:** 305 (and earlier papers cited therein).
17. DUNFIELD, L. G. & H. A. SCHERAGA. 1980. Macromolecules **13:** 1415.
18. WAKO, H., N. SAITÔ & H. A. SCHERAGA. 1983. J. Protein Chem. **2:** 221.
19. MOMANY, F. A., R. F. MCGUIRE, A. W. BURGESS & H. A. SCHERAGA. 1975. J. Phys. Chem. **79:** 2361.
20. NÉMETHY, G., M. S. POTTLE & H. A. SCHERAGA. 1983. J. Phys. Chem. **87:** 1883.
21. VÁSQUEZ, M. & H. A. SCHERAGA. 1985. Biopolymers **24:** 1437.
22. GŌ, N. & H. A. SCHERAGA. 1978. Macromolecules **11:** 552.
23. KINCAID, R. H. & H. A. SCHERAGA. 1982. J. Computational Chem. **3:** 525.
24. RAPAPORT, D. C. & H. A. SCHERAGA. 1981. Macromolecules **14:** 1238.
25. RAPAPORT, D. C. & H. A. SCHERAGA. 1982. J. Phys. Chem. **86:** 873.
26. PAINE, G. H. & H. A. SCHERAGA. 1985. Biopolymer **24:** 1391.
27. POTTLE, C., M. S. POTTLE, R. W. TUTTLE, R. J. KINCH & H. A. SCHERAGA. 1980. J. Computational Chem. **1:** 46.
28. GIBSON, K. D., S. CHIN, M. R. PINCUS, E. CLEMENTI & H. A. SCHERAGA. 1985. *In* Supercomputer Simulations in Chemistry, (symposium held in Montreal, August 25–27, 1985).
29. NÉMETHY, G. & H. A. SCHERAGA. 1962. J. Phys. Chem. **66:** 1773.
30. PATERSON, Y., G. NÉMETHY & H. A. SCHERAGA. 1982. J. Solution Chem. **11:** 831.
31. MEHROTRA, P. K., M. MEZEI & D. L. BEVERIDGE. 1981. *In* Computer Simulation of Organic and Biological Molecules. Proceedings of a NRCC Workshop: 63. LBL-12979, January 1981.
32. GIBSON, K. D. & H. A. SCHERAGA. 1967. Proc. Natl. Acad. Sci. USA **58:** 420.
33. HOPFINGER, A. J. 1971. Macromolecules **4:** 731.
34. HODES, Z. I., G. NÉMETHY & H. A. SCHERAGA. 1979. Biopolymers **18:** 1565.
35. OOI, T., R. A. SCOTT, G. VANDERKOOI & H. A. SCHERAGA. 1967. J. Chem. Phys. **46:** 4410.
36. YAN, J. F., G. VANDERKOOI & H. A. SCHERAGA. 1968. J. Chem. Phys. **49:** 2713.
37. YAN, J. F., F. A. MOMANY & H. A. SCHERAGA. 1970. J. Am. Chem. Soc. **92:** 1109.
38. CHOU, K. C., M. POTTLE, G. NÉMETHY, Y. UEDA & H. A. SCHERAGA. 1982. J. Mol. Biol. **162:** 89.
39. CHOU, K. C. & H. A. SCHERAGA. 1982. Proc. Natl. Acad. Sci. USA **79:** 7047.
40. SCHERAGA, H. A., K. C. CHOU & G. NÉMETHY. 1983. *In* Conformation in Biology. R. Srinivasan & R. H. Sarma, Eds.: 1. Adenine Press. Guilderland, NY.
41. CHOU, K. C., G. NÉMETHY & H. A. SCHERAGA. 1983. J. Mol. Biol. **168:** 389.
42. CHOU, K. C., G. NÉMETHY & H. A. SCHERAGA. 1983. Biochemistry. **22:** 6213.
43. HESSELINK, F. T., T. OOI & H. A. SCHERAGA. 1973. Macromolecule **6:** 541.
44. GŌ, M., F. T. HESSELINK, N. GŌ & H. A. SCHERAGA. 1974. Macromolecules **7:** 459.
45. GŌ, M. & H. A. SCHERAGA. 1984. Biopolymers **23:** 1961.
46. FU, Y. C., R. F. MCGUIRE & H. A. SCHERAGA. 1974. Macromolecules **7:** 468.
47. TANAKA, S. & H. A. SCHERAGA. 1975. Macromolecules **8:** 516.
48. CHOU, K. C., G. NÉMETHY & H. A. SCHERAGA. 1983. J. Phys. Chem. **87:** 2869.
49. CHOU, K. C., G. NÉMETHY & H. A. SCHERAGA. 1984. J. Am. Chem. Soc. **106:** 3161. Erratum: *ibid,* **107:** 2199.
50. CHOU, K. C., G. NÉMETHY, S. RUMSEY, R. W. TUTTLE & H. A. SCHERAGA. 1985. J. Mol. Biol. **186:** 591.
51. CHOU, K. C., G. NÉMETHY, S. RUMSEY, R. W. TUTTLE & H. A. SCHERAGA. 1986. J. Mol. Biol. **188:** 641.
52. MILLER, M. H. & H. A. SCHERAGA. 1976. J. Polymer Sci. Polymer. Symp. **54:** 171.
53. MILLER, M. H., G. NÉMETHY & H. A. SCHERAGA. 1980. Macromolecules **13:** 470.
54. MILLER, M. H., G. NÉMETHY & H. A. SCHERAGA. 1980. Macromolecules **13:** 910.
55. NÉMETHY, G., M. H. MILLER & H. A. SCHERAGA. 1980. Macromolecules **13:** 914.
56. NÉMETHY, G. 1983. Biopolymers **22:** 33.
57. NÉMETHY, G. & H. A. SCHERAGA. 1984. Biopolymers **23:** 2781. Erratum: *ibid.,* 1985. **24:** 581.
58. DYGERT, M., N. GŌ & H. A. SCHERAGA. 1975. Macromolecules **8:** 750.
59. NÉMETHY, G. & H. A. SCHERAGA. 1984. Biochem. Biophys. Res. Commun. **118:** 643.

60. ISOGAI, Y., G. NÉMETHY & H. A. SCHERAGA. 1977. Proc. Natl. Acad. Sci. USA **74:** 414.
61. PINCUS, M. R., R. D. KLAUSNER & H. A. SCHERAGA. 1982. Proc. Natl. Acad. Sci. USA **79:** 5107.
62. FITZWATER, S. & H. A. SCHERAGA. 1982. Proc. Natl. Acad. Sci. USA **79:** 2133.
63. WARME, P. K., F. A. MOMANY, S. V. RUMBALL, R. W. TUTTLE & H. A. SCHERAGA. 1974. Biochemistry **13:** 768.
64. SWENSON, M. K., A. W. BURGESS & H. A. SCHERAGA. 1978. *In* Frontiers in Physicochemical Biology. B. Pullman, Ed.: 115. Academic Press. New York, NY.
65. MEIROVITCH, H. & H. A. SCHERAGA. 1981. Macromolecules **14:** 1250.
66. MEIROVITCH, H. & H. A. SCHERAGA. 1981. Proc. Natl. Acad. Sci. USA **78:** 6584.
67. KIKUCHI, T., G. NÉMETHY & H. A. SCHERAGA. 1986. J. Computational Chem. **7:** 67.
68. WAKO, H. & H. A. SCHERAGA. 1982. J. Protein Chem. **1:** 5, 85.
69. RACKOVSKY, S. & H. A. SCHERAGA. 1984. Accts. Chem. Res. **17:** 209.
70. PINCUS, M. R. & H. A. SCHERAGA. 1981. Accts. Chem. Res. **14:** 299.
71. SMITH-GILL, S. J., J. A. RUPLEY, M. R. PINCUS, R. P. CARTY & H. A. SCHERAGA. 1984. Biochemistry **23:** 993.
72. PAINE, G. H. & H. A. SCHERAGA. 1986. Biopolymers. **25:** 1547.
73. METROPOLIS, N., A. W. ROSENBLUTH, M. N. ROSENBLUTH, A. H. TELLER & E. TELLER. 1953. J. Chem. Phys. **21:** 1087.
74. LEPAGE, G. P. 1978. J. Computational Phys. **27:** 192.
75. PAINE, G. H. & H. A. SCHERAGA. 1987. Biopolymers. Submitted for publication.
76. WOOD, W. W. 1968. *In* Physics of Simple Liquids, H. N. V. Temperley, J. S. Rowlinson & G. S. Rushbrooke, Eds. North-Holland. Amsterdam.
77. TORRIE, G. M. & J. P. VALLEAU. 1974. Chem. Phys. Lett. **28:** 578.
78. MEZEI, M., S. SWAMINATHAN & D. L. BEVERIDGE. 1978. J. Am. Chem. Soc. **100:** 3255.
79. JORGENSEN, W. L. & C. RAVIMOHAN. 1985. J. Chem. Phys. **83:** 3050.
80. PURISIMA, E. O. & H. A. SCHERAGA. 1987. J.Mol. Biol. Submitted for publication.

Computer Simulation of DNA Supercoiling[a]

WILMA K. OLSON AND JANET CICARIELLO

Department of Chemistry
Rutgers University
New Brunswick, New Jersey 08903

INTRODUCTION

Superimposed upon the right-hand coiling of the familiar DNA double helix is a higher order of coiling of the helix axis itself, called supercoiling.[1] This coiling is analogous to the tertiary folding of helical segments in protein structures, but is spread over a much larger molecular scale in the nucleic acids. In naturally occurring duplex DNA there is roughly one superhelical turn for every twenty local turns (e.g., 200 base pairs) of the approximately tenfold B-DNA helix.[1] In proteins, in contrast, a corresponding 360° turn of the polypeptide backbone can be accommodated by 10–20 chain residues.[2] Furthermore, unlike the proteins, which fold in both a right-handed and a left-handed manner,[3] the supercoiling of native DNAs is almost exclusively left-handed.[4] The two complementary strands of the resulting DNA duplex are underwound compared to the number of times they would intertwine around one another in a relaxed (nonsupercoiled) structure and are consequently subjected to a persistent internal strain tending to unwind local regions of the double helix. Not surprisingly, supercoiling is an important facet of biological processes that entail local helical unwinding or denaturation, such as replication, transcription, and recombination.[5–8]

Until now, virtually all theoretical understanding of DNA supercoiling has been based on phenomenological models that account for certain macroscopic properties of the double helix, but ignore the details of its chemical structure and environment. The DNA duplex has been idealized in a number of studies[9–18] as a symmetric, linearly elastic rod which, like a rubber hose, exhibits no preferential directions of bending and twisting. The forces opposing deformation of the rod are partitioned locally into independent harmonic contributions from the changes of curvature κ and torsion τ. The latter parameters are standard functions of the equations describing the spatial trajectory of the helix axis.

The DNA double helix, however, is well known to exhibit preferential modes of bending. The potential energy, for example, is considerably lower for the rolling of base pairs about their long axes, and hence into the major and minor grooves of the structure, than the tilting about their short (dyad) axes.[19–22] Moreover, according to various X-ray crystallographic models of oligonucleotide helices,[23–25] this rolling is

[a]This work was supported in part by Grant GM 34809 from the United States Public Health Service and by the Charles and Johanna Busch Memorial Fund.

base-sequence-dependent, with certain base pairs more likely to bend into the major groove and others more apt to bend into the minor groove. Steric interactions with adjacent residues are apparently responsible for the observed preferences.[27–31] Furthermore, the potential energy reported for a superhelical turn of B-DNA is not a simple additive function of chain curvature and torsion. The likely coupling of bending and twisting in the DNA duplex is said[33] to be responsible for the failure of elasticity theories to account for observed occurrences of toroidal structures (see below) at low ionic strengths.

Supercoil-induced changes of double helical conformation are also only understood at the phenomenological level. Traditional helix-coil transition theory for unconstrained macromolecules has been extended to the treatment of torsionally stressed DNA systems. The twist of the supercoiled helix is partitioned between ordered and melted regions of the chain and the melting profile is computed as a function of base sequence, chain length, degree of twisting, and so forth.[34–36] The observed melting of underwound helices is well accounted for by this approach. The possible stress-induced transitions of B-DNA can be examined in the same manner, as can the competition between multiple conformational transitions.[37–43] Detailed understanding of the structural response of the DNA double helix to supercoiling and the possible conformational pathways linking the B-type duplex to melted structures, alternate helices, or cruciform structures, however, is not apparent from such studies.

Detailed molecular treatments of DNA supercoiling have been avoided in the past because of the difficulties in constructing constrained macromolecular structures with conventional model building (e.g., matrix generator) techniques. The coordinates of a chemical structure not available in the X-ray crystallographic literature must be generated mathematically from the chemical bond lengths, valence bond angles, and internal torsions of the system.[44] Unfortunately, even when these parameters are confined to their energetically preferred domains, there is no assurance that the ends of the structure will fall at the appropriate distances and orientations required for chain cyclization or other specific geometries. Moreover, because of the stiff wormlike character of the polynucleotide duplex, the DNA is not able to adopt closed circular forms until it is rather long. Direct sampling studies are necessarily restricted to relatively small cyclic species by current limitations of computer space and time. These sorts of simulations are completely out of question for longer chains containing an appreciable degree of supercoiling.

In this paper we present an alternative differential geometric procedure to obtain detailed models of DNA folding. We deform the local helical axis of regular double helical (secondary) structures along preselected spatial trajectories. Our approach is a straightforward extension of methods currently used to describe the topological parameters (e.g., linking number, writhe, twist, curvature, torsion, and so forth) of a space curve.[45,46] We monitor the extent to which the regular linear duplex can be forced to bend and twist with standard semiempirical potential functions. We focus particular attention on the nonbonded interactions of adjacent base pairs and the long-range electrostatic interactions between the negatively charged phosphate groups. We introduce a modified Coulombic potential that reproduces the electrostatic interactions between charges on the surface of a dielectric cylinder immersed in salt water. We then compare the stabilities of different three-dimensional forms and examine the detailed conformational features of the deformed duplexes.

DIFFERENTIAL GEOMETRIC CHAIN FOLDING

The established differential geometric treatment of DNA supercoiling[45] assumes that the trajectory of the double helical axis follows the movement of the tangent $\mathbf{t} = \mathbf{X}'(s)$ of a given space curve. The parameter, s, is the arc length or distance traversed by curve $\mathbf{X}$. The twist and curvature are measured in terms of the principal normal $\mathbf{n}$, the normalized first derivative of the tangent vector given by $\mathbf{n} = \mathbf{t}'(s)/\kappa$, where the normalization factor κ is the curvature at the point of interest. The three orthogonal axes $\mathbf{n}$, $\mathbf{b}$, and $\mathbf{t}$, where $\mathbf{b}$ is the binormal given by the scalar product $\mathbf{t} \times \mathbf{n}$, are a moving coordinate frame along $\mathbf{X}(s)$ used in the computation of writhe, twist, and linking number. This frame of reference can also be used, as detailed below, to describe the spatial locations of individual residues at selected points along a supercoiled trajectory.

If the DNA base pairs are assumed to be perpendicular to the local helical axis as in B-DNA and thus exclusively in the $\mathbf{n}$, $\mathbf{b}$ reference plane, the local coordinates of atom a can be represented by the position vector η.

$$\eta = a_n \mathbf{n} + a_b \mathbf{b} \tag{1}$$

In this expression a_n and a_b are the respective projections of the atom along the $\mathbf{n}$ and $\mathbf{b}$ coordinate axes. The coordinates of the atom in the frame of reference of the DNA trajectory as a whole are then given by the vector sum $\mathbf{X} + \eta$.

The determination of molecular coordinates consequently reduces to the determination of the unit vectors $\mathbf{n}$, $\mathbf{b}$, and $\mathbf{t}$ at specific locations along a chosen spatial trajectory. The procedure is fairly straightforward for a superhelical pathway where

$$\mathbf{X}(s) = (r \cos (s/r), r \sin (s/r), sP/2\pi r) \tag{2}$$

and r and P are the radius and pitch, respectively, of the superstructure. The evaluation is more complex, however, for other trajectories which are not simple functions of arc length.

This complication is found, for example, in the treatment of a closed circular superhelical (e.g., toroidal) trajectory

$$\mathbf{X}(\theta) = (R \cos \theta - r \sin m \theta \cos \theta, R \sin \theta - r \sin m \theta \sin \theta, r \cos \theta) \tag{3}$$

Here r is the radius of the superhelix and R the radius of the closed circle. The parameter θ is used to describe the relative location of any point on the closed circle and m to indicate the number of superhelical turns around its perimeter. The arc length is a function of θ that cannot be evaluated directly. Specifically,

$$ds/d\theta = ((R - r \sin m\theta)^2 + r^2m^2)^{1/2} \tag{4}$$

where the integral of ds is an undetermined elliptic integral. The evaluation of $\mathbf{t}$, $\mathbf{n}$, and $\mathbf{b}$ accordingly must be carried out numerically or by application of the chain rule of differentiation (e.g., $\mathbf{t} = \mathbf{X}'(s) = \mathbf{X}'(\theta) \, d\theta/ds$;

$\mathbf{n} = \mathbf{t}'(s)/\kappa = [\mathbf{X}''(\theta) \, (d\theta/ds)^2 + \mathbf{X}'(\theta) \, d^2\theta/ds^2)]/\kappa$; and so forth).

A simple interwound geometry is given by a series of four expressions representing each of the two superhelical axes and each of the two hairpin turns of the structure

$$\mathbf{X}(\theta) = (r \cos \theta, \, r \sin \theta, \, P\theta/2\pi);$$

$$(\pm r \cos (\theta/2), \, \pm \xi r \sin \theta, \, P(m\pi + 2\xi \sin (\theta/2))/2\pi);$$

$$(\mp r \cos \theta, \, \pm r \sin \theta, \, P(m\pi - \theta)/2\pi);$$

$$(-r \cos (\theta/2), \, \xi r \sin \theta, \, -2\xi P \sin (\theta/2)/2\pi) \qquad (5)$$

This form is chosen to avoid the self-intersection of distant parts of the chain backbone. The parameter m is again the number of superhelical turns in the structure, while ξ is an adjustable term to accommodate the chosen range of spacing between residues in the hairpin loops (see below) and the selected proportions of the attached superhelices.

POTENTIAL ENERGIES

The conformational stabilities of various DNA pathways are estimated here in terms of the stacking interactions between adjacent base pairs and the electrostatic interactions between negatively charged phosphate groups. The base stacking potential V_{ST} is a standard sum of van der Waals repulsions, London attractions, charge-induced polarizability contributions, and Coulombic interactions. The energy is summed over all pairwise combinations of atoms on successive bases and reported as a residue average. The atomic parameters are the same as those used in earlier calculations of single-stranded chain stacking.[47,48] The base pairs are assigned rigid geometries consisting of standard purine and pyrimidine units[49] positioned in a common plane relative to the usual Watson-Crick dyad axis.[50] Successive base pair planes are located according to the differential geometric procedure detailed above. Phosphorus atoms are positioned relative to the base pair coordinate frame of each residue.

The phosphate interactions are summed over all pairs of groups in the superstructure. For simplicity, each group is represented by an extended atom of 3.0-Å van der Waals radius located at the phosphorus center. Each residue is assigned a partial charge δ of -0.25 esu to reflect the predicted effects[51] of counterion condensation. The polarizability and effective number of valence electrons are taken to be 3.75Å^3 and 32, respectively. These two parameters can be combined in the Slater-Kirkwood equation[52] to evaluate pairwise constants of the London dispersion energy.

The electrostatic interactions between phosphate groups are evaluated at three different levels of approximation. A simple Coulombic expression V_{COU} is employed assuming a dielectric constant ϵ of 80. This value is thought to provide a better description of the likely solvent between surface charges than the dielectric of 2–4 generally used to represent the chemical medium (e.g., sugar and base) of the DNA.

The precise variation of dielectric constant with distance along the chain backbone is uncertain. One of the best estimates of this dependence is the reported solution of the Poisson-Boltzmann equation for a discrete source charge located on the surface of a rodlike polyelectrolyte in aqueous salt solution.[53] As illustrated in FIGURE 1, the interactions between charges on the same side of such a cylinder (solid curve) are enhanced compared to those expected for a Debye-Hückel exponential decay, while those on the opposite sides of the rod (dashed curve) are lower than predicted. The Poisson-Boltzmann energies are also a function of vertical displacement on the cylinder with interactions dying off with increasing displacement. The smooth curves $H(\theta)$ are a best fit of data reported for two helical ribbons ($+$ and $-$ symbols) winding around a polyelectric cylinder and are of the form

$$H(\theta) = 1 \pm \exp(-A \cos^2 \theta) \exp(-B \sin^2 \theta) \cos \theta \qquad (6)$$

The A and B in this expression are determined by the decreasing amplitude of H reported at half integral turns of the ribbons ($A = H_1 + H_2 \theta/2\pi$; $B = H_3 + H_4(\theta - \pi)/2\pi$). The parameter θ is the cylindrical angular location of any point on the ribbon relative to a fixed charge.

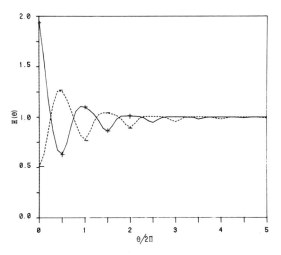

FIGURE 1. Ratio of electrostatic interactions $H(\theta)$ computed with the Poisson-Boltzmann equation and a simple Debye-Hückel screening term for two complementary helical phosphate trajectories around a rodlike polyelectrolyte[53] for the case $D^{-1} = 0$ Å^{-1}. The *solid* and *dashed curves* are fit to four successive data points using Eq. 6. $H(\theta)$ curves at alternate ionic strengths are obtained by interpolation between observed points where $D^{-1} = 0$ and 1 Å^{-1}. Integral values of $\theta/2\pi$ correspond to complete helical turns of the phosphate ribbons.

The electrostatic energy is then expressed as

$$V_{ELE}(d_{ij}, \theta) = (332 \, \delta_i \delta_j / \epsilon d_{ij}) H(\theta) \exp(-d_{ij}/D) \qquad (7)$$

where d_{ij} is the distance between residues i and j and D is the Debye radius characteristic of a particular ionic strength. The exponential term in this expression is the Debye-Hückel screening factor.

The interactions between phosphate and base are ignored in this preliminary study. The intervening sugar residues are also omitted in the calculations. The sugar geometry is severely deformed in the superstructure-generating procedure and must be refined by either energy minimization or constrained modeling techniques.[54]

RELAXED CLOSED CIRCLES

The relative stabilities of various closed circular trajectories of B-DNA are illustrated in FIGURE 2. The base pair separation is fixed at 3.4Å and the radii of the circle are adjusted (using $r = 3.4 \, N/2\pi$ Å) to accommodate increasing chain length (N). For simplicity, the chain sequence is represented by the homopolymer, $dG_N \cdot dC_N$. The average base stacking energy per residue $\overline{V}_{ST}$ is seen to drop off rapidly in value in small circles but to approach its limiting value (associated with the linear structure) more gradually at longer chain lengths. The computed energy is within 1 kcal/mol of the asymptote in rings as small as 50 residues (where adjacent base planes are oriented at an angle Λ of $360°/50 = 7.2°$), within 0.1 kcal/mol of this reference in circles of 120 residues (where Λ is equal to 3°), and within 0.01 kcal/mol of the limit in circles of 350 residues (where Λ is about 1°).

In contrast, the mean Coulombic energy per residue $\overline{V}_{COU}$ increases with chain length. While the overall variation of $\overline{V}_{COU}$ with N is less dramatic than that of $\overline{V}_{ST}$, the Coulombic effect persists at chain lengths where $\overline{V}_{ST}$ levels off. In rings of the size illustrated here, long-range interactions between phosphate pairs compete effectively, because of their large numbers and the $1/r$ dependence of the Coulombic potential, with those between close neighbors. Debye-Hückel screening depresses these contributions so that the average energy per residue $V_{D-H} = V_{COU} \exp(-d_{ij}/D)$ exhibits a local maximum at very short chain lengths and subsequently decreases with increase in N. The data reported in FIGURE 2 presume a Debye radius of 10Å. The $H(\theta)$ correction term increases the magnitude of the total Debye-Hückel energy contribution slightly ($\sim.04$ kcal/mol per residue), but has essentially no effect on the computed variation of

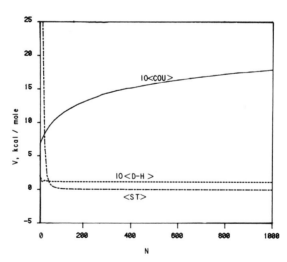

FIGURE 2. The computed variation in mean stacking energy $\overline{V}_{ST}$ between adjacent bases, average Coulombic energy per residue $\overline{V}_{COU}$, and average Debye-Hückel energy $\overline{V}_{D-H}$ (with D = 10Å) in relaxed closed circular B-DNA structures as a function of chain length N. Data were computed at integral numbers of helical turns and connected by smooth curves.

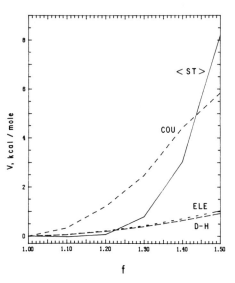

FIGURE 3. The computed variation in average base stacking energy and total phosphate-phosphate interactions as a function of the axis/residue spacing ratio f of closed elliptical trajectories of $dG_{200} \cdot dC_{200}$. Data are reported relative to a circular structure where f = 1.

electrostatic energy with chain length. For this reason only the $\overline{V}_{D-H}$ energies appear in FIGURE 2.

The ease of deforming relaxed B-DNA closed circles is described by the energy plots in FIGURE 3. A 200-residue poly dG · poly dC sequence is forced to conform to a series of elliptical trajectories of increasing axis ratio f and the various contributions to the potential energy are reported. The spacing between residues is a function of f in these structures with the maximum fluctuations in Δs from ideal 3.4Å spacing equal to 3.4 (f − 1)/(f + 1) Å. As f is increased in value, selected residues are forced together and the various energy contributions are accordingly increased. The most sensitive term is clearly the mean base stacking energy, which is more than 5 kcal/mol greater at f = 1.45 than the base stacking energy of a circular structure. The increase in Coulombic energy is much smaller over this range of f. The total change in V_{COU} for the cyclic species as a whole is comparable to the change in $\overline{V}_{ST}$ reported on a per-residue basis. Assuming D = 10Å, the net changes in V_{D-H} and V_{ELE} are even less over this range of f, their variations being 0.7 and 1.0 kcal/mol, respectively.

The electrostatic energy is far more sensitive to the formation of cyclic structures from linear pieces than to their local deformation. This process is approximated in FIGURE 4 by the condensation of a superhelical trajectory from a 200-residue B-DNA rod to a 200-residue circle. The rod only has pitch (e.g., r = 0 in Eq. 2), while the closed circle only has radius (e.g., P = 0). Intermediate structures on the pathway are described by the ratio $P/2\pi r$. Spacing between residues is fixed at 3.4Å by adjusting the value of r. Interatomic distances are decreased by the change in curvature between rod and circle and the energy is accordingly increased. Because the spacing is constant, $\overline{V}_{ST}$ is only slightly perturbed (0.06 kcal/mol) in each residue of the chain. The net change in Coulombic energy between rod and circle is somewhat greater than this on a per-residue basis (32.70/200 = 0.16 kcal/mol). The total increase in V_{D-H} or V_{ELE}, however, is only 0.4 kcal/mol assuming a Debye radius of 10Å.

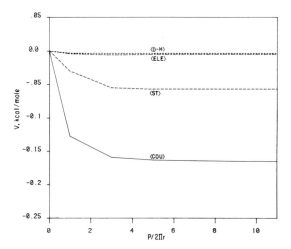

FIGURE 4. The dependence of local base stacking energy and overall phosphate interactions on the degree of condensation $P/2\pi r$ of a superhelical DNA chain. At $P/2\pi r = 0$ the chain is a relaxed closed circle and when this ratio is infinite, the chain is perfectly rodlike.

SUPERCOILED DNA

The energetic profile of closed circular toroidal DNA containing a single left-handed superhelical turn is described in FIGURE 5. These structures are described in terms of the parameter F, which specifies the range of spacing between adjacent residues in the chain. F is the ratio of the maximum and minimum arc lengths between residues and is taken here as $(3.4 + \delta\Delta s)/(3.4 - \delta\Delta s)$, where $\delta\Delta s$ is the allowed

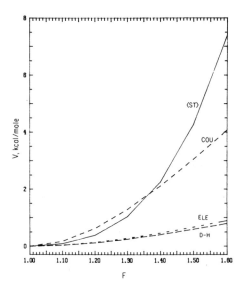

FIGURE 5. The computed variation in average base stacking energy and total phosphate-phosphate interactions as a function of the spacing ratio F in 200-residue toroidal forms of $dG_{200} \cdot dC_{200}$ containing one left-handed superhelical turn. Data reported relative to a relaxed closed circular trajectory where $F = 1$.

fluctuation in Δs. Both the radius of the superhelix and that of the circular pathway which it follows are determined by F. As F is increased in value, $\delta\Delta$s is likewise increased and selected residues are forced closer together. The energy is thereby increased with the most noticeable effects on base stacking terms. $\overline{V}_{ST}$ is raised at F = 1.55 more than 5.0 kcal/mol above the (per-residue) value observed for a 200-mer relaxed closed circular poly dG · poly dC. The net change in total Coulombic energy between phosphates in the toroidal structure as a whole, however, is only 3.0 kcal/mol over this range of F. The total variations in V_{D-H} and V_{ELE} are even smaller (0.6 − 0.7 kcal/mol) in the toroidal form with D = 10Å.

Alternative interwound arrangements containing a single left-handed superhelical turn of B-DNA are described in TABLE 1. The 200-residue structures are constructed according to Eq. 5 from a sequence of superhelical fragments and hairpin loops. The superhelices are described, as above, by the ratio $P/2\pi r$ and the hairpin turns by the spacing parameter F. The proportions of the interwound DNAs are altered by changing the number of residues in the superhelical and hairpin loop fragments (N_{SH} and N_{LP}, respectively), Energies are reported in the table relative to those found for a relaxed closed circular structure of the same chain length and base sequence. As expected, the various energy contributions are largest in the structure ($N_{SH} = N_P = 100$) with the broadest range of loop spacing (F = 1.62). In the most favored of the

TABLE 1. Relative Stabilities of Selected Interwound Structures

N_{SH}	N_{LP}	$P/2\pi r$	F	ξ	V_{COU} (kcal/mol)	V_{D-H} (kcal/mol)	V_{ELE} (kcal/mol)
140	60	1.80	1.03	0.22	34.3	−0.3	−0.6
120	80	1.63	1.30	0.38	45.4	0.9	0.8
100	100	1.58	1.62	0.62	56.4	1.9	1.9

three examples (where F = 1.03), the Debye-Hückel and electrostatic energies are more favorable than those found in a perfectly relaxed circular arrangement. The interwound geometries, however, are strongly disfavored compared to both circular and toroidal DNA by the Coulombic potential. This potential is expected to predominate in extremely dilute solutions (D → ∞) where toroidal forms of supercoiled DNA are found experimentally.[55,56] At higher ionic strengths, where D is finite, interwound forms are predicted to occur in greatest numbers. Such mixtures of toroidal and interwound structures have been reported in recent physicochemical studies of supercoiled DNA.[57,58]

The computed energies (V_{DH} and V_{ELE}) of a representative toroidal and interwound trajectory of supercoiled poly dG · poly dC are reported for selected values of $1/D$ in FIGURE 6. With a single left-handed superhelical turn, each of the 200-residue structures is characterized by a fixed linking number of $200/10 - 1 = 19$. The spatial trajectories of the two forms, however, are vastly different. These topological differences are illustrated in FIGURE 7 in terms of the distribution of interphosphate distances $W(d_{ij})$ in the two structures. The data are plotted with respects to d_{max}, the maximum extension of the 200-residue chain, here set equal to 680Å. As evident from the figure, most of the phosphate groups are distantly separated in the toroidal

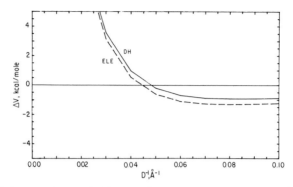

FIGURE 6. Relative Debye-Hückel V_{D-H} and electrostatic energies V_{ELE} of representative toroidal and interwound structures as a function of inverse Debye length D^{-1}. Both structures contain a single left-handed superhelical turn ($m = -1$) spread out over $200 \, G \cdot C$ base pairs. The toroidal form is characterized by a spacing ratio $F = 1.5$ with $r = 22.2$ Å and $R = 105.0$ Å. The interwound form is that described by the first entry in TABLE 1 with $r = 36.8$ Å and $P = 416.1$ Å.

arrangement, the maximum in $W(d_{ij})$ occurring at $d_{ij}/d_{max} \sim 1/\pi$ (a distance corresponding to the diameter of the toroidal closed circle). In contrast, the groups are, on the average, more closely spaced in the interwound model with the maximum in $W(d_{ij})$ occurring at $d_{ij} \sim 0.1 \, d_{max}$ (a distance corresponding to the diameter of the superhelical core of this structure). With only weak screening of long-range Coulombic interactions permitted (e.g., $1/D \sim 0$ Å^{-1}), the interwound form is disfavored over the toroidal arrangement by the high proportion of intermediate range phosphate-phosphate distances. As $1/D$ increases in magnitude and D correspondingly decreases in value, Coulombic interactions across the superhelical core of the interwound structure are also effectively screened. The relative stabilities of the two structures are then determined by short-range contacts, which are slightly more favorable for the interwound arrangement (i.e., there are fewer very close nonbonded contacts).

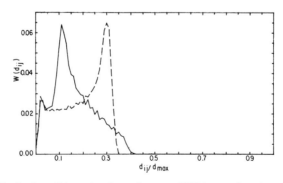

FIGURE 7. Distribution of interphosphate distances $W(d_{ij})$ as a function of relative chain extension d_{ij}/d_{max} in the toroidal and interwound structures described in FIGURE 6.

According to these calculations, the toroidal and interwound forms are roughly comparable in energy at $1/D = 0.05 \text{Å}^{-1}$.

DISCUSSION

The differential geometric approach set forth here provides a convenient way to generate detailed structural models of DNA supercoiling. The regular B-DNA duplex is easily deformed along a chosen spatial trajectory by taking advantage of the moving coordinate frame of the curve. The spacing between adjacent residues is not necessarily constant, but can be maintained within prescribed limits by an appropriate choice of overall chain dimensions. The parallel alignment of adjacent residues is, however, disrupted by the smooth bending of the duplex. The relative extent of rolling and tilting of successive units is then determined by the precise alignment of adjacent coordinate frames. The lateral twisting of adjacent residues is also affected by orientational variations of the moving coordinate frame. Noticeably large fluctuations in local twist have been previously reported in certain toroidal structures[59] and in the hairpin turn of interwound DNA.[22]

Structural details associated with these macroscopic changes are readily monitored by simple potential energy functions. The stacking interactions of adjacent bases are particularly sensitive to deformations which decrease the arc-length spacing along the helix axis. The elliptical perturbations of relaxed closed circular B-DNA are, for example, limited to structures with a maximum axis ratio of about 1.5 by the minimum close contacts of neighboring base planes. Toroidal DNAs are similarly prevented by variations in local spacing from adopting structures with superhelical radii of magnitudes comparable to the radii of the circles around which they wrap.

The electrostatic interactions between negatively charged phosphate groups are especially sensitive to the formation of cyclic structures from linear pieces and to the different three-dimensional ways of accommodating a given level of supercoiling. It costs roughly 30 kcal/mol of electrostatic energy to fold a 200-residue rodlike B-DNA into a relaxed closed circle with 3.4Å arc-length spacing. The phosphates are, on the average, more closely separated in the circle than in the rod so that the circle is disfavored by Coulombic interactions. The effect is less pronounced when long-range terms are screened by added salt (e.g., decreased Debye length), although the circle is still the disfavored form. A somewhat similar situation is found with toroidal and interwound supercoiled trajectories. The more compact toroidal arrangement is disfavored by greater Coulombic repulsions. Even when the residue charge is reduced to -0.25 esu and the dielectric constant of water is employed, the difference in energy found in an underwound 200-mer is in excess of 30 kcal/mol. Short-range contacts are, however, more favored in the interwound than in the toroidal form so that the relative structural preferences are switched at higher ionic strengths. This is apparently the case with natural DNA where the relative proportions of toroidal and interwound forms are ionic-strength-dependent.[55–58]

ACKNOWLEDGEMENT

Computer resources were supplied in part by the Center for Computer and Information Services of Rutgers University.

REFERENCES

1. BAUER, W. R. 1978. Ann. Rev. Biophys. Bioeng. **7:** 287–313.
2. RICHARDSON, J. S. 1981. Adv. Protein Chem. **34:** 167–339.
3. SIBANDA, B. L. & J. N. THORNTON. 1985. Nature **316:** 170–174.
4. BAUER, W. R., F. H. C. CRICK & J. H. WHITE. 1980. *Sci. Am.* **243:** 118–133.
5. COZZARELLI, N. R. 1980. Science **207:** 953–960.
6. SMITH, G. R. 1981. Cell **24:** 599–600.
7. LILLEY, D. M. J. 1983. Nature **305:** 276–277.
8. FISHER, L. M. 1984. Nature **307:** 686–687.
9. BENHAM, C. J. 1977. Proc. Natl. Acad. Sci. USA **74:** 2397–2401.
10. BENHAM, C. J. 1978. J. Mol. Biol. **123:** 361–370.
11. BENHAM, C. J. 1979. Biopolymers **18:** 609–623.
12. BENHAM, C. J. 1983. Biopolymers **22:** 2477–2495.
13. LEBRET, M. 1978. Biopolymers **17:** 1939–1955.
14. LEBRET, M. 1979. Biopolymers **18:** 1709–1725.
15. LEBRET, M. 1984. Biopolymers **23:** 1835–1867.
16. VOLOGODSKII, A. V., V. V. ANSHELEVICH, A. V. LUKASHIN & M. D. FRANK-KAMENETSKII. 1979. Nature **280:** 294–298.
17. SHIMADA, J. & H. YAMAKAWA. 1984. Macromolecules **17:** 609–698.
18. YAMAKAWA, H. 1985. J. Mol. Biol. **184:** 319–329.
19. ZHURKIN, V. B., YU P. LYSOV & V. I. IVANOV. 1979. Nucleic Acids Res. **6:** 1081–1096.
20. ZHURKIN, V. B., YU. P. LYSOV, V. L. FLORENTIEV & V. I. IVANOV. 1982. Nucleic Acids Res. **10:** 1811–1830.
21. ULYANOV, N. B. & V. B. ZHURKIN. 1984. Biomol. Str. Dynam. **2:** 361–385.
22. OLSON, W. K., A. R. SRINIVASAN, M. A. CUETO, R. TORRES, R. C. MAROUN, J. CICARIELLO & J. L. NAUSS. 1986. *In* Biomol Stereodynamics IV. R. H. SARMA & M. H. SARMA. Eds.: 75–100. Adenine Press. Guilderland, NY.
23. DICKERSON, R. E. & H. R. DREW 1981. J. Mol. Biol. **149:** 761–783.
24. FRATINI, A. V., M. L. KOPKA, H. R. DREW & R. E. DICKERSON. 1982. J. Biol. Chem. **257:** 14686–14707.
25. SHAKKAD, Z., D. RABINOVICH, O. KENNARD, W. B. T. CRUSE, S. A. SALISBURY & M. A. VISWAMITRA. 1983. J. Mol. Biol. **166:** 183–201.
26. CALLADINE, C. R. 1982. J. Mol. Biol. **161:** 343–352.
27. DICKERSON, R. E., M. L. KOPKA & H. R. DREW. 1982. *In* Conformation in Biology. R. Srinivasan & R. H. Sarma, Eds.: 227–257. Adenine Press. Guilderland, NY.
28. DICKERSON, R. E. 1983. J. Mol. Biol. **166:** 419–441.
29. CALLADINE, C. R. & H. R. DREW. 1984. J. Mol. Biol. **178:** 773–782.
30. HARVEY, S. C. 1984. Nucleic Acids Res. **11:** 4867–4878.
31. TUNG, C.-S. & S. C. HARVEY. 1984. Nucleic Acids Res. **12:** 3343–3356.
32. LEVITT, M. 1978. Proc. Natl. Acad. Sci. USA **75:** 640–644.
33. CALLADINE, C. R. 1980. Biopolymers **19:** 1705–1713.
34. ANSHELEVICH, V. V., A. V. VOLOGODSKII, A. V. LUKASHIN & M. D. FRANK-KAMENETSKII. 1979. Biopolymers **18:** 2733–2744.
35. VOLOGODSKII, A. V., A. V. LUKASHIN, V. D. ANSHELEVICH & M. D. FRANK-KAMENETSKII. 1979. Nucleic Acids Res. **6:** 967–982.
36. BENHAM, C. J. 1980. J. Chem. Phys. **72:** 3633–3639.
37. BENHAM, C. J. 1980. Nature **286:** 637–638.
38. BENHAM, C. J. 1982. Biopolymers **21:** 679–696.
39. BENHAM, C. J. 1981. J. Mol. Biol. **150:** 43–68.
40. BENHAM, C. J. 1982. Cold Spring Harbor Symp. Quant. Biol. **47:** 219–227.
41. VOLOGODSKII, A. V. & M. D. FRANK-KAMENETSKII. 1984. J. Biomol. Str. Dynam. **2:** 131–148.
42. MIYAZAWA, S. 1985. J. Chem. Phys. **83:** 859–883.
43. MAJUMDAR, R. & A. R. THAKUR. 1985. Nucleic Acids Res. **13:** 5883–5893.
44. OLSON, W. K. 1982. *In* Topics in Nucleic Acid Structure. Part II. S. Needle, Ed.: 297–314. Macmillan, London.
45. FULLER, W. B. 1971. Proc. Natl. Acad. Sci. USA **68:** 815–819.

46. CRICK, F. H. C. 1976. Proc. Natl. Acad. Sci. USA **73:** 2639–2643.
47. OLSON, W. K. 1978. Biopolymers **17:** 1015–1040.
48. TAYLOR, E. R. & W. K. OLSON. 1983. Biopolymers **22:** 2667–2702.
49. TAYLOR, R. & O. KENNARD. 1982. J. Am. Chem. Soc. **104:** 3209–3212.
50. ARNOTT, A. & D. W. L. HUKINS. 1972. Biochem. Biophys. Res. Commun. **47:** 1504–1509.
51. MANNING, G. S. 1979. Acc. Chem. Res. **12:** 443–449.
52. PITZER, K. S. 1959. Adv. Chem. Phys. **2:** 59–83.
53. SKOLNICK, J. & M. FIXMAN. 1978. Macromolecules **11:** 867–871.
54. CICARIELLO, J., A. R. SRINIVASAN & W. K. OLSON. Unpublished data.
55. GRAY, N. B. 1967. Biopolymers **5:** 1009–1019.
56. BRADY, G. W., D. B. FEIN, H. LAMBERTSON, V. GRASSIAN, D. FOOS & C. J. BENHAM. 1983. Proc. Natl. Acad. Sci. USA **80:** 741–744.
57. BRADY, G. W., D. FOOS & C. J. BENHAM. 1984. Biopolymers **23:** 2963–2966.
58. SHIBATA, J. H., J. WILCOXON, J. M. SHURR & Z. KNAUF. 1984. Biochemistry **23:** 1188–1194.
59. OLSON, W. K., N. L. MARKY, A. R. SRINIVASAN, K. D. DO & J. CICARIELLO. 1985. *In* Molecular Basis of Cancer, Part A: Macromolecular Structure, Carcinogens and Oncogenes. R. Rein, Ed.: 109–121. Alan R. Liss, New York, NY.

Calculation of Atomic Charges in Large Molecules[a]

S. SHANKAR,[b] W. J. MORTIER, AND S. K. GHOSH

Department of Chemistry
University of North Carolina
Chapel Hill, North Carolina 27514

A formalism that employs the principle of electronegativity equalization has been developed for the calculation of atomic charges in molecules that are connectivity- and geometry-dependent.[1,2] The effective electronegativity of an atom, in a molecule, which is equal to the molecular electronegativity, is given by

$$\chi_\alpha = (\chi_\alpha^\circ + \Delta\eta_\alpha) + 2(\eta_\alpha^\circ + \Delta\eta_\alpha)\, q_\alpha + \Sigma q_\beta / R_{\alpha\beta}$$

with χ_α° and η_α° the neutral-atom electronegativity and hardness respectively, q_α and q_β the charges on atoms α and β, and $R_{\alpha\beta}$ the internuclear distance. The parameters $\Delta\chi_\alpha$ and $\Delta\eta_\alpha$ are the corrections to the neutral-atom electronegativity and hardness that arise as a consequence of bonding. A rigorous theoretical justification for the above formulation is provided within the framework of density functional theory.[2] $\Delta\chi$ and $\Delta\eta$ are obtained from small-molecule calculations and are transferable and consistently usable for calculating charges in any molecule.

Our principal motive was to be able to calculate atomic charges in large biological molecules, for which *ab initio* calculations are extremely difficult. Besides serving as an index of chemical reactivity, these charges are essential for molecular-mechanics and molecular dynamics simulations. In view of this, we used *ab initio* STO-3G Mulliken charges[c] calculated for a large number of small molecules to parametrize $\Delta\chi$ and $\Delta\eta$. We then tested the use of these parameters for calculating charges on other small and large molecules. Charges have been obtained for four different configurations of the alanine dipeptide, for a representative deoxyribose and for several methyl-substituted cyclohexanes. In all cases the charges obtained by this method are extremely close to those obtained independently from *ab initio* calculations.[3] The flexibility of using a parametrizing scheme is exploited further to study atomic charges in interacting water dimers. Charges are calculated for linear, cyclic, and bifurcated dimers and correlation with *ab initio* calculations[4] is shown to be very good. FIGURES 1 and 2 present some illustrative correlations between charges obtained by electronegativity equalization method and those obtained from STO-3G *ab initio* MO calcula-

[a]This work was supported by grants from the National Institutes of Health and the National Science Foundation to the University of North Carolina.

[b]Present address: Department of Chemistry, University of Houston, University Park, Houston, Texas 77004.

[c]These calculations were done in a VAX 11/780 computer at the Chemistry Department, University of North Carolina, using the QCPE program #GAUSSIAN 82.

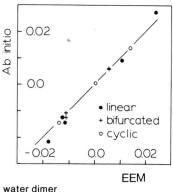

FIGURE 1. Correlation between STO-3G and EEM charges in water dimers.

tions. The time taken for these calculations is of the order of 30–60 sec on a VAX-11/780 computer.

We suggest the possibility of using this method in conjunction with molecular-mechanics or -dynamics calculations to recalculate the electrostatic force field as the configuration changes. This may be important especially for simulation of interactions with ions and ionic groups.

ACKNOWLEDGMENT

We are grateful to Professor Robert G. Parr for his valuable discussions and hospitality.

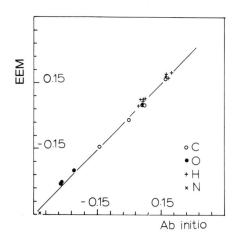

FIGURE 2. Correlation between STO-3G and EEM charges in 2 amino deoxyribose.

REFERENCES

1. MORTIER, W. J., K. VAN GENECHTEN & J. GASTEIGER. 1985. J. Am. Chem. Soc.
    107: 829–835.
2. SHANKAR, S., W. J. MORTIER & S. K. GHOSH. Personal communication.
3. WEINER, S. J., U. C. SINGH, T. J. O'DONNELL & P. A. KOLLMAN. 1984. J. Am. Chem. Soc.
    106: 6243–6254.
4. KOLLMAN, P. A. & L. C. ALLEN. 1969. J. Chem. Phys. **51:** 3286–3293.

A Vectorized Near-Neighbors Algorithm of Order N for Molecular Dynamics Simulations

S. G. LAMBRAKOS,[a] J. P. BORIS,[a] I. CHANDRASEKHAR[b]
AND B. GABER[b]

[a]Laboratory for Computational Physics and Fluid Dynamics
Code 4410, Naval Research Laboratory
[b]Bio/molecular Engineering Branch
Code 6190, Naval Research Laboratory
Washington, D.C. 20375-5000

The Laboratory for Computational Physics (LCP) is currently performing molecular dynamics simulations with substantial reductions in computation time by implementing a powerful new "location indexing" scheme, the Monotonic Logical Grid (MLG) algorithm.[1,2] Benchmark studies are underway for large macromolecular systems.

Efficient algorithms for keeping track of "close" objects[3,4] are based on the concept of a short range "cutoff" distance R_c. Efficiency obtains by limiting the computations to nearby objects, within the distance R_c. This approach, however, has two inherent features which limit efficiency. First, in order to determine which objects are within R_c, separation distances must be computed for many objects which are often not "close." The second is due to the required decision-making process, which interferes with efficient vectorization.

It is always possible to associate with a set of points in a 3D space, grid indices that are ordered according to their relative positions; this is described in FIGURE 1. Such an indexing scheme can be used to construct a grid where adjacent particles in space have close grid indices. A "monotonic" mapping of the positions of objects (X, Y, Z) onto the MLG indices (i, j, k) obtains if the conditions

$$X(i, j, k) < X(i + 1, j, k); \quad 1 < i < NX - 1$$

$$Y(i, j, k) < Y(i, j + 1, k); \quad 1 < j < NY - 1$$

$$Z(i, j, k) < Z(i, j, k + 1); \quad 1 < k < NZ - 1 \quad \text{(1)}$$

are satisfied, where the total number of objects equals $NX \times NY \times NZ$.

Using an MLG to index positions and attributes of particles in computer memory permits a tracking algorithm to be based on a "maximum indexing offset," N_c, rather than a short range "cutoff," R_c. N_c is selected such that no "close" object can have a grid index offset from the grid index of a "target object" which is larger than N_c. This condition removes the necessity of having to test distances. For each "target object" in the system, computations are performed for all objects with index offsets less than N_c. Although this approach can entail computing interactions for "far" objects, it still represents a substantial reduction in computational cost. Further, "close" objects will be indexed via contiguous memory, thus permitting efficient vectorization of computations and compact data bases.

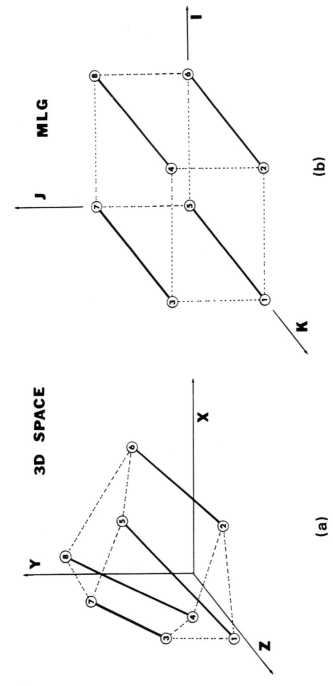

FIGURE 1. (a) Objects distributed randomly in 3D space; and (b) regular Monotonic Logical Grid (MLG) for indexing object positions in space. Comparison of (a) and (b) shows that the natural ordering of spatial coordinates defines a pattern that may be mapped onto an ordered grid structure. This MLG is optimum for representing the average separations of objects located randomly within some spatial domain. This means, for example, referring to FIGURE 1, that if one were to sample object positions over a sufficiently long time, the objects with MLG indices corresponding to, for example, "location" 8 would on average have the largest separation from objects with grid indices corresponding to "location" 1.

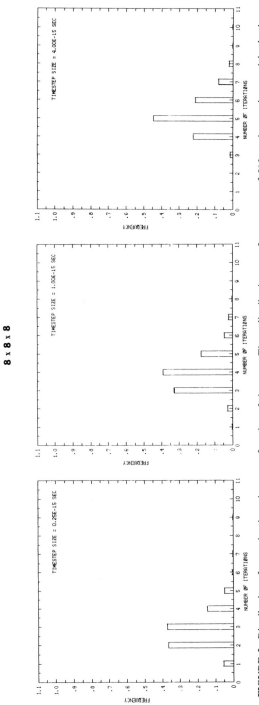

FIGURE 2. Distribution of swapping iterations as a function of timestep. These distributions are for a system of 512 noninteracting particles having random velocities.

In order to assure contiguous storage of positions and attributes the monotonic mapping of positions onto grid indices must be continuously updated. To maintain monotonicity, the MLG indices of any two adjacent objects are exchanged or "swapped" if they pass each other in real space relative to a chosen coordinate direction. The process of "swapping" is iterative and will depend on the extent to which the MLG indexing deviates or "distorts" from monotonicity. Even for large "MLG distortion," the "swapping" process converges rapidly (see FIGURE 2). The relative computational cost of "swapping," which is also vectorizable, is very small.

More than one type of MLG can be constructed for indexing objects within a given spatial domain. LCP is investigating the geometric properties of MLG's [4] to determine optimum grid configurations. The MLG described in FIGURE 1, constructed according to Eq. 1, is a "regular" MLG. One MLG being studied has a "skew periodic" structure, which makes it very efficient for indexing the positions of objects using vector memory.

REFERENCES

1. BORIS, J. 1985. A vectorized "nearest-neighbors" algorithm of order N using a monotonic logical grid. NRL Memo Report 5570. J. Comp. Phys. To be published.
2. LAMBRAKOS, S. G. & J. P. BORIS. Geometric properties of the monotonic logical grid algorithm for near neighbor calculations. J. Comp. Phys. In preparation.
3. HOCKNEY, R. W. & EASTWOOD, J. W. 1981. Computer Simulation Using Particles. Vol. 8: 267–309. McGraw-Hill. New York, NY.
4. VAN GUNSTEREN, W. F. et al. 1984. On searching neighbors in computer simulations of macromolecular systems. J. Comp. Phys. 5: 272.

A Comment on Hamiltonian Parameterization in Kirkwood Free Energy Calculations

ALBERT J. CROSS[a]

Department of Chemistry
University of California, San Diego
La Jolla, California 92093

The difference in Helmholtz free energy, ΔA, between two states of a molecular system can be computed by a technique attributable to Kirkwood sometimes called thermodynamic integration or Hamiltonian warping.[1-4] If a Hamiltonian $H(\mathbf{r}^N, \mathbf{p}^N, \xi)$ which depends on a parameter ξ is constructed so that the two states of interest are described by $H(\mathbf{r}^N, \mathbf{p}^N, \xi_0)$ and $H(\mathbf{r}^N, \mathbf{p}^N, \xi_1)$,

$$\Delta A = A(\xi_1) - A(\xi_0) = \int_{\xi_0}^{\xi_1} d\xi \left\langle \frac{\partial H(\mathbf{r}^N, \mathbf{p}^N, \xi)}{\partial \xi} \right\rangle_\xi \tag{1}$$

The brackets represent an average over a classical canonical ensemble, obtained from molecular dynamics or Monte Carlo at a fixed value of ξ. Since A is a state function, ΔA does not depend on the path from $H(\mathbf{r}^N, \mathbf{p}^N, \xi_0)$ to $H(\mathbf{r}^N, \mathbf{p}^N, \xi_1)$, but the choice of path can have a strong influence on the numerical convergence of approximations to Eq. 1.

FIGURE 1 illustrates an example of such a ΔA calculation for a one-dimensional system described by a Lennard-Jones potential. A parameterization of H which is linear in ξ is perhaps the simplest one, since this gives rise to a $\partial H(r, \xi)/\partial \xi$ which is ξ-independent. However, the ensemble average generates an integrand which becomes very large near zero; nearly half of the contribution to the integral is between ξ values of 0.0 and 10^{-3}.

The nonlinear parameterization of H consisting of scaling σ and ϵ linearly by ξ gives rise to an integrand which is nearly constant over the entire range of ξ, and thus requires much less dense sampling to obtain good convergence of approximations of ΔA. For this example, ΔA obtained by the average of the end points of the dashed curve differs by only 1% from the true ΔA. The faster numerical convergence of nonlinear parameterization should be particularly useful in calculating the nonbond contribution to ΔA for insertion of a particle into a dense environment such as a liquid or the interior of a macromolecule.

[a]Present address: BIOSYM Technologies, Inc., 9605 Scranton Rd., Suite 101, San Diego, California 92121.

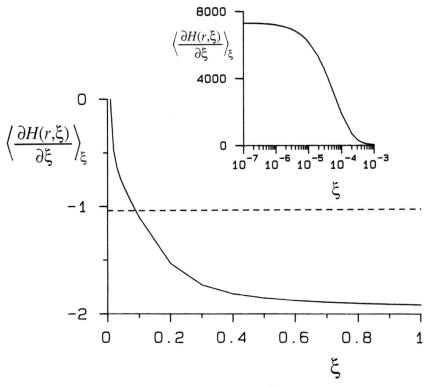

FIGURE 1. Integrands for Equation 1 using linear (—) and non-linear (– –) parameterizations of $H(r, \xi)$. The momentum contribution is omitted since it can generally be computed analytically. The integral of either curve is ΔA between systems with $\sigma = 1, \epsilon = 1$ and $\sigma = 2, \epsilon = 2$ at $k_B T = 10^{-1}$. Correct value of ΔA for this system is -1.03. *Inset* shows rapid increase of linearly parameterized integrand near $\xi = 0$.

REFERENCES

1. KIRKWOOD, J. G. 1935. J. Chem. Phys. **3:** 300–313.
2. MRUZIK, M. R., F. F. ABRAHAM & G. M. POUND. 1976. J. Chem. Phys. **64:** 481–491.
3. MEZEI, M., S. SWAMINATHAN & D. L. BEVERIDGE. 1978. J. Am. Chem. Soc. **100:** 3255–3256.
4. BERENS, P. H., D. H. J. MACKAY, G. M. WHITE & K. R. WILSON. 1983. J. Chem. Phys. **79:** 2375–2389.

Dynamics of Coordinated Water: A Comparison of Experiment and Simulation Results[a]

P. A. MADDEN

Physical Chemistry Laboratory
Oxford OX1 3QZ, England

R. W. IMPEY

Department of Chemistry
National Research Council of Canada
Ottawa K1A 0R6, Canada

INTRODUCTION

The modified behavior of the water molecules in the vicinity of an ion in aqueous solution and its influence on the macroscopic physical properties of the solution is one of the long-standing interests of physical chemistry.[1] Ions have been recognized as "structure-making" or "structure-breaking" in a phenomenological way, by comparing their observed effect upon the macroscopic properties with that predicted by the classical hydration shell models introduced by Frank and Gurney.[2,3] This categorization has proven a most useful guide to predicting trends in a wide range of experiments, but the establishment of a common quantitative description has proven more difficult. Attempts to substantiate these concepts by directly examining, at the molecular level, the state of the water molecules around an ion are relatively recent; most notable is the work of Hertz[4] (NMR) and Enderby and Nielson[5] (neutrons and X rays) and their coworkers. The aqueous environment around an ion is also discussed in the context of the interaction of ions with other molecules, as in electron-transfer reactions, or in their interaction with larger bodies, such as proteins, membranes, and electrodes.[6] Computer simulation studies have much to offer in helping to bring together the information that is available from these different strands of experimental investigation. In this article we will describe how the results of simulations of alkali and halide ions in water have been tied together with the results of molecular-level experiments on electrolyte solutions and used to relate the picture that emerges to the classical concept of hydration.

Thanks to the comprehensive neutron diffraction experiments, using the isotope difference technique, which have been performed on ionic solutions by Enderby and coworkers,[5,7] detailed information on the arrangement of the water molecules around an ion is available. The comparison of these data with simulation results has already

[a]This work was carried out while we were at the University of Cambridge and were supported by the Science and Engineering Research Council (U.K.).

been described in some detail and we will present only a brief survey here.[8] The results described below were obtained with the potentials devised by Clementi et al,[9] with the simulations performed at the experimental density. While it is clear that these potentials have limitations, their ability to reproduce the diffraction data is very good. Some examples are shown in FIGURE 1. Numerous other workers have compared

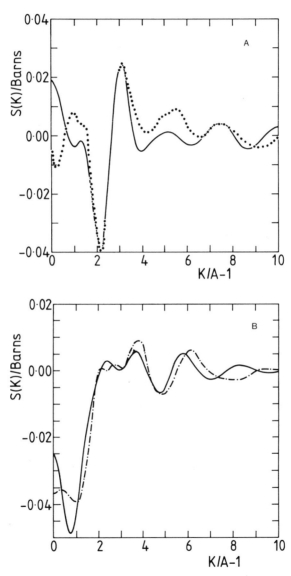

FIGURE 1. First-order difference neutron scattering functions for Li^+ (A) and Cl^- (B). The MD results (*solid line*) are compared with experimental data (*dotted line*) on a 3.57 *M* solution of LiCl (Refs. 5 and 7 and *loc. cit.*).

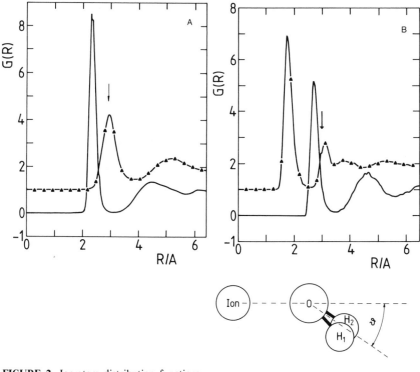

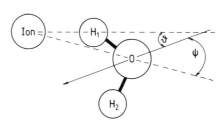

FIGURE 2. Ion-atom distribution functions. The *unbroken curves* show the ion-oxygen and the *curves with triangles* ion-hydrogen for Na^+ (**A**) and F^- (**B**); (**C**) the geometries of coordinated water, with *arrows* in the main diagram showing the expected H-atom positions if the angle θ is zero.

results obtained with different potentials with these data and have obtained similar levels of agreement.[10–12]

When the individual ion-atom distribution functions are viewed in real space, a clear picture of the arrangement of the water molecules emerges. In the first shell the water molecules around the cation and anion are as in FIGURES 2A and B, with the angles θ and ψ distributed over a range about zero. For the F^- ion the expected peak positions when the F^-—H—O "bond" is linear are indicated by arrows. The fact that the g_{NaO} function falls to zero shows the existence of a well-defined first coordination shell for the sodium ion. The coordination number of the sodium ion, defined as the integral of g_{NaO} up to this minimum, is found to be six. The number of hydrogen atoms within the first peak of g_{NaH} is only slightly more than 12, which shows that this peak is made up of the protons of the first shell water molecules, with little penetration from the bulk. The water molecules are appreciably ordered in the second shell, as shown by the relationship of the positions of the second maxima in the g_{NaO} and g_{NaH}, and (in larger simulation systems) a third shell is also found. Substantially the same picture is found for the other alkali atoms, although a much tighter coordination is evident for Li^+ from the narrowness of the peaks in the distribution functions. For K^+ the first minimum is appreciably shallower than for Na^+ and Li^+, suggesting a less well-defined first shell, and the coordination number of the first peak in g_{KH} is more than twice that of g_{KO}, showing appreciable penetration from the bulk. The same trend of weakening coordination with increasing ion size is found for the anions, although, of course, the organization of the water molecules within the shells is different from that for the cations.

The information on the ionic environment that emerges from these structural studies is consistent with the picture[1,13] established from the interpretation of macroscopic measurements, of quantities such as conductivities and solution entropies, on the basis of the Frank-Gurney model. Small (and highly charged) ions are "structure-makers," whereas large ions, such as ClO_4^-, are "structure-breakers." Despite this "structural" terminology it is the altered *dynamic* behavior of the water molecules in the vicinity of the ion that seems most directly related to the ionic effect on some of these observations. For example, data on the variation of ionic mobility with ionic size seem to be naturally interpreted by the concept of a "solventberg" of several molecules which move with the ion.[14] Bockris and Reddy[1] discuss the hydration number as the number of water molecules that "surrender their own translational degrees of freedom and remain with the ion." They show how this dynamic concept of *persisting coordination* may be used to describe thermodynamic data, such as solution entropies and compressibilities, as well as having an obvious relevance to transport properties.

It is with the characterization of the dynamic behavior of the water in the vicinity of the ion with which we shall be primarily concerned here. We have already shown how a measure of persisting coordination may be introduced in simulations and used to obtain hydration numbers that agree well with those deduced phenomenologically. We have also investigated how a measure of the fluidity of the first coordination shell may be used to interpret the variation of the ionic mobility with ion size. Our particular concern here will be the comparison of simulation results with experiments that probe the dynamic behavior of the water molecules at a molecular level—notably NMR, infrared, Raman, and dielectric experiments. In order to establish a basis for this comparison we begin by reviewing the relationship between these experiments and simulated correlation functions for pure water.

2. Molecular Motion in Pure Water

The motion of the water molecules is conveniently described with the orientational and velocity correlation functions. Various characteristics of these functions may be observed in NMR, Raman, infrared and neutron spectroscopies, and it is the relationship between the features observed in the spectra and the properties of the correlation functions that we will review here. More detailed comparisons have been given elsewhere.[15]

2.1 Orientational Correlation Functions

A convenient set of correlation functions for visualization of the molecular reorientation is given by

$$C_\ell^\alpha(t) = \langle P_\ell[{}^1e_\alpha(t) \cdot {}^1e_\alpha(o)] \rangle \qquad (1)$$

where P_ℓ is a Legendre polynomial of rank ℓ and ${}^1e_\alpha$ a unit vector that points along the αth direction (x, y or z) in the molecular coordinate frame of molecule 1, as indicated in FIGURE 3. The $\alpha = x$, y and z functions for $\ell = 2$ are shown in this figure. The remarkable feature of these functions, first noted by Rahman and Stillinger,[16] is the "glitch" at short times. This feature is seen only in fluids with hydrogen bonding; it is a

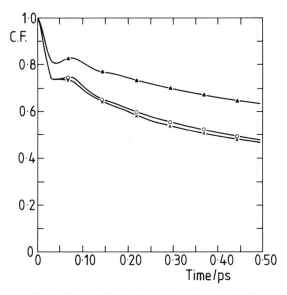

FIGURE 3. The second-rank ($\ell = 2$) orientational correlation functions (Eq. 1): C_2^X (X); C_2^Y (▲), and C_2^Z (O) for pure water at 286K.

signature of the librational motion of a water molecule in the potential of its neighbors. The amplitude of the librational displacement is ~25°. The short-time structure seen in the orientation correlation function may be related to oscillatory structure in the angular velocity correlation functions. The frequency of the oscillation decreases only slightly as the temperature is raised, although the amplitude of the glitch decreases somewhat more. At longer times the orientational correlation functions relax exponentially; the rapid transition from the glitch, which reflects the influence of specific intermolecular interactions, to featureless, diffusive decay is another extraordinary property of these functions. Although the same general shape is found with other potentials,[16,17] the close resemblance of the x and z functions and their decay rate relative to the y function is peculiar to the MCY model.

The orientational correlation functions influence the shape of the infrared and Raman spectra of water (in the "intermolecular" region below 1000 cm^{-1}) which are, in part, determined by the reorientation of the molecular dipoles and polarizabilities. The relevant contribution to the Raman spectrum is given by

$$I^{Ram}(\omega) \, \alpha \int_0^\infty dt \, e^{-i\omega t} \{\alpha_{aa}\alpha_{bb} \sum_{i,j} \langle P_2({}^i e_a(t) \cdot {}^j e_b(o)) \rangle\} \cong N \int_0^\infty dt e^{-i\omega t} \alpha_{aa}^2 \, C_2^a(t) \quad (2)$$

where the α_{aa} are the diagonal elements of the molecular polarizability tensor in the coordinate frame of FIGURE 3 and a and b are summed over x, y and z. The infrared absorption spectrum is given by

$$I^{ir}(\omega) \alpha B(\omega) \int_0^\infty dt \, e^{-i\omega t} \mu^2 \sum_{i,j} \langle P_1({}^i e_z(t) \cdot {}^j e_z(o)) \rangle \cong B(\omega)N \int_0^\infty dt \, e^{-i\omega t} \mu^2 C_1^z(t) \quad (3)$$

where $B(\omega)$ is a weighting factor determined solely by the temperature. We have written the expressions for I^{Ram} and I^{ir} in this form to emphasize the fact that they are both given by "collective" correlation functions, which involve the correlation between the orientations of the different molecules. The collective dipole correlation function involved in I^{ir} is also the one that determines the dielectric relaxation data. However, we have also indicated in these expressions the approximation of replacing the collective correlation function by a single particle one, which we shall actually use in discussing the origin of the librational features found in the infrared and Raman spectra. Note that the infrared spectrum involves only one of the orientational correlation functions—that of the dipole direction—whereas the Raman spectrum involves all three.

The proton NMR relaxation time of light water is determined by the intramolecular dipole-dipole coupling of the protons and thus by the reorientation of the interproton vector, i.e.,

$$T_1^{-1} \, \alpha \int_0^\infty dt \, C_2^y(t) \quad (4)$$

Note that these probes are sensitive to the motions of the water molecule about different axes and, as FIGURE 3 shows, these are not necessarily related.

The process of Fourier transformation of the orientational correlation functions to give the spectra is illustrated in FIGURE 4. The full orientational function may be resolved into an exponential and an oscillatory function which decays very rapidly, i.e.,

$$C_\ell^\alpha(t) = \phi_\ell^\alpha(t) + A_\ell^\alpha \exp(-t/\tau_\ell^\alpha) \quad (5)$$

The exponential gives rise to a Lorentzian feature centered at zero and the oscillating function, ϕ_ℓ^α associated with the glitch, transforms into a high-frequency peak. This peak may be regarded as being derived from the spectrum of the oscillations in the angular velocity correlation function.

The spectra calculated from the simulated correlation functions are in many respects in very good agreement with experimental Raman and infrared spectra. The high-frequency structure is indeed seen as the "librational" 400 cm^{-1}–900 cm^{-1} band systems.[18–20] Furthermore, the effect of temperature and isotopic substitution on the intensity and frequency of these librational bands is recovered. The sharp timescale separation between the transient glitch and subsequent diffusive relaxation is consistent with the form of the dielectric and far-infrared relaxation data.[21] These exhibit an almost perfect Debye relaxation up to quite high frequencies (~ 50 cm^{-1}) followed by a series of resonance peaks.

FIGURE 4 also illustrates how the NMR relaxation time is affected by the shape of the correlation function. T_1 is related to the transform at zero frequency (Eq. 4) to which only the diffusive Lorentzian contributes. However, T_1 is affected by the *proportion* of the orientational decay that is diffusive; consequently the NMR relaxation time is not simply related to the decay constant of the long-time exponential region of the correlation function, in fact (cf., Eqs. 4 and 5)

$$T_1^{-1} \alpha A_2^y \tau_2^y \qquad (6)$$

When the NMR relaxation times calculated in the simulation using the MCY

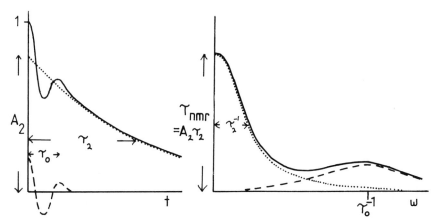

FIGURE 4. Schematic illustration of the Fourier transformation of a function of the shape of the orientational correlation functions of FIGURE 3, to illustrate the origin of the librational features in the infrared and Raman spectra, and the relationship of the NMR relaxation time τ_{NMR} to the decay constant of the exponential $\tau_2 \cdot \tau_0$ is the period of the librational oscillation.

potential are compared with experiment, they are found to be slightly too short (cf. 2.1 psec at 285 K versus 2.5 psec at 295 K).[4]

The comparison between the data on the reorientational motion obtained in these experiments and the behavior simulated with the MCY model is very satisfactory in the sense that the phenomena that are regarded as peculiar to hydrogen-bonded fluids are indeed recovered. (The same may be said of other potential models, for which relatively limited comparisons with experiment have been made.)[16,17] However, at a more detailed level the comparison reveals some shortcomings. In some cases this is because we have compared with experiment in a rather limited way. For example, the infrared and Raman spectra[18-20] show features at circa 60 cm^{-1} and 200 cm^{-1} that are not reproduced in the calculations described above. These arise (see below) because of dipoles and polarizability fluctuations induced by intermolecular interactions, the necessary ingredients for which have not been included in the comparison. In other cases the shortcomings are undoubtedly failures of the potential model. The most notable failures are that the calculated librational features occur at too low a frequency and the diffusive exponential decay of the orientational correlation functions is too rapid. Another significant shortcoming is that when the dielectric properties of the model are calculated,[15,22] the intermolecular contributions to collective dipole correlation function do not emerge correctly. The model fails to give the correct degree of orientational correlation between water dipoles; this means that the dielectric constant of the model is too low and also that the dielectric relaxation time is too short.[23] It is now known that some of these properties are better described by other potential models.

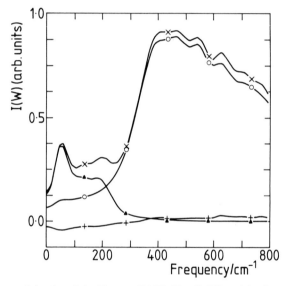

FIGURE 5. Spectral density of the H atom VACF (Eq. 7) (X) and its decomposition into a rotational (O) and center-of-mass translational (▲) component and a cross-term (+) (cf. Eq. 9).

2.2 Velocity Correlation Functions

The velocity correlation function (VCF) which may be examined in most detail is that of the hydrogen atom, i.e.,

$$C_H(t) = \langle V^H(t) \cdot V^H(0) \rangle \qquad (7)$$

where V^H is the velocity of a hydrogen atom. The spectrum of this function may be compared to the spectral density function pieced together from incoherent neutron-scattering data.[24] The time integral (or zero frequency spectrum) is the diffusion coefficient. This (simulated) spectrum is shown in FIGURE 5.

To interpret it, it is convenient to resolve $\mathbf{V}^H$ into a component due to center of mass motion ($\mathbf{V}^C$) and a rotational motion

$$\mathbf{V}^H = \mathbf{V}^C + \mathbf{\Omega} \times \mathbf{r}^H \tag{8}$$

where $\mathbf{\Omega}$ is the angular velocity of the molecule and $\mathbf{r}^H$ is the vector from the molecular center to the hydrogen atom. This gives (schematically)

$$C_H(t) = C_{COM}(t) + C_{ROT}(t) + C_{CROSS}(t) \tag{9}$$

where C_{COM} is the center-of-mass VCF, C_{ROT} is (the second time derivative of) an orientational correlation function and C_{CROSS} is the cross correlation between these two motions. The corresponding resolution of the spectrum is also shown in FIGURE 5. It can be seen that the strongest feature, the broad band extending upwards from about 300 cm^{-1}, is associated with C_{ROT}, i.e., with the same librational motion also seen in the infrared and Raman spectra in the same frequency regime. We have already remarked that such features reflect the spectrum of oscillations in the angular velocity correlation function and that these occur at too low a frequency with the MCY model. The new features at about 60 cm^{-1} and 200 cm^{-1} are associated with oscillations in the center of mass velocity correlation function. The position of these peaks agrees quite well with observed structure in the neutron data[24] and (as we shall see) is substantiated by infrared and Raman spectra. Since the oscillations in C_{COM} are influenced by the molecular mass, whereas those in C_{ROT} are influenced by the moment of inertia, the low-frequency peaks are insensitive to changing the hydrogen atom mass ("deuteration"), whereas the 400-cm^{-1} band is strongly shifted. This too is in accord with experiment.

To see in more detail the origin of the oscillations in C_{COM} we may examine the autocorrelation functions of the components of the center of mass velocity referred to the molecular frame of FIGURE 3. These are shown in FIGURE 6. The structure in the three functions may be described in terms of two damped oscillations of different frequency which beat together; the higher frequency oscillation is least prominent in the correlation function corresponding to the out-of-plane motion. The fact that there are only two frequencies, rather than the three that might be expected, for the rattling of a C_{2v} molecule and the relationship of the three functions may be artefacts of the MCY model. However, as we have noted, they do appear to give a reasonable agreement with experiment.

As we have remarked previously, the Raman and infrared spectra which were calculated from the simulation via equations failed to reproduce features at about 60 cm^{-1} and 200 cm^{-1} which are observed experimentally,[25,26] yet such frequencies are characteristic of oscillations associated with the center of mass motion. These motions may become spectrally active when the dipoles and polarizabilities induced by intermolecular interactions are included in the description of the spectrum. An important component of the interaction-induced dipole in a polar fluid is the dipole-induced dipole ($^{DID}\mu$)

$$^{DID}\mu^i = \alpha^i \cdot \sum_J \mathbf{T}(\mathbf{r}^{ij}) \cdot \mu^j \tag{10}$$

where α^i and μ^i are the molecular polarizability and dipole, respectively, and $\mathbf{T}(r)$ is the dipole field tensor

$$T_{\alpha\beta}(\mathbf{r}) = [3\, r_\alpha r_\beta - r^2\delta_{\alpha\beta}]r^{-5} \tag{11}$$

which gives the field at the origin due to a dipole at the point $\mathbf{r}$. Through the dependence of $\mathbf{T}(r)$ on the relative position of the centers of mass of two molecules, the correlation function of the *total* dipole density of the fluid (which is what is observed experimentally) may relax, in part, because of the molecular translation. Inclusion of this term, and the corresponding dipole-induced dipole polarizability, in the Raman case, leads to a qualitative improvement in the simulated infrared and Raman spectra.[27] Interaction-induced effects are important in determining the distribution of intensity in the spectra of non-hydrogen-bonded fluids,[28] but there they rarely produce

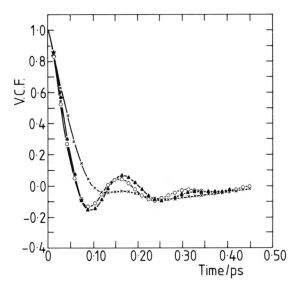

FIGURE 6. The autocorrelation functions of the components of the center-of-mass velocity referred to the molecule-fixed X (X), Y (▲), and Z (O) axes.

new spectral features, as seen in water (although the appearance of symmetry-forbidden transitions is an exception). In fact, in simulations of such fluids it is often found that the (relatively weak) oscillatory features in the angular velocity and COM velocity correlation functions occur at the same frequency.[29] From the viewpoint of simulation this must be regarded as an important characteristic difference between these fluids and water.

2.3 Summary of Results on Pure Water

The extensive comparison between the simulation results obtained with the MCY model and experimental data on molecular motion has shown that potentials of this type are capable of reproducing the distinctive spectral features associated with a hydrogen-bonded liquid. In this way an interpretation of these features in terms of properties of correlation functions is obtained, which is directly comparable with the interpretation of the spectra of other fluids. This description is at least complementary to the traditional description of water spectra in terms of equilibria between several special structures.[6]

It should be clear, even leaving aside the shortcomings of the intermolecular potential, that the comparison of simulations with dynamic experiments is not usually straightforward. Effects such as the interaction-induced terms, which we have mentioned above, and the fact that observed correlation functions are usually collective in nature and not necessarily related to single-particle motion (which we have ignored) make detailed comparison a matter of considerable complexity. Quantitative comparisons of both intensities and line positions have not been made for water, although, in our opinion, sufficient is now known about the properties of other model potentials for this to be a reasonable undertaking. A good deal of successful work has been done on comparing experiment and simulation at a quantitative level on unassociated liquids and this should serve as a good guide.[30]

3. The Dynamics of Water Molecules Around an Ion

As we have shown, the simulated dynamic behavior of water molecules in the pure fluid is consistent with experimental observations; further, the structure of the water molecules in the simulations of ions in water is consistent with the experimental diffraction data. We therefore approach with some hope the comparison of the simulated dynamic behavior in the vicinity of an ion with such experiments as probe this behavior. The objectives of this comparison are twofold: On the one hand, the extraction of information on coordinated water from spectroscopic experiments on electrolyte solutions is a much more complicated task than in the pure fluid and we might hope to help unravel these observations with insights provided by the simulation. In addition, we hope to show that, insofar as they may be judged against existing experimental information, these simulations, with relatively simple model potentials, are providing a good description of coordinated water so that results obtained from such simulations on quantities for which no experimental probe exists may be accepted with some confidence.

The potentials used in these simulations were those devised by Kistenmacher[9b] et al., and the MCY[9a] model is used for the water-water interactions. The simulations involve one ion in 64 water molecules and the Ewald summation was used to describe the Coulombic interactions. Other details have been described in References 8 and 31.

In this work we are attempting to mimic, as closely as presently feasible, the environment of the real alkali and halide ions. Complementary to our study is that of Geiger,[32] who, by using model potentials in which the ion-water interaction parameters were varied systematically, investigated the origin of the ionic effects on the structure and motion of coordinated water. In particular his work illuminates the nature of the "structure-breaking" ions, whereas the ions we have studied are "structure-making" or neutral.

3.1 Residence Times

A crucial parameter in discussing the interpretation of experiments which probe the dynamic behavior of water molecules in aqueous solutions is the residence time, τ_R. τ_R is the mean time for which a given water molecule remains in the first coordination shell of an ion before returning to the bulk. If τ_R is short compared to the characteristic timescale associated with an experimental observation, then the experiment will see only the average behavior of the water molecule in the bulk and coordination shell environments.

For solutions of diamagnetic ions, such as the alkali and halide ions, residence times cannot be directly measured using existing techniques.[4] For divalent paramagnetic ions, residence times have been measured by NMR and these times have been confirmed in quasielastic neutron studies.[7] On the basis of NMR measurements, it has been inferred that the residence time for the lithium ion is about 30 psec[33]; this is expected to be the longest value of the τ_R for the alkali and halide series.

Residence times may be conveniently measured in the course of a simulation.[8] A characteristic radius of the first coordination shell may be established from the position of the first minimum of the ion-oxygen atomic distribution function and a water molecule is considered coordinated if its oxygen atom is within a spherical shell of this radius. We may define a quantity which characterizes the exchange of water molecules with the bulk as

$$n_{ion}(t) = \frac{1}{N_t} \sum_{n=1}^{N_t} \sum_j Q_j(t_n, t; t^*) \tag{12}$$

Here $Q_j(t, t_n; t^*)$ is a property of water molecule j; it takes the value 1 if molecule j lies within the shell at both timesteps t_n and $t + t_n$ and has not left the shell for longer than t^* in the interim; otherwise it is zero. The sum over n is over the times steps of the simulation; it follows immediately that $n_{ion}(o)$ is just the coordination number of the ion, whereas $n_{ion}(t)$ gives the mean number of molecules that remain within the shell after the elapsed time t. The introduction of t^* is an attempt to improve upon our geometrical definition of coordination by allowing a molecule to temporarily leave the shell. It was given the value of 2 psec, which is characteristic of the timescale on which water molecules exchange neighbors in the bulk. The function $n_{ion}(t)$ has the same general form for all the alkali and halide ions, but at short times it decays exponentially

$$n_{ion}(t) = n_{ion}(o) \exp(-t/\tau_R) \tag{13}$$

As this expression shows, we regard the decay time of the exponential as a measure of the residence time.

Values for the residence times obtained in this way are shown in TABLE 1. The order of times corresponds to the order of increasing strength of coordination, inferred from the structural studies, and with expectations based upon the strength of the coulombic interactions between ion and water molecule. The result for Li$^+$ is consistent with the NMR-based estimates.

Apart from their role in the interpretation of spectroscopic experiments, which we shall discuss below, the residence times that we have defined are clearly relevant to the concept of persisting coordination which is at the heart of the idea of hydration, as discussed by Bockris.[1] We have explored this relationship elsewhere[8] and shown how the residence times may be used to give a plausible way of estimating a hydration number, values of which agree well with experiment. We have also discussed how the residence times may be used to rationalize ionic mobility data in a way that avoids the introduction of structural notions, such as the "solventberg."

3.2 A Survey of the Experimental Situation

The residence times are much shorter than the timescales of the various NMR experiments which may be made on the nuclei of a water molecule, and its isotopic variants, in diamagnetic electrolyte solutions. For the most part[4] these experiments probe the relaxation induced by intramolecular interactions, such as the dipole-dipole coupling between protons, which are modulated by the tumbling motion of the water molecules. Because of the relative timescales, the NMR experiments see a weighted average of the tumbling behavior of water in the first coordination shell and in the bulk; the rotational correlation time associated with coordinated water may be extracted by varying the ionic concentration.

TABLE 1. Residence Times Calculated in the Simulations from Equations 12 and 13

Ion	T (K)	π_R (psec)
Li^+	278	33
Li^+	368	6.0
Na^+	282	10
K^+	274	4.8
F^-	278	20
Cl^-	287	4.5
H_2O	286	4.5

Information about the long-time reorientation of the dipole axis is contained in dielectric relaxation data; even in ionic solutions the polarization caused by the fluctuations in the molecular dipole orientation dominates the observation.[21,34] However, it must be borne in mind that the relaxation time measured in dielectric experiments is that of a collective orientational correlation function (Eq. 3) and that for pure water the difference between this time and the corresponding single-particle relaxation time is considerable.[21] This difference may be attributed to the short-range orientational correlation between dipoles, as measured by the Kirkwood g-factor.[23] For this reason inferences about the effect of adding ions on the single molecule dynamics must be treated with reserve. It is equally plausible that the influence on the observed relaxation time is due to a change in the static orientational correlation induced by the presence of ions.

A somewhat more direct handle on the ion water interactions exists in solutions containing $^{19}F^-$ and $^7Li^+$, for which the ionic nuclei themselves have appreciable magnetic dipoles.[4] The ion-proton dipole-dipole-induced relaxation of either the proton or ionic nuclear spins may then be observed; this is determined by the motion of the vector between the ion and the proton, which may be appreciably different from the tumbling sensed by the intramolecular relaxation. A similar intermolecular motion may be studied in yet another way: $^7Li^+$, $^{35}Cl^-$ and $^{23}Na^+$ have appreciable nuclear quadrupoles and the NMR relaxation of these nuclei is caused, in part, by the fluctuations in the electric field gradient at the nucleus caused by interactions with water molecules.[4] The nature of this interaction has been characterized in *ab initio* electronic structure calculations[35] so that it is now possible to compare measured relaxation times of these quadrupolar nuclei with calculations.[36]

As we have seen in the last section, the NMR relaxation times are sensitive to the long-time diffusive behavior of the water molecules. The short-time librational and oscillatory translational features seen in the angular velocity and velocity correlation functions of the water molecules affect infrared, Raman, and inelastic neutron-scattering spectra.[6,18–20] The timescale situation in these experiments should be in the opposite limit to that found in NMR. The characteristic observation time should be related to the dephasing time of the oscillatory motions. From the widths of the observed bands and from ideas gleaned on simpler fluids one would expect this time to be no longer than 10^{-13} seconds, which is comfortably shorter than the residence times. If the librational and translational oscillations of the water molecules were appreciably shifted by moving into the coordination shell from the bulk this should lead to new bands. However, for these to be observable it is also necessary for the shift to be significant compared to the bandwidth found in pure water, otherwise they will be obscured by the bulk signal. Another motion that should, in principle, be observable is the oscillatory translational motion of the ion relative to the surrounding water molecules.[20] This should appear in an inelastic neutron study of a strongly incoherent

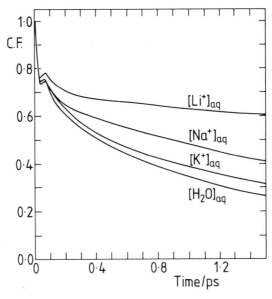

FIGURE 7. The C_2^z correlation function for water molecules in the first coordination shell of the cations and for bulk water.

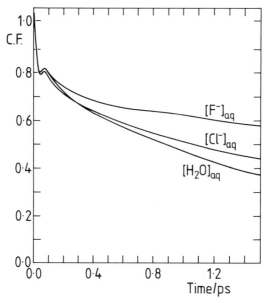

FIGURE 8. The C_2^z correlation function for water molecules in the first coordination shell of the anions and for bulk water.

scattering ion in D_2O (for which the incoherent scattering is low), but unfortunately no such experiment appears to have been done. However, as we have mentioned in connection with experiments performed on bulk water, interaction-induced dipoles and polarizabilities provide a mechanism by which translational motions may also influence infrared and Raman spectra.

3.3 Water Reorientation in the First Coordination Shell

The reorientational motion will be discussed in terms of the same orientational correlation functions $C_\ell^\alpha(t)$ discussed in Section 2.2. However, only water molecules that are members of the first coordination shell of the ion are included. The same geometric definition of a coordination shell is used as in the last section and time correlation is only continued while the molecule is still a member, formally:

$$C_\ell^\alpha(t) = \frac{1}{N_t} \sum_{n=1}^{N_t} \sum_J P_\ell({}^j e_\alpha(t + t_n) \cdot {}^j e_\alpha(t_n)) \times Q_j(t, t_n; 0) \qquad (14)$$

The orientational correlation functions of the dipole axis $C_2^z(t)$ for the three cation systems Li^+, Na^+ and K^+ are shown in FIGURE 7, together with the corresponding function for bulk water. FIGURE 8 shows the same set of functions for the anions. Very similar plots are found for the C_ℓ^y and C_ℓ^z.

The most remarkable feature of these figures is that the glitch at short times is almost unchanged from the bulk water case. This similarity of the short-time behavior suggests that the infrared, Raman, and inelastic neutron spectra of the electrolyte solutions, in the librational regime, should resemble those in pure water. We shall discuss the comparison with these experiments in Section 3.5, where we also discuss the short-time translational motion.

In contrast, the diffusive, exponentially decaying region of the correlation functions displays marked differences in the various systems. We therefore look to the NMR relaxation data on the intramolecular dipole-dipole coupling for confirmation of these effects. The relaxation time measured in this way is actually the time integral of $C_2^x(t)$. In TABLE 2 we give values for the time integrals of the x, y and z correlation functions for $\ell = 2$, for Li$^+$, Na$^+$, K$^+$, F$^-$ and Cl$^-$, and we include the bulk water values for comparison. In all cases these times are longer than in pure water, indicating that the rotational motion of the coordinated water molecules about any axis is hindered. The three cation systems show a well-defined trend in the calculated reorientation times. The water molecules in the first coordination shell of the lithium ion are strongly hindered, particularly with regard to the reorientation of the z-axis, (the dipole direction), which can be understood in terms of the geometry of the coordinated water (FIG. 2). The ion-water complex is far less stiff for Na$^+$, while the reorientation times

TABLE 2. Calculated $\ell = 2$ Reorientation Times[a]

Ion	Temperature (K)		Time (psec)		
			X	Y[b]	Z
Li$^+$	297	$A_2\tau_2$	3.7	5.8	8.0
Na$^+$	298	$A_2\tau_2$	1.4	2.3	2.1
K$^+$	274	$A_2\tau_2$	1.1	2.1	1.4
F$^-$	278	$A_2\tau_2$	3.9	4.8	5.2
Cl$^-$	287	$A_2\tau_2$	1.1	2.1	1.4
H$_2$O	286	$A_2\tau_2$	1.03	2.01	1.07

[a]Defined as the integral of the normalized correlation function.
[b]The set labeled Y is related to the NMR intramolecular dipole-dipole relaxation times.

for the K$^+$ system are very similar to those of bulk water. In the anion systems the reorientational motion is appreciably hindered for the F$^-$ ion, for which the fluidity of the coordination shell seems lower than for Na$^+$, but for Cl$^-$ the reorientation rates are almost the same as in pure water.

As we have indicated, the intramolecular relaxation times for the protons in the first coordination shell are extracted from the measured relaxation time by calculating the proportion of water molecules in the coordination shell at a given ionic concentration.[4] We may use the coordination numbers determined by neutron scattering and by simulation to remove one degree of uncertainty in the results given by Hertz. We then find ratios of NMR reorientation times for coordinated to bulk water, which we may conveniently compare with the same ratio (for the C_2^x correlation functions) obtained from simulation. This is done in TABLE 3; the two sets of results agree extremely well, with the possible exception of the Na$^+$ ion.

Our ability to directly compare with the dielectric data is somewhat limited; we have not attempted to calculate the collective dipole correlation function whose properties are actually observed. However, we can present data on the single-particle

TABLE 3. Ratio of NMR Relaxation Time for Coordinated Water to That in the Bulk (from Simulation) Compared with the Experimental Ratio Given by Hertz[4]

	Simulation	Experiment
Li^+	2.9	2.4
Na^+	1.2	1.6
K^+	1.0	0.9
F^-	2.4	2.3
Cl^-	1.0	0.9

relaxation time of the dipole axis for water molecules in the coordination shell and compare them with the behavior of this quantity which has been deduced from dielectric relaxation measurements.[34] These data are given in TABLE 4; they are the relaxation times that characterize the exponential region of $C_1^z(t)$. (A complete set of relaxation times is given in Reference 31.) We note that all the dipole relaxation times are slowed with respect to pure water; this is particularly true of the cation-coordinated water, for which the dipole points towards the ion. We also note that the dipole correlation times are very similar to the residence times (which is suggestive of a mechanism for the coordination shell break-up). This further suggests that the dielectric relaxation should take place on the same timescale as the exchange of coordinated water with the bulk so that a distinctive relaxation feature for coordinated water should not appear. This is in accord with what is observed.[34] The relaxation is basically Debye-like (i.e., almost semicircular Cole-Cole plot), although less so than for pure water.

Pottel[34] has noted some important differences between the behavior of the dielectric relaxation time and the NMR times measured by Hertz et al.[4] For anions and for large cations the data are consistent and the dielectric data may be interpreted by considering the altered rate of molecular tumbling in the coordination shell. However, for the small cations Li^+ and Na^+ it is found that the dielectric relaxation rates increase with respect to pure water in apparent contradiction to the NMR results and our own findings. It has been suggested that this result may be understood if the water dipoles in the first coordination shell do not contribute to the dielectric polarization of the solution. The data on the static dielectric permittivity may be taken to show that up to 10% of water molecules in a $1M$ NaCl solution are in this category.[21,34] It has been argued that this is because the dipole of a Na^+ or Li^+ coordinated water molecule is tightly constrained to point towards the ion and cannot be reoriented by an external field.[6,21] Such a water molecule could still give NMR

TABLE 4. Reorientation Times of the Dipole Axis Defined as the Relaxation Time Characterizing the Exponential Regime of $C_1z_{(t)}$

	Temperature (K)	$\tau_1 z$ (psec)
Li^+	297	32
Na^+	282	11
K^+	274	6
F^-	278	14
Cl^-	287	3.8
H_2O	286	3.8

relaxation since it could tumble about the dipole axis; for anions the dipoles do not point towards the ion and they may reorient by tumbling about the X^-—H—O bond, so that the same effect should not occur. However, this "single-molecule" interpretation of the findings does not hold up, as our results show that a molecular dipole *does* reorient in space and therefore, by linear response theory, can be oriented by a suitable external field. A more tractable explanation is to note (again) the collective nature of the dielectric experiment and the fact that the water molecules in the first coordination shell are mutually arranged so that their *net* dipole is very small (it would be zero for perfect tetrahedral or octahedral coordination). The total dipole correlation function must therefore contain interference terms between the dipoles of different molecules within the shell such that the contribution of a shell to the total dipole density correlation function comes only from the *fluctuations* about the perfectly coordinated structure. It is these *collective* fluctuations that are suppressed by the coulomb field in Na^+ and Li^+.

This idea has been built into an attractive model of the dielectric relaxation data by Giese.[37] He considered the effect on the polarization response of allowing water molecules to enter and leave a coordination shell in which the polarizability was low. The characteristic timescales for this process, which he deduced by examining the experimental data with his model, agree very well with the residence times we have obtained in the simulations.

3.4 Intermolecular NMR Relaxation

The NMR experiments that directly probe ion-water interactions are affected by both the rotational and translational motion of a water molecule with respect to the ion. The magnetic dipole-magnetic dipole relaxation measures the correlation time[4]

$$\tau_{DD} = \int_0^\infty dt \, \langle T_{zz}[\mathbf{r}_i^H(t)] \, T_{zz} \, [\mathbf{r}_i^H(o)] \rangle \qquad (15)$$

where T is given in Eq. 11 and $\mathbf{r}_i^H$ denotes the vector joining proton i to the ionic center.

For cationic coordination this is sensitive to the reorientation of the (rigid) coordination complex as a whole, to the exchange of molecular positions within the complex, diffusion of a water molecule out of the complex and, to a limited extent, to rotation of a water molecule about the O—M^+ bond. By comparing the rate of relaxation of protons in $^7Li^+$- and $^6Li^+$-containing solutions[38] and of the $^7Li^+$ nucleus in H_2O and D_2O[39] solutions, values of τ_{DD} are obtained. The values should be equal, but in fact the latter is some 40% shorter. The somewhat different rate of water reorientation in H_2O and D_2O solutions could contribute to this discrepancy. The values quoted by Hertz[4] were obtained using a coordination number of four and a Li^+-proton distance of 2.7 Å. They should therefore be corrected to be consistent with the coordination geometry determined by neutron studies (5.5 and 2.57 Å) and confirmed in the simulations. The corrected value (from the proton relaxation time) indicates a value of the order of 14 psec for τ_{DD}. The fact that this time is substantially shorter than both the residence time and the estimated reorientation time of the rigid coordination complex indicates that an important contribution comes from motion of water molecules within the coordination complex.

The dipole-dipole relaxation time may also be measured for the interaction of an F^- ion with the protons on a water molecule. The geometry of the coordination complex found in the simulation studies supports the illlustration of FIGURE 2b, with a coordination number of 5.8 and a distance between the ion and nearest proton of 1.73 Å.[8] If we combine this information with the NMR relaxation time quoted by Hertz we find τ_{DD} = 4 psec. This time is considerably shorter than the residence time found in the simulations, suggesting that intrashell motions make an important contribution. We note that this value for τ_{DD} is quite similar to the reorientation times of the molecular x and y axes, as quoted in TABLE 2. It could be that reorientation of a water molecule about the dipole direction, so as to interchange the identities of the nearest neighbor protons, is an important effect.

It is clear that the interpretation of the dipolar relaxation rates could be improved by the direct calculation of τ_{DD} (Eq. 15) in a simulation. This is not a difficult task, but it has not been done. Fortunately, for the other intermolecular mechanism, the quadrupole relaxation of $^{23}Na^+$, $^{35}Cl^-$ and $^7Li^+$, direct calculations of the relaxation time have been attempted. These have been described in detail in a paper by Engström et al.[36] and so we shall only summarize the salient points briefly here.

The electric field gradient at the nucleus (**F**) was obtained from *ab initio* electronic structure calculations[35] and "parameterized" into a two-body interaction between the ion and a water molecule. For Li^+ **F** is caused by overlap interactions between the ion and water molecule, but for more polarizable ions it is proportional to the field gradient (**E**) caused by the permanent charge distribution of the water molecules

$$\mathbf{B} = (1 + \gamma)\mathbf{E} \tag{16}$$

γ is known as the Sternheimer shielding factor. The calculated relaxation times for the ionic nuclei were in good agreement with experiment for $^{23}Na^+$ and $^{35}Cl^-$ (approximately 50% too large), but about a factor of 4 too small for Li^+. The time correlation functions of F_{ZZ} were calculated and found to exhibit a quite complex decay with a very rapid initial relaxation (on a timescale of order 0.1 psec) followed by a relatively slow decay on a timescale of the order of 1 psec. This behavior was not anticipated by the theories of quadrupole relaxation, which predict single exponential decay.[40,41]

To illustrate what may be learned from the quadrupole relaxation it is convenient to consider the case of those ions for which Equation 16 holds and to further simplify by imagining that the sources of **E** are simply the water molecule dipole moments. We then have

$$\mathbf{E} = \sum_{J} \nabla \mathbf{T} (\mathbf{r}_j) \cdot \mu^j \tag{17}$$

where $\mathbf{r}_j$ is the vector from the ion to the center of the water molecule j and **T** is given by Eq. 11. The NMR relaxation time is given by

$$T_1^{-1} \alpha \int_0^\infty dt \langle F_{ZZ}(t)F_{ZZ}(o) \rangle$$

$$= (1 + \gamma)^2 \int_0^\infty dt \sum_{j,k} \langle \nabla_Z T_{Z\alpha}[r_j(t)] \nabla_Z T_{Z\beta}[r_k(o)] \mu_{j,\alpha}(t)\mu_{k,\beta}(o) \rangle \tag{18}$$

The important thing to notice about this correlation function is that it is a *collective* correlation function, i.e., it involves the correlation between the dipoles on different molecules; in this it is quite unlike the correlation function that determines τ_{DD}. In fact, as we have already seen in the case of the similar collective dipole correlation function observed in dielectric relaxation, the $j \neq k$ terms are likely to make an important contribution to the relaxation behavior. In the simulations[36] the importance of these terms was brought out and the lack of correspondence between the two characteristic times observed and any of the single molecule relaxation times calculated in the simulation was noted.

3.5 Velocity Correlation Functions and the Infrared and Raman Spectra

The center of mass velocity correlation functions for the water molecules in the first coordination shell were calculated from

$$C_{COM}(t) = \frac{1}{N_t} \sum_{n=1}^{N_t} \sum_j V_j(t + t_n) \cdot V_j(t_n) Q_j(t, t_n; o) \tag{19}$$

For the cation systems these functions are shown in FIGURE 9 and for the anions in FIGURE 10. As with the reorientational functions, the remarkable feature is the close resemblance to the corresponding functions in pure water, on this short timescale. We recall that in pure water the structure may be described as the result of two characteristic oscillations with frequencies of ~ 60 cm^{-1} and ~ 200 cm^{-1}. If anything, the first minima in the most strongly coordinated cases (Li$^+$ and F$^-$) occur slightly earlier than in the bulk.

The close similarity of both the orientational and velocity correlation functions to those of pure water is consistent with what is observed in the librational and translational region of the inelastic neutron on Raman and infrared spectra of alkali halide solutions.[18-20] Numerous studies have confirmed the absence of new features associated with coordinated water despite the anticipated favorable situation of the experimental timescale with respect to the residence times noted in Section 3.2.

Systematic shifts of both the librational and translational bands to lower frequencies are observed, with the largest shifts being obtained for the most strongly hydrated ions. However, the size of these shifts is in every case considerably smaller than the widths of the bands. The asymmetry of the bands also increases. The data are therefore consistent with the simulation finding that the systematic shift in the librational and translational frequencies is well within the spread of frequencies found in pure water, as reflected in the spectral bandwidths.

The most notable change in the infrared and Raman spectra is the change in the intensity of the bands.[20,42] In the infrared spectra the same effect (an increase in intensity) is found for anions and cations, but in Raman spectra anions cause a marked increase in intensity while cations cause a decrease. The origin of these changes is quite complex. We recall that the infrared and Raman experiments actually measure a collective quantity so that intermolecular correlation effects may be involved. Also, a change of intensity may reflect a change in the amplitude of a particular motion. Alternatively, interaction-induced effects may alter band intensities. In view of the distinctive difference detected by infrared and Raman spectroscopy this must be playing a significant role.

Although no new characteristic frequencies are found that characterize the coordinated water, oscillatory structure does appear in the velocity correlation function of the ions (see FIGURE 11). We have discussed the origin of this structure elsewhere[8] (see also Ref. 43). In the tightly hydrated ions the frequency of this oscillation is quite different from the natural frequencies of water. Induced dipoles and polarizabilities caused by the ion-water interaction will depend upon the distance between the ion and the water molecule. Consequently, we may hope that the oscillatory structure of the ion VCF may appear in the solution's infrared and Raman spectra, by virtue of the same argument which accounts for the appearance of translational bands in pure water.[27]

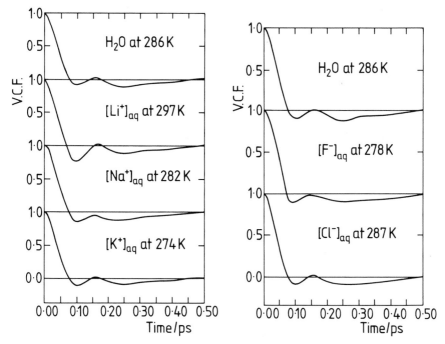

FIGURE 9. The VACFs for water molecules in the first coordination shell of the cations and for bulk water.

FIGURE 10. The VACFs for water molecules in the first coordination shell of the anions and for bulk water.

The new frequencies which appear in the ionic VCFs appear at ~ 400 cm^{-1} for Li$^+$, at ~ 260 cm^{-1} for F$^-$ and at ~ 260 cm^{-1} for Na$^+$; for K$^+$ and Cl$^-$ the structure in the VCFs gives the same frequencies as for a water molecule. Some support for these findings comes from the Raman studies of Nash et al.[44] In solutions containing Li$^+$ they report a band at ~ 380 cm^{-1}, which they assign to the "F$_2$ mode of the Li$^+$ ion in its tetrahedral cage" and a band at ~ 440 cm^{-1}, which they assign to the A$_1$ mode of the coordination complex (i.e., the breathing vibration of the water). If one accepts these assignments then one should compare the former with the simulation frequency. It might be expected that this type of oscillation should show up better in infrared spectra. Bands at about the right frequencies to correspond to these ionic oscillations are seen in infrared spectra of the alkali ions when these are dissolved in aprotic solvents.[20] It could be that they are simply obscured by the water translation bands in aqueous solutions.

4. CONCLUSION

We have surveyed the comparison of computer simulation and experimental information on the motion of water molecules in the first coordination shell of the alkali and halide ions. We have only presented properties gleaned from simulated orientational and velocity correlation functions, and in some cases the interpretation of experimental data could have been sharpened by calculating the correlation functions that are directly observed, such as the collective dipole correlation function observed in

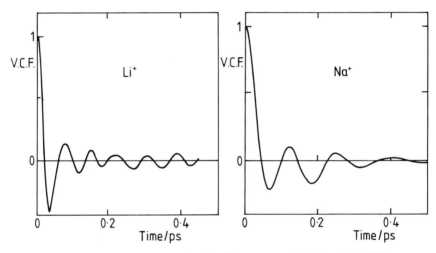

FIGURE 11. The VACFs of Li^+ and Na^+ in aqueous solution (see also Ref. 8).

dielectric experiments or the intermolecular properties observed in some NMR experiments. Further, we have used only the intermolecular potentials devised by Kistenmacher *et al.*[9b] and the MCY potential[9a] for water (although the dielectric and librational properties of this model are incorrect), and we have not evaluated other potentials. Despite these limitations, which could easily be removed in future work, the comparison of simulation and experiment has proven remarkably successful. For all the experiments the simulations have mirrored the observed behavior. In some cases, notably those experiments for which a detailed numerical analysis is available (e.g., NMR), the agreement is quantitative. In dealing with intermolecular infrared and Raman spectra the simulation has reproduced the insensitivity of the experiments to replacement of water molecules by ions, which seems extraordinary when the differences between the intermolecular potentials involved is considered.

One area that emerges from our survey as particularly meriting further attention is the distinction between the relaxation of collective correlation functions, as in the dielectric experiments and in the NMR quadrupole relaxation, and the corresponding single-particle functions. An understanding of these effects is important as they bear upon the nature of the "dielectric saturation" experienced by water molecules in the ionic field. This phenomenon is known to play an important role in many biophysical problems.[6]

SUMMARY

Simulations are of great value in clarifying the interpretation of spectroscopic observations. They may help in unravelling from what is usually a complex observation simpler quantities which directly reflect molecular motion. Also, simulations may be used to enhance the information that may be derived from experiment by helping us to get an improved "handle" on molecular behavior. These principles are illustrated here with examples relevant to NMR, infrared, Raman, neutron and dielectric experiments on aqueous solutions of alkali and halide ions. The objective of these experiments is to probe the altered dynamic behavior of the water molecules in the coordination shell of the ions. The connection between this altered behavior and the hydration number of the ion, as perceived through macroscopic measurements on ionic solutions, is loosely sketched.

ACKNOWLEDGEMENT

We are particularly grateful to I. R. McDonald for helping to get these calculations undertaken.

NOTES AND REFERENCES

1. BOCKRIS, J. O'M. & A. K. N. REDDY. 1973. Modern Electrochemistry. Plenum. New York, NY.
2. FRANK, H. S. & M. J. EVANS. 1945. J. Chem. Phys. **13:** 507.
3. GURNEY, R. W. 1953. Ionic Processes in Solution. McGraw-Hill. New York, NY.
4. HERTZ, H. G. In Water: A Comprehensive Treatise, Vol. 3. F. Franks, Ed.: 301. Plenum. New York, NY.
5. ENDERBY, J. E. & G. W. NIELSON. 1981. Rep. Prog. Phys. **44:** 593.
6. CONWAY, B. E. 1981. Ionic Hydration in Chemistry and Biophysics. Elsevier. New York, NY.
7. ENDERBY, J. E. 1983. Annu. Rev. Phys. Chem. **34:** 155.
8. IMPEY, R. W., P. A. MADDEN & I. R. MCDONALD. 1983. J. Phys. Chem. **87:** 5071.
9a. MATSUOKA, O., E. CLEMENTI & M. J. YOSHIMINE. 1976. J. Chem. Phys. **64:** 1351.
9b. KISTENMACHER, H., H. POPKIE & E. CLEMENTI. 1973. J. Chem. Phys. **58:** 5627, 5842.
10. SZÁSZ, G. I., K. HEINZINGER & W. O. REIDE. 1981. Z. Naturforsch Teil A **36:** 1067.
11. MEZEI, M. & D. L. BEVERIDGE. 1981. J. Chem. Phys. **74:** 6902.
12. BOUNDS, D. L. 1985. Molec. Phys. **54:** 1335.
13. FRANKS, F., Ed. 1973. Water: A Comprehensive Treatise, Vol. 3. Plenum. New York, NY.
14. KAY, R. L. In Franks,[13] p. 173.
15. IMPEY, R. W., P. A. MADDEN & I. R. MCDONALD. 1982. Molec. Phys. **46:** 513.

16. RAHMAN, A. & F. H. STILLINGER. 1976. J. Chem. Phys. **55:** 3336; 1972. J. Chem. Phys. **57:** 1281.
17. GEIGER, A., A. RAHMAN & F. H. STILLINGER. 1979. J. Chem. Phys. **70:** 263.
18. WALRAFEN, G. E. 1972. *In* Water: A Comprehensive Treatise, Vol. 1. F. Franks, Ed. Plenum. New York, NY.
19. Articles by T. H. Lilley (p. 265) and R. E. Verrall (p. 211) in Franks.[13]
20. DESNOYERS, J. E. & C. JOLICOEUR. 1983. *In* Comprehensive Treatise in Electrochemistry, Vol. 5. B. E. Conway *et al.,* Eds.: 1. Plenum. New York, NY.
21. HASTED, J. B. 1972. *In* Water: A Comprehensive Treatise, Vol. 1. F. Franks, Ed. Plenum. New York, NY.
22. NEUMANN, M. 1985. J. Chem. Phys. **82:** 5663.
23. MADDEN, P. A. & D. KIVELSON. 1984. Adv. Chem. Phys. **56:** 467.
24. PAGE, D. *In* Franks (c.f. Ref. 21).
25. HASTED, J. B., S. K. HUSAIN, F. A. M. FRESCURA & J. R. BIRCH. 1985. Chem. Phys. Lett. **118:** 622.
26. KRISHNAMURTHY, S., R. BANSIL & J. WIAFE-AKENTEN. 1983. J. Chem. Phys. **79:** 5863.
27. MADDEN, P. A. & R. W. IMPEY 1986. Chem. Phys. Lett. **123:** 502.
28. MADDEN, P. A. 1984. *In* Molecular Liquids. A. Barnes *et al.* Eds. Reidel. Dordrecht; and *in* Phenomena Induced by Intermolecular Interactions. G. Bimbaum, Ed. Plenum. New York, NY.
29. MADDEN, P. A. & D. TILDESLEY. 1983. Molec. Phys. **48:** 129.
30. MADDEN, P. A. *In* Computer Simulation, Enrico Fermi Summer School in Physics 96. Italian Physical Society. Bologna, Italy. In press.
31. IMPEY, R. W. 1982. Ph.D. thesis, Cambridge University.
32. GEIGER, A. 1981. Ber. Bunsenges. Phys. Chem. **85:** 52.
33. HUNT, J. P. & H. L. FRIEDMAN. 1983. Prog. Inorg. Chem. **30:** 359.
34. POTTEL, R. *In* Franks.[13]
35. ENGSTRÖM, S. & B. JÖNSSON. 1981. Molec. Phys. **43:** 1235.
36. ENGSTRÖM, S., B. JÖNSSON & R. W. IMPEY. 1984. J. Chem. Phys. **80:** 5481.
37. GIESE, K. 1972. Ber. Bunsenges. Phys. Chem. **76:** 495.
38. FABRICAND, B. P. & S. GOLDBERG. 1967. Molec. Phys. **13:** 323.
39. HERTZ, H. G., R. TUSCH & H. VERSMOLD. 1971. Ber. Bunsenges, Phys. Chem. **75:** 1177.
40. HERTZ, H. G. 1973. Ber. Bunsenges. Phys. Chem. **77:** 531, 668.
41. HYNES, J. T. & P. G. WOLYNES. 1981. J. Chem. Phys. **75:** 395.
42. ARMISHAW, R. F. & D. W. JAMES. 1976. J. Phys. Chem. **80:** 501.
43. NGUYEN, H. L. & S. A. ADELMAN. 1984. J. Chem. Phys. **81:** 4564.
44. NASH, C. P., T. C. DONNELLY & P. A. ROCK. 1977. Solution Chem. **6:** 663.

Hydrophobic and Ionic
Hydration Phenomena[a]

PETER J. ROSSKY[b]

Department of Chemistry
University of Texas at Austin
Austin, Texas 78712

I. INTRODUCTION

Computer simulation studies now provide a powerful approach for the elucidation of the molecular details of solution structure. The productivity of these methods has recently been enhanced by the rapid growth of computational facilities available to researchers. At the same time, new approaches to computer simulation have opened avenues for the study of previously completely inaccessible problems. Perhaps foremost in this area are recent path integral techniques[1-6] which permit the simulation of the structure associated with degrees of freedom that have substantial quantum-mechanical character. A few examples of such simulations in chemical systems are now available,[3-7] and we discuss one of these here.

Of particular importance in deriving physical insight from simulation studies is the ability of the theoretician to study well-defined models which may be either basically a caricature of nature or a realistic representation of a chemical system. The former class of studies allows the investigator to focus on the impact of specific elements of the laws for molecular interaction and hence to obtain relatively unclouded conclusions as to their significance. The latter class focuses on developing experimentally inaccessible structural information.

We consider an example of each type here. In particular, we consider, first, the nature of hydrophobic hydration; that is, the structural consequences of aqueous solvation of nonpolar solutes. The cases we consider are the hydration of a small nonpolar spherical solute[8] and that of the contrasting extreme case of a semi-infinite nonpolar solid that shares a planar interface with the aqueous solvent.[9] These are prototypical of the behavior that may be expected for small solutes, on the one hand, and macromolecular assemblies on the other. Particular attention is paid to the solvent's orientational structure and to the influence of the solute on the solvent-solvent hydrogen bonding.

We then consider a particular case of ionic hydration, but for a charged solute of a special nature, namely an excess electron.[7] This structural study has been carried out

[a]This work was supported by grants from the National Institute of General Medical Sciences and the Robert A. Welch Foundation, and by a grant of computer time from Cray Research, Inc.

[b]Alfred P. Sloan Foundation Fellow and recipient of an NSF Presidential Young Investigator Award, a Dreyfus Foundation Teacher-Scholar Award, and an NIH Research Career Development Award from the National Cancer Institute.

115

using path integral simulation techniques to treat the highly quantum-mechanical solute; these methods are outlined briefly in the next section, along with a description of the electron-water pseudopotential employed. In the analysis, we also focus on the solvation structure of the (localized) hydrated electron, and consider the orientational and radial distribution of solvent. By comparing the results for a very compact charge distribution, similar to that of an atomic ion, with that characteristic of the excess electron, we can evaluate the degree of similarity between the classical charged solute and the quantum case.

In Section II we briefly describe the models used and some elements of the simulation methods. Section III focuses on hydrophobic hydration, while Section IV describes results for charge solvation. The conclusions are presented in Section V.

II. MODELS AND METHODS

All simulation studies reported here are for systems near room temperature. Because of the limitations of space, we do not attempt to describe the simulations considered here in detail, but refer to the literature reports.[7-9] Rather, we focus on the results obtained and the insights into solvation structure that can be obtained from them.

The models we use to describe the interactions among water molecules have the now familiar form of a spherical short-ranged Lennard-Jones interaction and a set of embedded partial electrostatic charges arranged on the nuclear framework.[10] For the studies of hydrophobic hydration we have used the four-point charge ST2 model,[11] whereas for the studies of electron solvation we have employed the somewhat simpler three-charge SPC model.[10,12] Both models provide reasonable descriptions of bulk water properties, although the former model is somewhat more thoroughly studied.

The apolar materials interact with the water via a purely Lennard-Jones type of interaction. For the small spherical-type solute, the parameterization is typical of a krypton-like atom, while the apolar surface acts as a semi-infinite continuum material with a perfectly planar surface exerting a potential, u, on the solvent of the form

$$u = A/z^9 - B/z^3$$

where z is the normal distance from the surface to the solvent center, and the parameters A and B are chosen to be characteristic of typical paraffinic materials.[9]

The interaction potential of the water with the excess electron is more subtle. We use a pseudopotential constructed in analogy to those in studies of electron-molecule scattering[13,14] and of the solid state.[15] The details of the potential will be published elsewhere as part of a complete report of this work.[7] Basically, the pseudopotential is founded on the interaction of an electron with a water molecule which is assumed to be fixed in the ground-state electronic structure of the isolated molecule.

The interaction potential includes three contributions. The first is a purely electrostatic term which is taken to be that produced by the charge distribution of the SPC water model. This part of the potential is in reasonable accord with that calculated directly from an *ab initio* molecular wavefunction, except quite close to the nuclei.

The second term is a spherically symmetric polarization term, referred to the oxygen nucleus, taken from electron-molecule scattering technology. It is of the form

$$\frac{\alpha_0}{2r^4} (1 - \exp [(-r/r_0)^6])$$

where α_0 is the isotropic part of the molecular polarizability and r_0 is a cutoff parameter. For r_0, we use the sum of the OH bond length and the Bohr radius.

The third and most subtle term is an effective repulsive potential included to account for the requirements of orthogonality between the one-electron wavefunction describing the excess electron and those describing the water molecular wavefunction. We use a pseudopotential approach from solid-state physics,[15] analogous to that used some time ago to describe the electron-helium interaction.[14] Under reasonable assumptions of smoothness of the excess electron wavefunction, one can approximate this core term in a local form, in close analogy to the Slater-type of local exchange approximations.[16] The final form of this part of the potential is evaluated with the s-type basis functions of a double-ζ multicenter *ab initio* molecular wavefunction. The

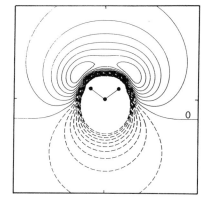

FIGURE 1. Electron-water pseudopotential in the solvent molecular plane; the solvent nuclear framework is indicated, with the oxygen at the center of the 10 × 10Å region shown. The contours are separated by 20 kJ/mol, starting from zero, with positive potential indicated by *dashed lines* and negative potential by *solid lines*. The zero potential line is labeled.

final repulsive core potential used consists of nine exponential terms, two centered at each hydrogen nucleus and five centered at the oxygen nucleus. Exchange terms were estimated in a local density approximation and found to be quite small compared to other uncertainties in the potential; such effects are therefore omitted.

The final potential used is shown in FIGURE 1 as a contour plot in the water molecular plane. The general features are not surprising. It is seen that the potential exhibits minima at bond-oriented positions with an optimal attractive potential of about -35 kcal/mol (~1.5 eV).

The computer simulations described here all use molecular-dynamics techniques to sample configurations, a particularly efficient approach for sampling the solvent. In the case of the solvated electron problem, we also periodically (about every 0.5 psec) reassign velocities from a Boltzmann distribution.

While the simulation techniques for the purely classical systems are quite standard, that appropriate for a quantum system is rather recent and deserves a brief description. More detail is given in the literature.[4,5]

The quantity of interest is the thermal distribution of electron and solvent positions which follows from the Hamiltonian described above. It has been shown[1,2] that for this case, the distribution is equivalent to the results obtained for a completely *classical* system in which the electron is replaced, in the simulation, by a cyclic chain polymer consisting of P (pseudo)atoms, each connected by harmonic springs to its two nearest neighbors and interacting with the solvent via the potential described above, but reduced by a multiplicative factor of P^{-1}. The harmonic force constant within the chain is given by $mP/\beta\hbar^2$ where m is the quantum particle mass, $2\pi\hbar$ is Planck's constant, and β is the inverse of the product of Boltzmann's constant and the temperature.

A simulation of this classical system provides the desired spatial distributions if one interprets the polymer pseudoatom distribution as the thermal electronic density, i.e., the spatial diagonal elements of the one-electron density matrix. The simulations have been carried out for P = 24, 200 and 1000, using 300 SPC water molecules in truncated

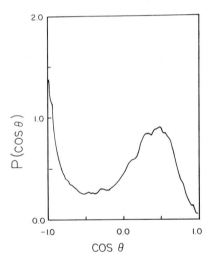

FIGURE 2. Relative distribution of solvation shell solvent OH bond directions with respect to the inward pointing surface normal of an atomic apolar solute; $\cos\theta = -1$, at the left, corresponds to OH bonds, which point away from the solute.

octahedral boundary conditions,[17] with all interactions calculated in the minimum image prescription.

III. HYDROPHOBIC HYDRATION

We consider here some results pertinent to hydrophobic hydration. We emphasize the variation in solvent structure which can result from the presence of an apolar surface in solution, and the attempt to maintain optimal hydrogen bonding among solvent molecules.

The case of small apolar solutes has been considered by several investigators.[8] It is generally accepted that the solvent tends to organize so that the hydrogen bonding groups preferentially avoid directions toward the apolar surface. The distribution of water molecule OH bond directions with respect to the center of an apolar spherical solute is shown in FIGURE 2. Such a distribution is representative of a clathrate-like geometry.

A detailed analysis of the structure[8] shows that the immediate environment of each solvent molecule which is proximal to the solute is best characterized as one that completely retains the hydrogen bonding neighbors present in the bulk. Although the total number of solvent neighbors is reduced from the bulk, the number in hydrogen bonding directions is essentially unchanged; the loss can be associated with weakly interacting solvent pairs.

Of most interest here is the contrasting behavior exhibited by a large planar apolar surface.[9] Although this case is idealized, it is reasonable to anticipate corresponding behavior for any relatively large apolar surface including significant regions with small (or negative) curvature, as may occur, for example, in protein subunits.

The structure observed in the simulation[9] is characterized phenomenologically by an expected loss of solvent coordination of individual solvent molecules as these approach the apolar surface; in the immediate neighborhood of the surface (within about one molecular diameter), the coordination number drops to about half of the bulk value, as expected. In contrast to the behavior for small apolar solutes, the number of hydrogen bonds per solvent molecule is *not* maintained, but decreases near the surface.

However, the frequency of hydrogen bonding does not drop in proportion to the loss in coordination number, as would be expected in the absence of structural reorganization. One observes that the frequency of intermolecular hydrogen bonding in the immediate vicinity of the surface is, in fact, 75% of the bulk value.

The structural interpretation of this phenomenon is pictured in an *idealized* structure in FIGURE 3a. The structure shown is an oriented segment of ice I (only oxygen atoms are shown), in which the apolar surface would be present at the bottom of FIGURE 3b. This structure satisfies the requirement that the coordination number drops while retaining three-quarters of the bonding capacity. This is accomplished by a molecular orientation in the immediate vicinity of the surface which projects one hydrogen bonding group directly *into* the surface, leaving three such directions pointing generally toward the bulk. The "sacrifice" of one such bonding direction maximizes hydrogen bonding in this severely restrictive situation, as the very different structure observed for apolar sphere solvation does in that case.

The structure in FIGURE 3 implies more than has been said above. First, one should observe broad density maxima in the solvent distribution in the direction normal to the surface; these maxima are separated by about 3.7Å (the separation of the positions labeled 2 and 6 in FIGURE 3b). This is, in fact, found in the simulation.[9] Further, one would also expect distinct orientational patterns as a function of distance as one proceeds outward from the surface.

If one uses the structure pictured in FIGURE 3 to align the simulated system and the idealized structure according to the expected positions for peak densities, the corresponding orientational predictions can be tested. This was carried out by dividing the simulated liquid into spatial slices at various distances measured outward along the surface normal. These slices (about 0.9Å thick) are indicated by label at the right of FIGURE 3b, with increasing index corresponding to increasing distance from the surface.

The orientational distribution of solvent OH bond directions with respect to the

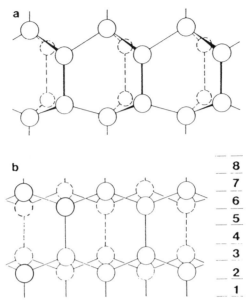

FIGURE 3. Schematic illustration of the structure of ice I and of idealized interfacial solvent structure. The three-dimensional ice I structure is shown in (**a**), and a two-dimensional projection is shown in (**b**). The slice intervals used in the orientational analysis are also shown in (**b**).

surface normal (appropriately weighted by the available solid angle) is shown in FIGURE 4. The distribution corresponds for this case to that shown in FIGURE 2 for the case of solvation of a small apolar sphere, except that in FIGURE 2 no spatial subdivision of the solvation shell into different radial distances is made, nor is any needed.[8]

The orientational preferences apparent in the figure are consistent with that expected from the idealized structure in FIGURE 3. At the surface (panel 1), one OH direction points inward, opposing the surface normal, while about 2Å further away (panel 3), the preference is reversed.

Clearly, the structure in FIGURE 3 is highly idealized, and the orientational preference is strongly dissipated by about 6Å from the surface. However, the essential point here is that the interfacial structural preferences manifest are substantial, are driven by the attempt to maintain maximal hydrogen bonding, and are clearly *inverted* from those familiar from numerous studies of small solutes.

Hence, it is clear that, at the least, caution is indicated in the general extrapolation of the results of small molecule solvation studies to macromolecular assemblies.

IV. IONIC SOLVATION OF AN EXCESS ELECTRON

In this section, we present some illustrative results obtained via the path integral simulation of an excess electron in bulk liquid water at 298 K. We focus here on the solvent orientational and radial distribution with respect to the localized quantum electron.

It is perhaps of first interest to display a typical electron "path" (a contribution to the electronic density), and this is shown in FIGURE 5. Here, we have displayed the 1000 points (P) of the electron path sampled in this case, each connected to its neighbors in the "polymer" by a straight line. The edges of the boundary confining the 300 solvent molecules is indicated. The electron is clearly localized and roughly spherical, with a radius of about 3Å (see below), in general agreement with earlier estimates.[18] However, the electronic distribution clearly manifests irregularities and diffuse regions at its surface.

The solvent orientational structure can be analyzed in much the same way as for ordinary atomic solutes, as shown in FIGURE 6. There we show the distribution of solvent dipole directions (solid lines) and OH bond directions (dashed lines) with respect to the electron position, plotted as the cosine of the angle with respect to the oxygen to electron vector (cf. FIG. 2) and normalized so that unity is the random orientational value. The upper frame includes all electronic positions on the path and hence is relatively diffuse. The lower frame corresponds to an electronic reference

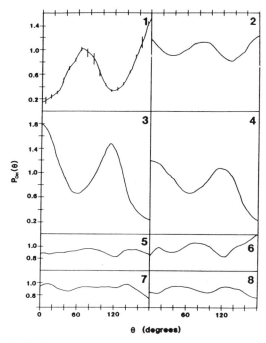

FIGURE 4. Relative distribution of angles between solvent OH bond vectors and outward pointing surface normal for an infinite apolar planar surface; $\theta = 0$, at the left, corresponds to OH bonds which point away from the surface. The eight panels correspond to the spatial solvent slices indicated in FIGURE 4. The bars in panel 1 illustrate error estimates.

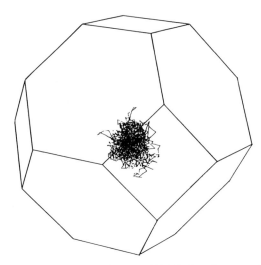

FIGURE 5. Representative electron path for P = 1000. The boundary containing the 300 water molecules is also indicated.

point at the center of mass of the electronic distribution, and corresponds to the view familiar from analysis of atomic ion solvation.[19,20]

As is clear from the figure, the solvent is bond-oriented like it is around a negative atomic ion.[19,20] Except for a somewhat increased breadth in the peaks, the distribution is, in fact, exactly that which has been observed in earlier negative ion solvation studies.

The radial distribution of solvent is shown in FIGURE 7. In FIGURE 7a, we show the result for P = 24, while in FIGURE 7b, we show the result for P = 1000. In each case, the upper set of curves corresponds to all points in the electronic distribution, whereas the lower curve describes the distribution with respect to the electronic center of mass. In all cases, the dashed line is the solvent hydrogen distribution and the solid line is that of the oxygen. As above, we focus on the lower set of curves.

For P = 24, the electronic distribution is too compact compared to the more accurate description obtained with a larger P. We include the result here as a point of reference, since the radial distributions manifest are in good accord with that expected for an atomic ion.[19,20]

For P = 1000 (FIG. 7b), the radial correlations are clearly quite different in first appearance from the ion-like behavior shown in FIGURE 7a. Although the bond orientation of the solvent is still apparent (hydrogen approaching by ~1Å closer than does oxygen), both hydrogen and oxygen peaks are strongly broadened. Nevertheless, the electron is solvent-coordinated in an ionic-like manner. The coordination number obtained from integration of the oxygen radial correlation function shown at the bottom in FIGURE 7b is between five and six, depending on the choice of radial position used to define the radius of the first solvation shell.

Considering the well-defined orientational correlations found and the reasonable, and small, coordination number observed, it is reasonable to attribute the apparent diffuse nature of the radial correlations to fluctuations in the shape and radius of the excess electron, rather than a lack of structure. An examination of a sequence of paths analogous to that shown in FIGURE 5 supports this view. It is perhaps in this respect that the excess electron differs most from a simple ion. The electron exhibits the solvation structure expected of an ion which has the additional freedom to fluctuate in size and shape, while remaining compact and roughly spherical.

Finally, it is of interest to compare, as far as possible, the calculated structure to experimental results. While inferences have been drawn from a variety of measurements,[18] one direct structural analysis has been carried out.[21] Using electron spin echo measurements to determine a set of solvent proton populations and distances in an aqueous glass, Kevan has proposed a specific idealized structure for the hydrated electron. Keeping in mind that the glass is formed at high salt concentration, we can compare the liquid state results described above to the experimental assessment. Kevan finds that his data are best fit by a (glassy) solvent with six nearest-neighbor protons at a distance of 2.1Å and a second set of six protons at a distance of 3.5Å from the electronic center, implying an electron-oxygen distance of 3.1Å and bond-oriented solvent.

The present results are in very good accord with these results. We find a roughly six-coordinate bond-oriented solvent, with six nearest protons centered at 2.3Å and oxygen atoms centered at 3.3Å. When we consider the difference in systems, the limited information that can be extracted from the experiment, and the uncertainties in our present potential, the agreement in structural results is quite striking. Not surprisingly, the present results clearly show a substantial degree of dispersity in the detailed solvation structure, and it is now of substantial interest to explore the degree to which the experimental probe is sensitive to such fluctuations.

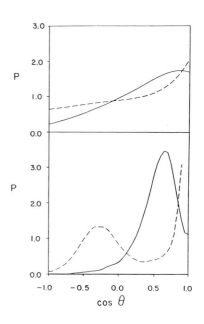

FIGURE 6. Relative distribution of solvation shell solvent OH bond (*dashed lines*) and dipole (*solid lines*) directions with respect to the solvent oxygen to electron position vector; $\cos \theta = -1$, at the left, corresponds to a solvent direction that points away from the electron. The *upper curves* include all electron positions on the path; the *lower curves* refer to the electronic center of mass.

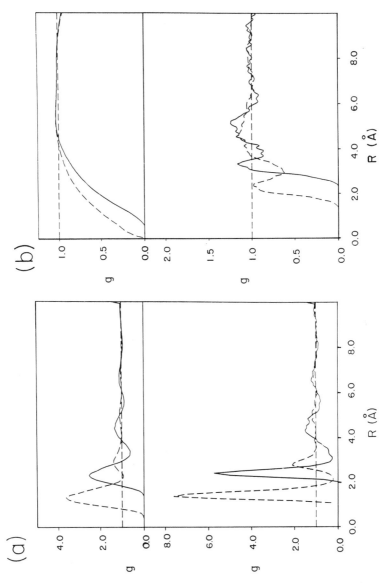

FIGURE 7. Electron-solvent radial correlation functions, g_{eH} (*dashed lines*) and g_{eO} (*solid lines*) for (**a**) P = 24 and (**b**) P = 1000. *Upper* and *lower curves* include all electron positions on the path or only the electronic center of mass, respectively.

V. CONCLUSIONS

Simulation methods can provide insight into quite complex aqueous systems that is presently not available via direct experimental probe. This ability is clearly increasing rapidly, facilitated by new techniques and computers.

The present examples show that although both detail and insight are emerging from such studies, many elements of aqueous system behavior remain to be explored. The phenomenon of hydrophobic interaction is a major example, when we consider the questions raised by the observed dependence of hydration structure on solute shape.

The area of quantum simulations is quite new, and the study of electron solvation is only a first step in a detailed understanding of such complex chemical processes as molecular ionization and electron transfer in liquids. These more demanding cases will no doubt be accessible to comparable analysis in the near future.

SUMMARY

Computer simulation offers an increasing opportunity to examine the details of solvation structure for well-defined solution models. One can characterize idealized models that exemplify specific aspects of solution behavior or those that realistically model systems experimentally difficult to access. As an example from the first category, results are presented for the aqueous solvation structure and solvent hydrogen bonding in the presence of a small apolar solute, and these are contrasted with the quite different behavior observed for a large apolar surface. As an example from the second category, structural results are presented for hydration of an excess electron in bulk water. These latter results, obtained via recent path integral simulation methods, manifest significant solvent structural disruption compared to the case of a simple ion, due to the irregular and fluctuating shape of the quantum solute.

ACKNOWLEDGMENTS

I am indebted to C.Y. Lee, J.A. McCammon, D.A. Zichi, and J. Schnitker for their scientific collaboration in the research reported in this paper.

REFERENCES

1. FEYNMAN, R. 1972. Statistical Mechanics. Benjamin. Reading, MA.
2. CHANDLER, D. & P. G. WOLYNES. 1981. J. Chem. Phys. **74:** 4078–4095.
3. THIRUMALAI, D., R. W. HALL & B. J. BERNE. 1984. J. Chem. Phys. **81:** 2523–2527.
4. PARRINELLO, M. & A. RAHMAN. 1984. J. Chem. Phys. **80:** 860–867.
5. DeRAEDT, B., M. SPRIK & M. L. KLEIN. 1984. J. Chem. Phys. **80:** 5719–5724.
6. KUHARSKI, R. A. & P. J. ROSSKY. 1985. J. Chem. Phys. **82:** 5164–5177.
7. SCHNITKER, J. & P. J. ROSSKY. Manuscript in preparation.
8. ZICHI, D. A. & P. J. ROSSKY. 1985. J. Chem. Phys. **83:** 797–808 and references therein.
9. LEE, C. Y., J. A. McCAMMON & P. J. ROSSKY. 1984. J. Chem. Phys. **80:** 4448–4455.
10. JORGENSEN, W. L., J. CHANDRASEKHAR, J. D. MADURA, R. W. IMPEY & M. L. KLEIN. 1983. J. Chem. Phys. **79:** 926–935.
11. STILLINGER, F. H. & A. RAHMAN. 1974. J. Chem. Phys. **60:** 1545–1557.

12. BERENDSEN, H. J. C., J. P. M. POSTMA, W. F. VAN GUNSTEREN & J. HERMANS. 1981. *In* Intermolecular Forces. B. Pullman, Ed. Reidel. Dordrecht, The Netherlands.
13. TRUHLAR, D. G., K. ONDA, R. A. EADES & D. A. DIXON. 1979. Int. J. Quant. Chem. Symp. **13:** 601–632.
14. JORTNER, J., N. R. KESTNER, S. A. RICE & M. H. COHEN. 1965. J. Chem. Phys. **43:** 2614–2624.
15. HEINE, V. 1970. Solid State Phys. **24:** 1–36.
16. HARA, S. 1967. J. Phys. Soc. Jpn. **22:** 710–718.
17. VAN GUNSTEREN, W. F., H. J. C. BERENDSEN, F. COLONNA, D. PERAHIA, J. P. HOLLENBERG & D. LELLOUCH. 1984. J. Comp. Chem. **5:** 272–279.
18. KESTNER, N. R. 1976. *In* Electron-Solvent and Anion-Solvent Interactions. L. Kevan & B. C. Webster, Eds. Elsevier. New York, NY.
19. MEZEI, M. & D. L. BEVERIDGE. 1981. J. Chem. Phys. **74:** 6902–6910.
20. CHANDRASEKHAR, J., D. C. SPELLMEYER & W. L. JORGENSEN. 1984. J. Am. Chem. Soc. **106:** 903–910.
21. KEVAN, L. 1981. J. Phys. Chem. **85:** 1628–1636.

Neutron Diffraction Studies of Aqueous Solutions of Molecules of Biological Importance: An Approach to Liquid-State Structural Chemistry

J.L. FINNEY AND J. TURNER[a]

Crystallography Department
Birkbeck College
London University
London WC1E 7HX, England

INTRODUCTION

Diffraction methods are perhaps unique in allowing the *direct* determination of spatial correlations in liquids and solutions. The pair correlation function is related to the measured structure factor through a simple Fourier transform relationship. Experimental data so obtained can thus give us reliable information on structure both for modeling the liquid itself, and for testing potential functions used in computer simulation calculations. It is particularly important that such tests be rigorously made if our simulations are to relate to the real, rather than a hypothetical, world.

Similar tests of potential functions are possible using crystal hydrate data (e.g., amino acids, nucleic acid components, coenzymes, proteins), and work by ourselves and others[1-5] has already demonstrated the sensitivity of predicted structure to assumed potential function. Some aspects of this approach are discussed later in this volume by Goodfellow.[6] However, the relatively high degree of order in crystals—without which data of adequate quality could not be obtained reliably—of necessity restricts such tests to the lower-energy regions of the potential surface. In contrast, the use of liquid-state structural data allows us to probe the potential surface away from the minima, and to explore regions that are likely to be sampled in a real biomolecular system *in vivo*.

A single diffraction experiment on an *n*-component system yields a weighted superposition of the Σ_{k-1}^n k partial pair-correlation functions, and hence will be extremely difficult to interpret usefully. For aqueous solutions, the total pair correlation function will be totally dominated by the water signal, making the extraction of hydration information perilous. With the development of the isotope-substitution neutron diffraction method by Enderby and coworkers[7] it became possible for the first time to "difference out" the dominant water contribution, and so obtain *direct* information on the water coordination of a specific atom in a solution. As a result of the application of the method to aqueous electrolytes,[8-10] our understanding of ionic solutions has been revolutionized, and a large experimental database has been

[a]Recipient of a Research Studentship from the Science and Engineering Research Council.

established against which to test theoretical models, including potential functions used in computer simulations. Although such a statement may be contentious, we would argue that the experimental results in general have yet to be reproduced satisfactorily by simulations.

Recent improvements in neutron sources and instrumentation, amounting to more than two orders of magnitude in effective flux, have allowed us in recent years to extend the method to examine both *hydration* and *conformation in solution* of *molecules* of biological importance. Thus, we are now able *for the first time* to observe *directly, without making any interpretive assumptions,* the water coordination of charged, polar, and apolar moieties in *molecules.* The method allows us to build up an unambiguous picture of the structural consequences of solvent interactions with charged, polar, and apolar groups *in solution.* In addition to extending our understanding of these basic, chemically and biologically important, interactions, the results can provide strong tests of potential functions used in simulations of chemical and biomolecular systems.

In this paper, our initial experiments on a series of amides are presented, to illustrate the application of the technique and the kinds of results obtained. Additional problems of interpretation are introduced by the asymmetry of the molecules, and further experiments designed to overcome these will be discussed. Initial comparisons with simulation results suggest that the potentials used are inadequate to explain the experimental data. Finally, the prospects for the technique, as being developed in our current program at the Institute Laue-Langevin and in the Spallation Neutron Source at the Rutherford Appleton Laboratory, are summarized to illustrate the possible development of what is, in effect, the application of *liquid-state structural chemistry* to biomolecular systems.

BACKGROUND

General Formalism

The structurally significant part of the scattering from a solution, $F(\mathbf{k})$, can be written

$$F(\mathbf{k}) = \sum_i \sum_j c_i c_j b_i b_j (S_{ij}(\mathbf{k}) - 1) \qquad (1)$$

where c_i is the atomic concentration of atom i (solute or solvent), whose neutron scattering length is b_i. $S_{ij}(\mathbf{k})$ is the partial structure factor relating to atom types i and j, and is related to the real-space partial pair correlation function through a simple Fourier transformation. $\mathbf{k}$ is the usual scattering vector,

$$|\mathbf{k}| = \frac{4\pi \sin \theta}{\lambda}$$

Consider now two solutions, in one of which one of the atoms M has been substituted by a different isotope M' of the same element. The difference between the two $F(\mathbf{k})s$ can be written

$$\Delta_M(\mathbf{k}) = \Sigma \, A_{Mi}(S_{Mi}(k) - 1) \qquad (2)$$

where the coefficients A are given by

$$2c_M c_i b_i (b_M - b_{M'}) \qquad (i \neq M)$$

and

$$c_M^2 (b_M^2 - b_{M'}^2) \qquad (i = M) \qquad (3)$$

We see from Equation 2 that the differencing has removed all contributions from those correlations *not* involving atom M.

The Fourier transform of the difference with respect to atom M can be written as:

$$\overline{G}_M(r) = \Sigma \, A_{Mi}(g_{Mi}(r) - 1) \qquad (4)$$

Thus, the first difference technique effectively allows us to "sit on" the atom M, and in principle enables us to determine a concentration and scattering length weighted linear combination of all those partial pair correlations which involve atom M. The full theoretical background is given in Ref. 7.

Application to Aqueous Solutions of Biologically Important Molecules

The ability to extract useful information from $\overline{G}_M(r)$ depends upon several factors. Clearly, isotopes must be available that have significantly different scattering lengths. Of the nonhydrogen atoms commonly occurring in biomolecules, only nitrogen, with scattering lengths of 0.94 (^{14}N) and 0.64 (^{15}N) $\times \, 10^{-12}$ cm is of obvious use. Because of improvements in neutron sources and instrumentation, however, it is likely that carbon substitution will become a feasible technique in the reasonably near future. Secondly, the atomic concentration of the substituted isotope must be "reasonable," the dilutions accessible clearly depending upon the magnitude of the isotope difference (Equation 3). Consider, for example, the case of urea in D_2O. Using subscripts m and w to denote oxygen or deuterium atoms in the solute and solvent (water) respectively, Equation 4 expands to

$$\overline{G}_N(r) = A_{NN}g_{NN}(r) + A_{NC}g_{NC}(r) + A_{NO_m}g_{NO_m}(r) + A_{ND_m}g_{ND_m}(r)$$
$$+ A_{NO_w}g_{NO_w}(r) + A_{ND_w}g_{ND_w}(r) - B \qquad (5)$$

where B is the sum of all the A coefficients. This difference function contains both intramolecular and intermolecular (urea-water) terms. Provided the A coefficients are sufficiently large, and that some way of separating out the intra- and intermolecular contributions can be found (trivial for simple ionic solutions of single ions or isotropic molecular ions), direct information both on *the structure of the molecule in solution* and its *hydration* is obtainable.

TABLE 1 shows the A coefficients for a 2-molal solution of urea in D_2O, equivalent to 25 water molecules per urea. Two points might be noted. First, $\overline{G}_N(r)$ is dominated by the nitrogen-water correlations, the A coefficients for ND_w and NO_w being an order of magnitude greater than those weighting the intramolecular contributions. Secondly, the solvent values are an order of magnitude smaller than those for the original 4.41-molal $NiCl_2$ solution used in the initial demonstration experiment of Enderby and Neilson,[11] a comparison that illustrates the progress that has been made in neutron sources and instrumentation over the past decade. This progress is even more

ANNALS NEW YORK ACADEMY OF SCIENCES

TABLE 1. Coefficients (in millibarns) Weighting the Partial Contributions to $\overline{G}_N$ (r)

	A_{NN}	A_{NC}	A_{NO_m}	A_{ND_m}	A_{NH}	A_{NO_w}	A_{ND_w}
2.0-molal urea	0.26	0.11	0.10	0.44	—	2.37	5.46
1.9-molal formamide	0.06	0.05	0.04	0.10	−0.03	1.20	2.76
1.6-molal acetamide	0.04	0.07	0.03	0.18	—	0.97	2.24
	molecule				water		

NOTE: The values are calculated using scattering lengths appropriate to natural abundance nitrogen and 99% ^{15}N.

impressive if the intramolecular contributions, *weighted only by one- or two-tenths of millibarns,* can be detected in the urea experiment.

EXPERIMENTAL WORK

Our initial experiments were designed to examine the hydration of amides, as models of the peptide group in proteins. Data have been obtained at the Institut Laue-Langevin, Grenoble, and C.E.N. Saclay, France, on approximately 2-molal solutions in D_2O of urea, formamide, and acetamide. For each molecule, two isotopically distinct solutions containing natural-abundance nitrogen and 99% ^{15}N were used.

In all cases, the exchangeable hydrogens were replaced by repeated dissolving in D_2O (at 99.8%D) followed by drying (either freeze-drying or distillation). In acetamide, in order to reduce the incoherent background, the methyl hydrogens were deuterated ($CD_3 \cdot COND_2$); in formamide, the nonexchangeable hydrogen remained a proton ($HCO \cdot ND_2$). Care was taken to obtain samples with <0.4% light water content, and to equalize within 0.1% the hydrogen content of each pair of samples. Cylindrical geometry was used, with ~3 ml of solution in the beam in each case. Data were corrected for multiple scattering[12] and absorption[13] and normalized with reference to a vanadium standard of the same dimensions as the sample.[14] Full experimental details are given elsewhere.[15]

RESULTS

FIGURE 1 shows the k-space difference function $\Delta_N(\mathbf{k})$, the real space $\overline{G}_N(r)$, together with the urea molecule as seen by neutron analysis of the crystal structure.[16] The statistics of $\Delta_N(\mathbf{k})$ are extremely good, indicating immediately that it is possible to go to lower nitrogen concentrations (e.g., lower concentrations of urea, and/or use of other molecules with a lower nitrogen concentration). FIGURE 2 shows the $\Delta_N(\mathbf{k})$ curves for 1.9-molal formamide and 1.6-molal acetamide. As is evident from TABLE 1, the A coefficients for these latter two molecules are about half those for urea, reflecting the lower nitrogen concentration in the total system. Nevertheless, the statistics are excellent for formamide, the greater scatter in the acetamide case resulting from a shorter data collection time.

Molecular Structures

One advantage of the technique is that, provided the molecular structure is known, the form of $\overline{G}_N(r)$ in the intramolecular region can be used as an additional check on the results. Despite the very low weighting factors for the intramolecular contributions (TABLE 1), the molecular structure is clearly visible in $\overline{G}_N(r)$, which has peaks (FIG. 1b) at 0.95 Å, 1.38 Å, and 2.27 Å. When this is compared with the intramolecular distances obtained from the crystal structure (TABLE 2), the first two peaks can be provisionally identified with the N-D1 (FIG. 1c) and N-C distances, whereas the third broad feature is at the position that would be expected for N-O, N-N, and N-D2. These assignments are confirmed by integration, the first two (incompletely separated) peaks giving a total of 3.1 ± 0.1 atoms (expected 2 + 1), while the second peak area of

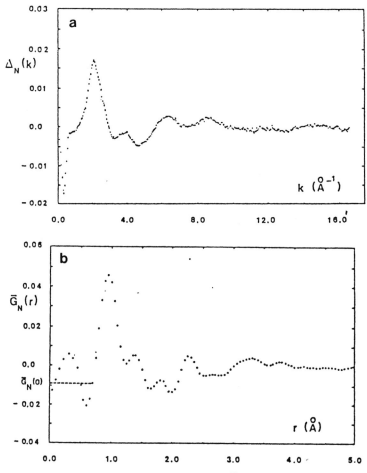

FIGURE 1. (a) First-order difference $\Delta_N(\mathbf{k})$ and (b) $\overline{G}_N(r)$ for 2.0-molal urea in D_2O.

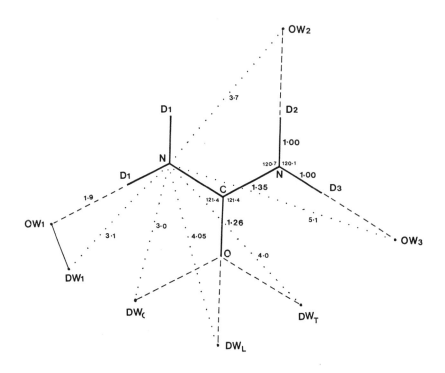

FIGURE 1c. Molecular geometry from the crystal coordinates, together with distances from the nitrogen of hypothetical water oxygen and deuterium positions assuming idealized hydrogen bonding.

2.8 ± 0.2 atoms is entirely consistent with the 3 atoms (O, N, D_2) expected. These comparisons are summarized in TABLE 3. Confirmation of the molecular structure can also be obtained by least-squares refining the molecular structure against the high k data in the experimental k-space difference. In addition to confirming quantitatively the solution's molecular structure, subtraction of the coherent distinct scattering of the molecule from the experimental difference should result in the *inter*molecular scattering only, with its Fourier transform giving only the *water* structure around the urea nitrogen atom. The geometrical parameters of the model resulting from this procedure (R ∼ 18%) are compared with the crystal values in TABLE 2. The data show that the refined solution is, within the estimated uncertainties quoted, very close to that in the crystal. The only possible discrepancy is in the N-D distance, which may be slightly shorter in solution, as might be expected for the expected weaker hydrogen bonding in solution. In real space, the Fourier transforms of $\Delta_N(k)$ and of the calculated intramolecular function are *very* close, except for the first peak position, which is susceptible to any Placzek "droop" in the data that might result from slight differences in the hydrogen content between the two samples. This agreement would also seem to confirm the absence of any nearest-neighbor urea-urea interactions, consistent with indirect data.[17,18] The 1.81-Å peak in $\overline{G}_N(r)$ (FIG. 1b) is also confirmed as a truncation ripple through this procedure.

The molecular structures of formamide and acetamide are also confirmed, by our results, although for a variety of reasons the errors are greater. For example, a known problem with the formamide experiment was a small inequality in light water content between the two samples, whereas the statistical accuracy of the acetamide difference was less good than that for the other two molecules (FIG. 2b). A point of particular interest in the formamide $\overline{G}_N(r)$ (FIG. 3) is a *negative* feature at 2.04 Å; this can be identified with the expected N-H distance (2.06 Å in the crystal), the negativity reflecting the negative scattering length of the proton ($b_H = -0.374$).

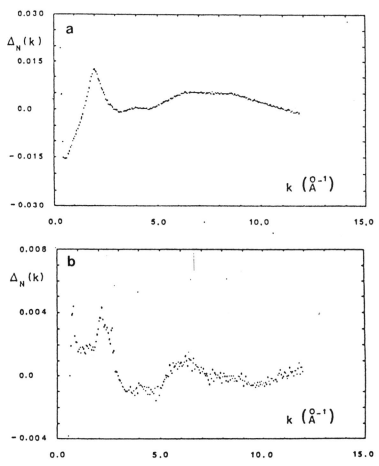

FIGURE 2. First-order differences $\Delta_N(k)$ for **(a)** 1.9-molal formamide and **(b)** 1.6-molal deuterated acetamide in D_2O.

TABLE 2. Intramolecular Distances (Å) and Angles for Crystalline Urea, and Comparisons with Refinement from Solution Scattering

	ND1	NC	NO	NN	ND2	ND3	NCN angle
Crystal[a]	1.00;0.99	1.35	2.29	2.30	2.50	3.21	117.2(1)°
Solution	0.97(1)	1.34(2)	2.31(1)	—	—	—	118.5(2.8)°

[a]Crystal data were obtained by room temperature extrapolation of low-temperature data of Ref. 16. Expected error is in the third decimal place, and therefore not quoted.

HYDRATION

Apart from the N-D3 distance of 3.21 Å, the $\overline{G}_N(r)$ for urea (FIG. 1b) beyond about 2.6 Å gives information on the nitrogen-water correlations. There are three apparent peaks at 2.67 Å, 3.25 Å, and 3.75 Å; above about 4 Å no significant structure is evident. Integration of the 2.67-Å feature gives a value consistent with 1.1 oxygen atoms at this distance; integrating out to 4.00 Å gives 7.1 ± 0.5 D_2O molecules after subtraction of D3 (FIG. 1c).

In contrast to results in strongly hydrating ions,[8,10,11] the solvent region $\overline{G}_N(r)$ is relatively featureless. Compare, for example, the $G_{Ni}(r)$ for aqueous $NiCl_2$, where separate peaks are clearly assignable to the deuterium and oxygen atoms of strongly hydrating water molecules (see, for example, Figure 5 in Ref. 8). It might be tempting to conclude, therefore, that the results presented here support a weak hydration model for urea. Further thought, however, suggests that this may well not be the case, and suggests additional experiments to clarify the situation, as discussed below.

Referring to FIG. 1c, we note the anisotropy of the molecule as seen from the nitrogen atom. Moreover, in addition to water oxygen atoms at about 3Å possibly accepting hydrogens from the ND groups (O_{W1} in FIG. 1c), the nitrogen will also see any water deuteriums that donate to the carbonyl oxygen *at about the same distance*. For discussion purposes, we take the hydrogen-bonded O . . . D distance to be 1.9Å. On this assumption, O_{W1} will occur at 2.9Å (or less for a nonlinear hydrogen bond), while the deuterium of a water approaching the CO vector at 120° "*cis*" to the nitrogen (D_{WC} in FIG. 1c) will be at about 3Å. "Bisecting" and "*trans*" water deuteriums (D_{WL} and D_{WT} in FIG. 1c) will be more distant at about 4Å; approach angles intermediate between the "*cis*" and bisecting positions will fall in the 3–4 Å distance range from the nitrogen. Also within this range would fall the oxygen from the water linearly

TABLE 3. Intepretation of $\overline{G}_N(r)$ for 2.0-Molal Urea

Peak Position (Å)	Peak Assignment	Number of Atoms at r	
		Measured	Predicted
0.95	2 × D1	2.2 } $3.1 \pm .1$	2 }
1.38	C	0.9	1 }
2.27	O, N, D2	2.8 ± 0.2	3
2.67	0 (water)	1.1 ± 0.2	—
2.55–4.00	D3	1.0	
	n × D_2O	7.1 ± 0.5	8 ± 0.5[20]

hydrogen-bonding to the other nitrogen (D2 at about 3.7Å for the "ideal" O_{W2} position in FIG. 1c). The deuterons complicate matters even further. Assuming a minimum D-D contact distance of 2.2Å,[19] the minimum ND_{W1} distance is about 3.1Å, again in this 3–4 Å range.

Even assuming a hydration model in which the water molecules are strongly restricted, both positionally and orientationally, both ND and NO distances would be expected to contribute to the $\overline{G}_N(r)$ up to 4Å. Considering the positional and orientational freedom likely even for a strong hydration model, peaks corresponding to these ideal distances will begin to smear out, with the result that a relatively featureless $\overline{G}_N(r)$ in this region is to be expected. To make further progress in interpreting the hydration region, a first step would be to separate out the hydrogen and oxygen contributions to $\overline{G}_N(r)$. Fortunately, the power of neutron techniques is such that this

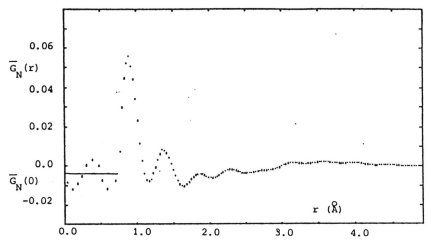

FIGURE 3. $\overline{G}_N(r)$ for 1.9-molal formamide, showing a negative peak at about 2.04Å corresponding of the intramolecular N-H distance.

can in principle be done by performing an experiment in either pure H_2O or an H_2O/D_2O mixture. The resulting $\overline{G}_N(r)$ would have the ND contribution weighted differently and could thus allow the separation of both NO and ND solvent contributions. Formally, this can be considered as a second difference experiment,[8] but one involving the solvent rather than the solute; subtraction of the two $\overline{G}_N(r)s$ would in principle give the g_{NO} contribution alone. A second possibility would be to utilize the opposite signs of b_H (-0.374) and b_D ($+0.667$) to prepare a "null hydrogen" H_2O/D_2O mixture, the resulting $\overline{G}_N(r)$ having no hydrogen contribution at all. Such experiments raise significant, if not severe, problems of sample characterization and multiple scattering corrections, but should be possible with care. They are currently being attempted.

COMPARISON WITH THEORETICAL MODELS

The experimental $\overline{G}_N(r)$ results provide a test of potential functions used in simulation calculations: even though the data themselves cannot be used without further experiments to set up a hydration model, they can act as a strong filter of potentials when the model is compared with computed $\overline{G}_N(r)$ distributions. Presently, MD simulations in urea have been published by Tanaka et al.[20] and by Kuharski and Rossky[21] using the MCY and ST2 water potentials, respectively. Formamide simulations have been published by Jorgensen and Swenson,[22] and by the Beveridge group.[23]

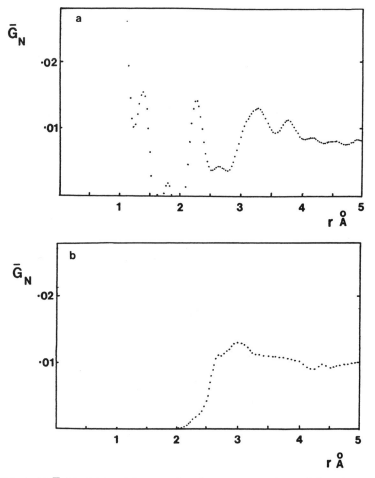

FIGURE 4. (a) $\overline{G}_N(r)$ of 2.0-molal urea, enlarged to ease comparisons of the solvent region ($r \gtrsim 2.5$Å) with simulation results. (b), (c), and (d) are computed solvent $G_N(r)$s for simulations using (b) MCY,[20] (c) TIPS2,[15] and (d) ST2[15] models for the water molecule. The ordinate for (b) is scaled only approximately with respect to the other curves; the curve has been smoothed by interpolation.

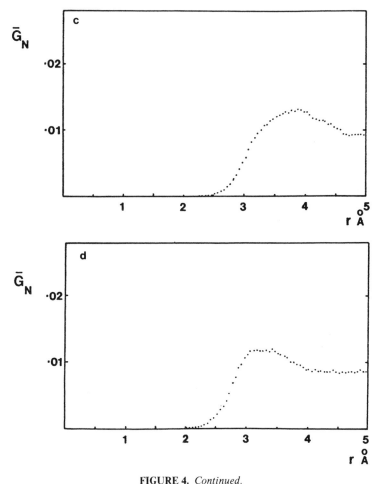

FIGURE 4. *Continued.*

Additionally, we have performed a urea simulation using ST2 and TIPS2 water models.[15]

Two major problems arise in comparing these simulation results with experiment. First, the simulations are all of solutions of lower concentration than the experimental 2-molal ones; the rate of decay of the nitrogen-centered distance correlations in the simulations does, however, suggest that the concentration differences are unlikely to seriously affect the conclusions drawn. This is also a problem in many previous comparisons of ionic solution simulations with the results from neutron experiments. Secondly, most of the published data are not in a form that allows a meaningful comparison to be made. Initially, the $\overline{G}_N(r)$ needs to be constructed from the simulation data; this requires separate $g_{NO}(r)$ and $g_{ND}(r)$ partial correlation functions, which need then to be suitably weighted by the appropriate concentration and scattering length factors (e.g., Equation 5). Because the g_{NO} and g_{NH} correlations are not given in the other published work, we are presently limited in making quantitative comparisons with our own simulations and with those of Tanaka et al.[20]

A relatively weak test can be made, however, with several simulations for urea and formamide by comparing the number of water molecules out to the distance at which the $\overline{G}_N(r)$ becomes effectively flat. Experimentally, 7.1 ± 0.5 and 10.1 (±1) D_2O molecules are found out to 4.00Å and 4.25Å for urea and formamide, respectively. These compare well with estimates of 8 ± 0.5 and 10 (±2) extracted from the Tanaka[20] data and relating to[24] the Marchese et al.[23] simulations, respectively. However, this agreement perhaps tells us little more than that the simulations and experiments are performed at the same density.

FIGURE 4a shows the solvent region of the experimental $\overline{G}_N(r)$, whereas computed functions from three simulations are given in FIGURES 4b–d, using MCY, TIPS2, and ST2 water, respectively. The urea-water potential parameters were determined from STO-3G dimer calculations in the work of Tanaka et al.,[20] which uses MCY water, whereas partial charges for urea were estimated using CNDO/2 calculations for the ST2 and TIPS2 calculations.[15]

The following points can be made in comparing the experimental data with the simulation results.

(1) The experimental function shows stronger oscillations than any of the three calculations. We should beware of concluding too much from this because it is not clear how much these oscillations might be due to truncation problems in Fourier transforming, and possibly other errors such as residual noise or inelastic effects, or errors in normalization. Because this structure is retained when the truncation point is changed, however, it seems likely to be real.

(2) The main experimental maximum occurs at about 3.3Å. Positions in the three simulations are at about 3.0Å, 3.7Å, and between 3.0 and 3.5Å for the MCY, TIPS2, and ST2 calculations, respectively.

(3) Neither of the Birkbeck simulations reproduces the small peak in the experimental distribution at about 2.7Å, although the possibility of this feature's being a truncation ripple should be borne in mind. The Tanaka results do show a bump in the computed $\overline{G}_N(r)$ (FIG. 4b) between 2.6 and 2.7Å, which results from a peak in g_{NO}.

(4) Both the MCY and TIPS2 correlation functions rise to the first peak more slowly than do the experimental functions, TIPS2 notably so. Although the ST2 function rises too soon, its rate of rise parallels that of experiment.

Because of truncation ripple problems, it is difficult to know how to weight items (1) and (3) above. The discrepancies between all the simulations and experiment noted in (2) and (4) do, however, seem significant. All seem to show basic disagreement in the first peak's position and shape, even allowing for the truncation and other reservations made above. Thus, we conclude that neither the Tanaka simulation[20] nor our Monte Carlo calculations give water distributions around the urea nitrogen in adequate agreement with experiment. Clearly, improved theoretical models are needed to explain the experimental data even at this level of comparison, where D and O are undifferentiated.

Once deuterium and oxygen contributions to $\overline{G}_N(r)$ can be separated by the experiments currently in progress (see above), the tests of simulation results will be much more severe. Even for our ST2 and TIPS2 simulations, which use a common model for urea, the deuterium and oxygen contributions are significantly different, implying different orientational behavior in the two cases. In the TIPS2 case (FIG. 5a), the oxygen and hydrogen curves peak at about the same first peak position, while in ST2, where we expect stronger orientational restrictions,[25] the oxygen and hydrogen curves are approximately out of phase. The Tanaka distributions show similar

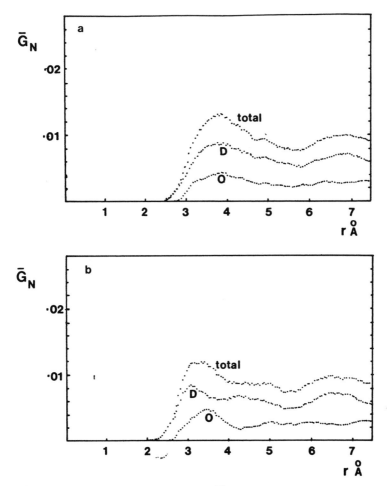

FIGURE 5. Total, oxygen, and hydrogen solvent $\overline{G}_N(r)s$ for (**a**) TIPS2 and (**b**) ST2 urea simulations.

out-of-phase behavior, although there are clear differences from the ST2 calculation, in which the NH distribution peaks *before* the NO. An examination of instantaneous configurations shows unexpected N-H . . . O-H hydrogen bond geometry, which might be traced to an inadequate partial charge on the urea hydrogens. Thus, the relatively featureless total curve results from a partial cancellation of peaks and troughs in the ND and NO distributions, underlining the point that a relatively featureless $\overline{G}_N(r)$ is not necessarily indicative of weak hydration. The separate H and O contributions are required before the weak hydration statement can be made confidently.

One final point to note from FIGURE 5 is that although the nitrogen-oxygen correlations have died out, the nitrogen-hydrogen curve has a significant broad maximum at 6.5Å. The work by Tanaka *et al.* shows, if anything, the opposite, with the nitrogen-oxygen correlation stronger than the nitrogen-hydrogen in this region. The experimental data are flat in this region, although a direct comparison here is difficult because of the higher urea concentration in the experimental system. Experiments on 0.5-molal urea (about the concentration of our simulations) are feasible, and should be able to clarify the reality or otherwise of this predicted long-range correlation.

SUMMARY AND PROSPECTS

Neutron diffraction isotope substitution methods have already revolutionized our understanding of aqueous electrolyte solutions, and provided data that have been difficult to reproduce adequately by simulation methods. Sources and instrumentation have developed sufficiently over the past few years to allow extension of the technique to aqueous solutions of *molecules* of basic chemical and biological interest. The urea work described here has established the feasibility of the method for direct determination of both the molecular conformation in solution, and the molecular hydration.

The negative scattering length of the hydrogen atom is a powerful tool in low-resolution contrast variation work, which is now a standard technique for low-angle neutron scatterers. It promises to be of great help also in separating out the water oxygen and deuterium contributions to the direct space difference function $\overline{G}_N(r)$. Even without this separation, the total $\overline{G}_N(r)$ for urea is not reproduced by any of the three simulations discussed using MCY, TIPS2, and ST2 models of water. The main peak is generally of incorrect shape, and also at the wrong position, indicating at least a scaling problem for the potentials, but probably also an incorrect handling of orientational structure. Two of the three simulations show N-O and N-D correlations that tend to be out of phase, underlining the point—similar to that now recognized in molten salts—that a relatively featureless $\overline{G}_N(r)$ does not necessarily imply weak hydration. In fact, the experimental $\overline{G}_N(r)$ shows more structure than any of the three simulations discussed, at least one of which represents a strong hydration situation. A further point to note is that the three simulations show different behavior in both total solvent $\overline{G}_N(r)s$, and the O and D contributions. That all are also distinguishable from experiment and each other underlines the power even of the total solvent $\overline{G}_N(r)$ as a test of model calculations.

With the recent start-up of the Spallation Neutron Source at the Rutherford Appleton Laboratory in England, the possibilities of the technique are further enhanced. Lower concentrations and larger molecules can now be tackled, as well as suitably chosen multicomponent systems. The interaction of water with polar, charged, and apolar groups on molecules of basic chemical and biological interest can now be tackled *directly,* and its perturbation by the addition of additional components monitored. Higher effective fluxes make carbon substitution a real possibility, as well as the use of second difference methods to probe intermolecular correlations. Again, similarly to the electrolyte solution work,[26] quasielastic scattering may also be used to obtain dynamic information on the water in biologically important systems.

We now have available a very powerful technique which is capable of giving direct, model-independent structural information on a variety of *liquid*-state systems of fundamental chemical and biological importance. Given adequate provision of neutron facilities, the potential of this technique of liquid-state structural chemistry is immense.

ACKNOWLEDGMENTS

This work was performed in collaboration with G.W. Neilson, S. Cummings, J. Bouillot, M.-C. Bellissent, and J.P. Bouquiere. Use of neutron facilities at the Institut Laue-Langevin, Grenoble, and at C.E.N. Saclay is gratefully acknowledged.

REFERENCES

1. GOODFELLOW, J. M., J. L. FINNEY & P. BARNES. 1982. Proc. Roy. Soc. Lond. Ser. B **214**: 213–228.
2. FINNEY, J. L., J. M. GOODFELLOW & P. L. POOLE. 1982. *In* Structural Molecular Biology: Techniques and Applications. D. B. Davies, W. Saenger & S. S. Danyluk, Eds.: 387–426. Plenum Press. New York, NY.
3. GOODFELLOW, J. M. 1984. J. Theor. Biol. **107**: 261–274.
4. VOVELLE, F., J. M. GOODFELLOW, H. F. J. SAVAGE, P. BARNES & J. L. FINNEY. 1985. Eur. Biophys. J. **11**: 225–237.
5. GOODFELLOW, J. M., P. L. HOWELL & R. ELLIOT. 1986. Proceedings of NATO ASI on Crystallography of Molecular Biology, Strasbourg, September 1985. In press.
6. GOODFELLOW, J. M. 1986. Ann. N.Y. Acad. Sci. This volume.
7. SOPER, A. K., G. W. NEILSON, J. E. ENDERBY & R. A. HOWE. 1977. J. Phys. **C10**: 1793–1801.
8. ENDERBY, J. E. & G. W. NEILSON. 1979. *In* Water: A Comprehensive Treatise, vol. 6. F. Franks, Ed. Plenum Press. New York, NY.
9. ENDERBY, J. E. & G. W. NEILSON. 1981. Rep. Prog. Phys. **44**: 593–653.
10. NEILSON, G. W. 1984. J. Phys. (Paris) **45** (C7, Suppl. 9): C7-119–129.
11. NEILSON, G. W. & J. E. ENDERBY. 1978. J. Phys. **C11**: L625–L628.
12. BLECH, I. A. & B. L. AVERBACH. 1965. Phys. Rev. **A137**: 1113–1116.
13. PAALMAN, H. H. & C. J. PINGS. 1962. J. Appl. Phys. **33**: 2635–2639.
14. ENDERBY, J. E. 1968. *In* Physics of Simple Liquids. H. N. V. Temperly, J. S. Rowlinson & G. S. Rushbrooke, Eds. North Holland. Amsterdam.
15. TURNER, J. 1986. Ph.D. thesis, University of London.
16. SWAMINATHAN, S., B. M. CRAVEN & R. K. McMULLAN. 1984. Acta Crystallogr. **B40**: 300–306.
17. FINER, E. G., F. FRANKS & M. J. TAIT. 1972. J. Am. Chem. Soc. **94**: 4424–4429.
18. HAMMES, C. G. & P. R. SCHIMMEL. 1967. J. Am. Chem. Soc. **89**: 442–446.
19. SAVAGE, H. F. J. 1983. Ph.D. thesis, University of London.
20. TANAKA, H., H. TOUKARA, K. NAKANISHI & N. WATANABE. 1984. J. Chem. Phys. **80**: 5170–5186.
21. KUHARSKI, R. A. & P. J. ROSSKY. 1984. J. Am. Chem. Soc. **106**: 5786–5793.
22. JORGENSEN, W. L. & C. J. SWENSON. 1985. J. Am. Chem. Soc. **107**: 1489–1496.
23. MARCHESE, F. T., P. K. MEHROTRA & D. L. BEVERIDGE. 1984. J. Phys. Chem. **88**: 5692–5702.

24. BEVERIDGE, D. L. Personal communication.
25. FINNEY, J. L., J. E. QUINN & J. O. BAUM. 1985. *In* Water Science Reviews, vol. 1. F. Franks, Ed. Cambridge University Press. Cambridge, England. In press.
26. HEWISH, N. A., J. E. ENDERBY & W. S. HOWELLS. 1983. J. Phys. **C16:** 1777–1791.

The Born Model of Ion Solvation Revisited

ALEXANDER A. RASHIN[a] AND BARRY HONIG[b]

[a]*Department of Physiology and Biophysics*
Mount Sinai School of Medicine
New York, New York 10029

[b]*Department of Biochemistry and Molecular Biophysics*
Columbia University
New York, New York 10032

We demonstrate that the 65-year-old Born theory of ionic solvation[1] can provide a very accurate means of calculating the solvation energies of ions. The consistent rederivation of the well-known Born equation,[1] $\Delta G_s = -(q^2/2r)(1 - 1/D)$, where r is a radius and D is the dielectric constant, shows that r represents the radius of the cavity formed by the ion in a particular solvent, and thus that different radii could be needed for

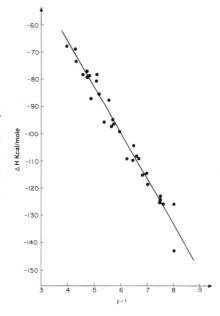

FIGURE 1. Comparison of experimental (*dots*) and predicted (*solid line*) enthalpies of ion hydration. The enthalpies are given per unit charge. The experimental values are given for the following ions: Li^+, Na^+, K^+, Rb^+, Cs^+, F^-, Cl^-, Br^-, I^-, Cu^+, Ag^+, Cu^{+2}, Mg^{+2}, Ca^{+2}, Sr^{+2}, Ba^{+2}, Zn^{+2}, Cd^{+2}, Hg^{+2}, Al^{+3}, Sc^{+3}, Y^{+3}, La^{+3}, Ce^{+3}, Ce^{+4}, Ga^{+3}, In^{+3}, NH_4^+, OH^-, S^{-2}, SH^-.

calculations in different solvents. We attribute the known failures of the Born theory[2] to the use of ionic radii rather than cavity radii and show, on the basis of electron density profiles in ionic crystals, that ionic radii are a reasonable measure of the cavity size for anions but that covalent rather than ionic radii are the consistent choice for cations. Cavity radii for complex spherically symmetric ions for which covalent radii

143

are not defined (e.g., ammonium or tetramethyl ammonium) can be calculated from the closest distance between the ion and the surrounding solvent molecules.

The introduction of a single correction factor, 7% increase in all radii, leads to an excellent agreement between the predicted and experimental hydration enthalpies of alkali halide salts. The absolute heat of hydration of a proton calculated with these corrected radii from the relative heats of hydration of alkali and halide ions does not depend on their radii. This provides another corroboration of the consistency of the model and allows us to obtain hydration enthalpies of individual ions from their experimental relative heats of hydration.

The Born formula, used with the radii described above, accurately predicts experimental heats of solvation for more than 30 ions of different size and charge (FIG. 1). The need for 7% correction in radii may be due in part to our neglect of dielectric saturation effects. However, it appears that dielectric saturation does not constitute a large contribution to the solvation enthalpy since the model works well for polyvalent ions where saturation effects should be largest.

The success of the Born model suggests that any detailed simulation that successfully reproduces short-range interactions between the ion and the molecules of a high dielectric solvent can lead to a successful prediction of solvation enthalpies as long as it predicts high enough (and not even necessarily correct) value of the dielectric constant.

REFERENCES

1. BORN, M. 1920. Z. Phys. **1:** 45.
2. BOCKRIS, J. O'M. & A. K. N. REDDY. 1977. Modern Electrochemistry, Vol. 1. Plenum. New York, NY.
3. RASHIN, A. A. & B. HONIG. 1985. J. Phys. Chem. **89:** 5588–5593.

The Structure, Energy, Entropy, and Dynamics of Peptide Crystals[a]

D. H. KITSON,[b] F. AVBELJ,[b] D. S. EGGLESTON,[c] AND
A. T. HAGLER[b]

[b]The Agouron Institute
La Jolla, California 92037

[c]Smith Kline and French Laboratories
Philadelphia, Pennsylvania 19101

INTRODUCTION

Because of their fundamental role in living organisms, peptides and proteins have been an intensively studied class of molecules. Their function is intimately related to their structural and dynamic properties and a variety of experimental methods, including NMR spectroscopy[1-3] and diffraction[4] have been used to study these properties. Many important questions are, however, inaccessible to these experimental methods. For example, what forces drive a peptide to adopt the conformation that it takes up? Can we predict what this conformation will be, given only the sequence of amino acids in the peptide? Why does a peptide pack in the observed packing mode in a crystal? How much strain is imposed on the molecule by the crystal lattice? In an attempt to answer such questions, a variety of theoretical methods have been used.[5-7] We have been applying several of these methods, including energy minimization and molecular dynamics, to study the structural, energetic, entropic, and dynamic properties of peptides in the crystal environment.

In this paper we summarize extensive theoretical studies of three peptide crystal systems: N-formyl-Met-NMePhe-t-butyl ester, cyclo-(Ala-Pro-D-Phe)$_2$ · 8H$_2$O and cyclo-(Gly-Pro-Gly)$_2$ · 4H$_2$O. We have carried out molecular dynamics simulations of these crystals, quenched, or minimized, the energy at points along the trajectory, calculated the entropy of the peptides in isolated and crystalline environments, and calculated their vibrational spectra. These studies were carried out with two overall objectives in mind. The first was to answer fundamental questions relating to peptide structure and dynamics and especially the effect of environment on these properties. The second was to take advantage of the excellent opportunity these systems provide to test the reliability of the theoretical methods that we employ. When energy minimization or molecular dynamics techniques are used for modeling peptide or protein systems, one obviously needs to have confidence that the results obtained accurately reflect the properties of the real system. This requirement can best be satisfied by attempting to calculate properties that are accessible experimentally, and peptide

[a]This work was supported by Grant No. PCM 8204908 from the National Science Foundation, Grant No. GM 30793 from the National Institutes of Health, Smith Kline and French Laboratories, and Abbott Laboratories.

crystal structures constitute a very accurate source of structural information which can be used to evaluate the force fields used for simulations.[5,8–12]

In addition to studying the properties of the peptides themselves, we are also interested in the properties of the solvent in the peptide crystals. Proteins, as well as peptides, form rather open arrays in the crystalline state. These arrays contain water molecules and often there are only a few close contacts between proteins or peptides. The ordering of water molecules in the protein and peptide crystals ranges from tightly bound, fully ordered molecules to those in a state similar to that of bulk water.[13,14] These crystals can thus provide excellent systems for studying the varieties of interactions between protein and peptide molecules and solvent.

DEVELOPMENTS IN COMPUTER SIMULATIONS OF PEPTIDES AND PROTEINS

Early Computer Simulations

The first application of computers to the calculation of energetic properties of molecules came in 1961 when Hendrickson showed that one could use a computer to

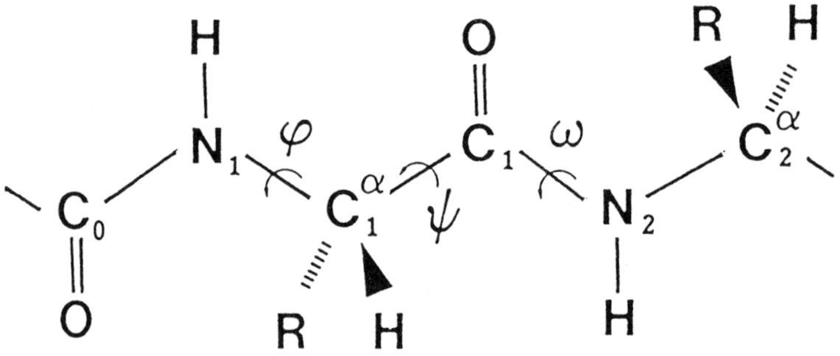

FIGURE 1. The structure of the peptide backbone, illustrating the ϕ, ψ and ω torsion angles.

calculate the energy of small cycloalkanes.[15] Calculations were first carried out on peptide models when Ramachandran et al. calculated so-called Ramachandran maps for various blocked amino acids.[16] These calculations involved the mapping of allowed combinations of the ϕ and ψ backbone torsion angles (FIG. 1) by searching for those combinations for which no overlap of atoms took place. This type of mapping was later combined with the use of energy functions to produce maps of the conformational energy as a function of the ϕ and ψ torsion angles.[17–19]

Minimization

The next advance in the simulation of peptides came when minimization algorithms were used to minimize the energy of a molecule as a function of the ϕ, ψ and χ torsion angles. In these "rigid-geometry" calculations all other conformational variables (bond lengths, valence angles, and ω torsion angles) were kept fixed. It thus became possible to compare different minimum energy conformations of a peptide in order to explore the conformational preferences of the molecule.

The bonds and valence angles in molecules are not rigid, however, and so the technique of "flexible-geometry" minimization was later introduced in which the energy is minimized with respect to all degrees of freedom of the molecule.

Evaluation of Entropy

In all of the cases that we have discussed so far, only the potential energy of a system has been considered. The entropy also contributes, however, to the free energy and therefore to the relative stabilities of different conformations of a molecule. Hagler *et al.*[7] showed that it is important to take this entropic contribution into account in an analysis of the energetic properties of a series of oligopeptides. In comparing the relative energies of α-helical ($\phi_i, \psi_i \approx -60°, -60°$) and C_7 ($\phi_i, \psi_i \approx -80°, +80°$) conformations of a series of host-guest hexapeptides, they found in all cases that the potential energy favored the α-helical conformation. The vibrational entropy, however, favoured the C_7 conformation and in one case (Boc-Met$_3$-Gly-Met$_2$-OMe) was sufficient to reverse the order of stability predicted from consideration of the potential energy alone ($E_\alpha - E_{C_7} = -2.35$ kcal/mol, $T(S_\alpha - S_{C_7}) = -4.13$ kcal/mol at 298°, therefore $A_\alpha - A_{C_7} = +1.83$ kcal/mol).

The entropy of a system depends on the extent of the phase space available to the system. Except for a very simple system with a limited number of degrees of freedom, there is no feasible method by which to rigorously explore the entire phase space of a molecule. Nevertheless, several methods have been developed for estimating entropy.

In the harmonic approximation (which was used in the study of oligopeptides described above) it is assumed that the system is constrained to a harmonic potential well and is vibrating around an average structure corresponding to a minimum energy conformation of the system. The entropy can then be obtained from the matrix of second derivatives of the energy with respect to the coordinates.[5] The behavior of a real molecule is, however, more complex than this simple model describes and techniques have therefore been developed for estimating entropy from molecular dynamics or Monte Carlo simulations (see below), which are able to give a better approximation to the dynamic behavior of the system. For cases where the system fluctuates around a single average structure, analogous to a harmonic oscillator (e.g., a native protein conformation), direct integration over the phase space can be used.[20,21] For systems where a conformational change is taking place (or, for example, an event such as the binding of a ligand is occurring), umbrella sampling methods[22-26] and perturbation methods[25,27,28] have been developed.

Monte Carlo and Molecular Dynamics

The conformational behavior of peptides and proteins is far more complex than is revealed in the static pictures given by energy minimization, and techniques that were developed to study the statistical thermodynamic properties of liquids have been recently applied to model the dynamic behavior of these system. These techniques include molecular dynamics[6,29-31] and Monte Carlo simulations.[13,32,33]

Monte Carlo Simulations of Peptide and Protein Crystals

Applications of the Monte Carlo method to peptides or proteins in the solid state include a simulation of the triclinic lysozyme crystal.[13] It was shown in this study that the behavior of water molecules in different environments in the crystal varies widely. Near the protein, and particularly near charges, the waters were found to be highly ordered, whereas far from the protein the water molecules can travel much larger

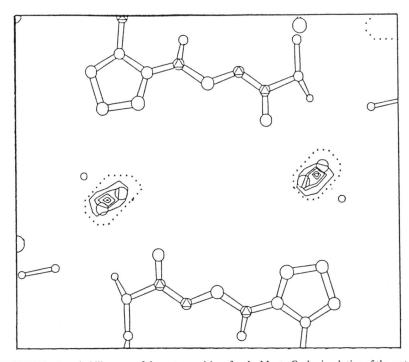

FIGURE 2. A probability map of the water positions for the Monte Carlo simulation of the water structure in the crystal of cyclo-(Ala-Pro-D-Phe)$_2$. *Dotted contours* represent the zero probability level. Solid contours are in steps of 0.56 water molecules per cubic angstrom above this. This map shows the density for two water molecules which each have two alternative positions ≈1Å apart. The X-ray positions of the waters are represented by *circles* within the density. The simulated density spans the X-ray positions, indicating that the water molecules move between these alternate positions during the simulation. (From Hagler *et al.*[32] Reprinted by permission.)

distances and represent essentially disordered waters. This variation in order is in good agreement with that observed crystallographically.[34] It had usually been assumed that all waters in the first hydration shell (within 3.5Å of the protein) have similar properties. This Monte Carlo simulation of the water structure in the lysozyme crystal showed that this picture is not correct and that the degree of ordering within this region varies widely. This arises from the very different environment of the water molecules within this layer.

The Monte Carlo method was also applied to study the interactions of a small peptide with water and the water structure in the crystal of cyclo-(Ala-Pro-D-Phe)$_2$ · 8H$_2$O.[32] This system is an attractive one for study for several reasons. First, there are only a few water molecules per unit cell (in comparison with a protein crystal where there are very many) and so it should be possible to understand precisely the role of each water molecule in the crystal. Second, the water structure had been characterized precisely in an X-ray crystal structure[35] and so this study provided an ideal opportunity to test the methods used. Several properties of the water molecules were investigated, including their average positions and root mean square movements from these positions. The averaged positions of the ordered water molecules, excluding one position that was found to be unoccupied, were in good agreement with experimental data (RMS ≤ 0.5Å). Since one of the ordered water positions was found to be vacant during the simulation, the water molecule having migrated to one of the cavities occupied by disordered waters (which were assumed in the experimental study to contain three waters), and since it was found that the energies of waters in the cavities were lower when there were four water molecules per cavity rather than three, it was proposed that the measured density of the crystal was not correct and that the formula should be cyclo-(Ala-Pro-D-Phe)$_2$ · 9H$_2$O. As an example of the type of detailed information that can be simulated by this technique, we consider the waters that were observed crystallographically to occur in two alternative, partially occupied locations (there are two such waters per unit cell). As shown in FIGURE 2, the simulation was able to reproduce the motion of the water between these two positions. Thus this study showed that it is possible to simulate both the positions and the behavior of individual water molecules in the crystal of a small peptide with a high degree of accuracy.

Molecular Dynamics Simulations of Peptide Crystals

In these Monte Carlo studies of peptide and protein crystals only the solvent is allowed to move, while the peptide or protein molecules are kept fixed. Through the use of the molecular dynamics method, as described here, one can also study the dynamic behavior of the peptide or protein.

METHODS

The potential energy, V, of a system is represented by a function involving the internal coordinates and the nonbonded interactions within the system (Equation 1).

$$
\begin{aligned}
V = \Sigma \{ D_b [1 - e^{-\alpha(b-b_0)}]^2 - D_b \} &+ \tfrac{1}{2} \Sigma H_\theta (\theta - \theta_0)^2 \\
&+ \tfrac{1}{2} \Sigma H_\phi (1 + s \cos n\phi) \\
&+ \tfrac{1}{2} \Sigma H_\chi \chi^2 + \Sigma\Sigma F_{bb'} (b - b_0)(b' - b_0') \\
&+ \Sigma\Sigma F_{\theta\theta'} (\theta - \theta_0)(\theta' - \theta_0') + \Sigma\Sigma F_{b\theta}(b - b_0)(\theta - \theta_0) \\
&+ \Sigma F_{\phi\theta\theta'} \cos \phi (\theta - \theta_0)(\theta' - \theta_0') + \Sigma\Sigma F_{\chi\chi'} \chi\chi' \\
&+ \Sigma\epsilon[2(r^*/r)^{12} - 3(r^*/r)^6] + \Sigma q_i q_j / r
\end{aligned}
\tag{1}
$$

There are two types of quantities represented in the equation: constants character-istic of the energy required to deform a given internal coordinate ($D_b \ldots F_{bb}, F_{\theta\theta} \ldots$) or characteristic of the strength of a given interatomic nonbonded interaction (r^*, ϵ, q); and the internal coordinates specifying the geometry of the molecule (bond lengths, b; bond angles, θ; dihedral angles, ϕ; out-of-plane angles, χ; and nonbonded interatomic distances, r). From the energy surface, we can obtain the forces on the individual atoms by calculating the negative of the first derivatives of the energy with respect to the Cartesian coordinates of the atoms. By solving for the coordinates at which all of the forces are zero, we can obtain the minimum energy conformation of the system. Since we know the masses of the atoms, by substituting the forces into Newton's equation of motion ($F = ma$) we can integrate the accelerations forward in time to follow the trajectory of the atoms, giving us the molecular dynamics of the system.

For our simulation all hydrogen atoms were explicitly included in the system. The Berendsen SPC water potential[36] was employed to model the water-water interactions. The parameters used for the peptide-peptide interactions will be described elsewhere.[37] Peptide-water parameters were taken as the geometric mean of the appropriate water and peptide parameters. The experimental translational symmetry of the system was used to generate the crystal environment through the use of periodic boundary conditions[38] and a cutoff was applied to nonbond interactions. Typically, in the past, a cutoff of 6–8Å has been used for this type of calculation.[39,40] When a cutoff is used, one must be sure that all of the interactions that are being excluded are negligible. It has been shown that a cutoff of at least 12Å is necessary to give $\approx 97\%$ of the total nonbond energy for a crystal system[8,9,41] and for the studies described here, we have chosen to use a 15Å cutoff.

CONFORMATIONAL PROPERTIES, ENERGETICS, AND DYNAMICS OF N-FORMYL-MET-NMEPHE-T-BUTYL ESTER

The crystal structure of N-formyl-Met-NMePhe-t-butyl ester (FIG. 3) was recently described by one of us[42] and provides a number of challenging features that we can attempt to reproduce and rationalize in terms of the energetics of the system. The amide bond between the Met and NMePhe residues is *cis* in the crystal, whereas fluxional *cis-trans* equilibrium is observed in solution, and the terminal -S-CH$_3$ of the Met residue occurs in two conformational states in the crystal. The disorder in the Met side chain arises mainly through a rotation of the χ_3 torsion angle, and the two resulting conformations of the peptide are illustrated in FIGURE 4.

Models of peptide crystals containing each of the conformations (FIG. 4A and B) were minimized, using periodic boundary conditions and a cutoff of 15Å. The minimized energies obtained were -27.0 and -24.7 kcal/mol for conformations A and B, respectively. Thus the predominant conformational isomer observed in the crystal is of lower energy, although from the relative fractions of roughly 65% A, 35% B observed in the crystal we would expect a smaller energy difference.

We were also interested in investigating the barrier to interconversion between the two conformations, and thus we calculated the energy of the system as a function of the Met χ_3 torsion angle; the result is shown in FIGURE 5. This plot shows that the barrier for conversion of B to A is very low (≈ 0.06 kcal/mol).

Beginning at the minimized structures, we carried out short molecular dynamics simulations starting from each of the two conformations A and B. FIGURE 6 shows the history of the χ_3 and χ_4 torsion angles of Met during the simulation that was started from conformation A. χ_3 shows considerable flexibility, with a large fluctuation at ≈ 1.5 psec, although no transitions to conformation B are seen. The methyl group is clearly very free to rotate, as shown by the rapid changes in χ_4, and as might be expected from the large free space near the methionine side chain in the crystal.

Given the low energy barrier for the conversion of conformation B to A, and the higher energy of conformation B, one would expect conformation B to change very readily to conformation A during a simulation started from conformation B and this

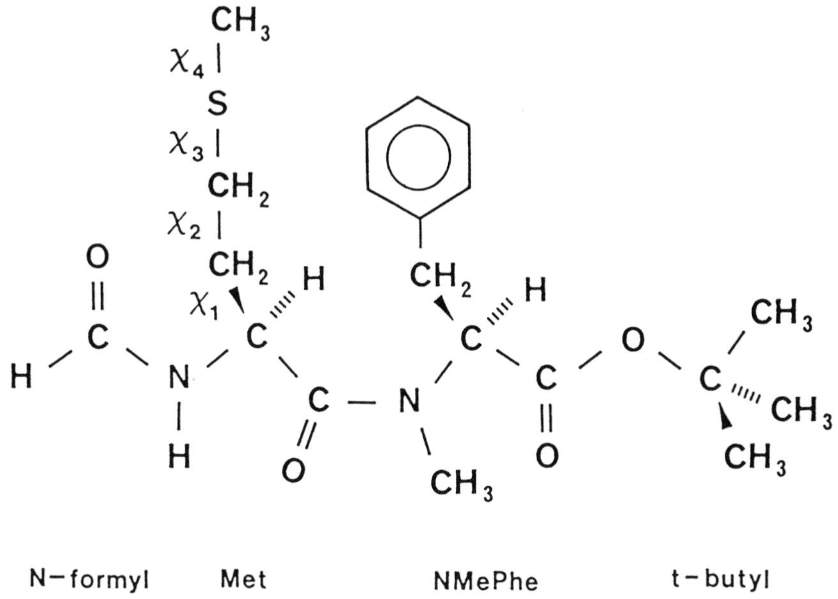

FIGURE 3. *N*-formyl-Met-NMePhe-*t*-butyl ester.

indeed occurs, as shown in FIGURE 7. By 0.5 psec the molecule has undergone a conformational transition to A, $\chi_3 = -171°$, about which it fluctuates throughout the rest of the simulation.

CYCLO-(ALA-PRO-D-PHE)₂

As we stated in the introduction, there were two main reasons for undertaking this type of study, namely, to explore questions related to the nature of the crystal structure and to test the reliability of the simulation techniques that we employ. To fulfill these objectives we clearly need to study a variety of different peptide crystals. We have

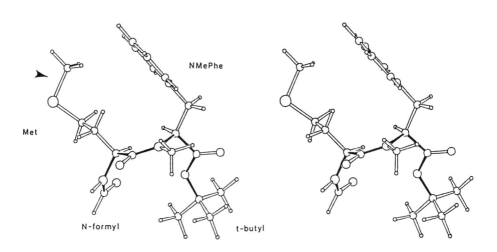

Conformation A

Conformation B

FIGURE 4. Stereo views of the two conformations of *N*-formyl-Met-NMePhe-*t*-butyl ester which are observed in the crystal. These conformations differ principally in the conformational state of the terminal —S—CH₃ of the Met residue (indicated by *arrows*). This disorder arises mainly through a rotation of the Met χ_3 torsion angle. Approximately 65% of the peptides in the crystal are in conformation A and approximately 35% are in conformation B.

previously studied the water structure in the cyclo-(Ala-Pro-D-Phe)$_2$ crystal by Monte Carlo *methods*[32] and now have returned to this system to compare these results with those obtained by molecular dynamics and to assess the effects of relaxing the peptide.

Cyclo-(Ala-Pro-D-Phe)$_2$ crystallizes in an orthorhombic space group with 2 peptide molecules and 16 water molecules per unit cell.[35] The backbone of the peptides is in a double type II β turn configuration with L-Pro and D-Phe residues at the corners of the turns ($\phi_2, \psi_2 = \phi_5, \psi_5 = -60, +122°$; $\phi_3, \psi_3 = \phi_6, \psi_6 = +79, +9°$). The X-ray study indicated that the water molecules range from fully ordered to fully disorder-

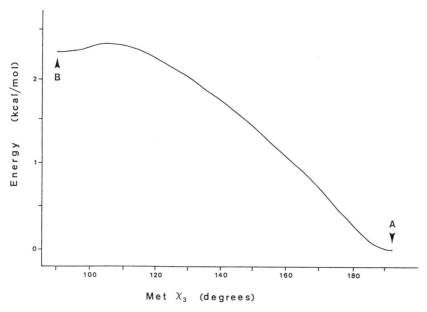

FIGURE 5. Relative energy as a function of the Met χ_3 torsion angle in the crystal of *N*-formyl-Met-NMePhe-*t*-butyl ester. The difference between the energies of conformations A and B is 2.3 kcal/mol. The barrier to the rotation of the torsion angle from conformation B to A is only 0.06 kcal/mol.

ed.[32,35] We chose to place the waters in random (but sterically reasonable) locations in the unit cell because one of our objectives is to assess the ability of the simulation to reproduce experimental properties. By starting with the waters in random locations, we are not causing them to be biased to the observed positions by the starting configuration and we thus provide a far more stringent test of the technique. In addition to performing an energy minimization and a 28-psec molecular dynamics simulation of the crystal system, we also minimized an isolated peptide and ran a 60-psec dynamics simulation of this isolated system. By comparing the results of the minimizations and dynamics simulations of the two systems, we can investigate the effects of the crystal forces on the structural, energetic, and dynamic properties of the peptide.

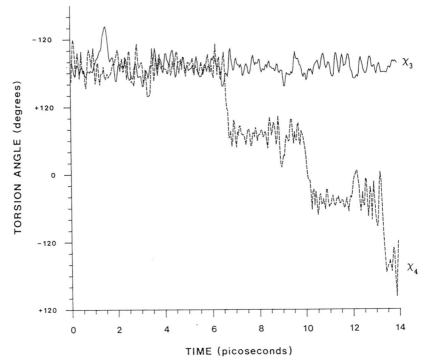

FIGURE 6. The history of the Met χ_3 (*solid line*) and χ_4 (*broken line*) torsion angles during a 14-psec simulation of the crystal of *N*-formyl-Met-NMePhe-*t*-butyl ester started from conformation A. These torsion angles were found to be very flexible, reflecting the large free space near the Met side chain (this is also reflected in the small barrier for rotation about χ_3, as shown in FIGURE 5). This flexibility can be seen most markedly in the χ_4 torsion angle which undergoes several transitions (corresponding to rotations through 60° of the methyl group on the end of the side chain) during the course of the simulation.

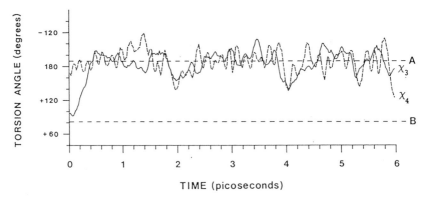

FIGURE 7. The history of the methionine χ_3 (*solid line*) and χ_4 (*broken line*) torsion angles during a 6-psec simulation of the crystal of *N*-formyl-Met-NMePhe-*t*-butyl ester started from conformation B. The χ_3 values for conformations A ($-171°$) and B ($82°$) are indicated by *dashed lines* on the figure. After 0.5 psec the molecule has changed from the higher-energy conformation B to the lower-energy conformation A, which is consistent with the low energy barrier for this transition, but does not reflect the two states observed in the crystal.

Ability of the Minimization to Reproduce the Experimental Structure

From a comparison of the torsion angles of the experimental and minimized crystal structures (TABLE 1), we can see that the minimization has not introduced any substantial changes in the conformation of the peptide, the biggest deviation being about 21° (for ψ of Pro^5, molecule 1), due to a slight change in the orientation of the amide group between the Pro and Phe residues. Other than this, there were no changes in torsion angles of more than about 10°, and the overall shape, including the type II β turns, has been maintained. Thus the peptide conformation is well reproduced in its crystal environment.

Effect of Lattice Forces on the Structure and Dynamics of the Peptide

The effects of lattice forces on the structure of the peptide can be seen by comparing the torsion angles of the minimized crystal system with those of the lowest

TABLE 1. Torsion Angles (in degrees) of Experimental, Minimized Crystal and Minimized Isolated Structures of Cyclo-(Ala-Pro-D-Phe)$_2$

Residue	Torsion Angle	Experimental[a]	Minimized Crystal Structure		Minimized Isolated Structure
			Molecule 1	Molecule 2	
Ala^1	ϕ	−156.6	−150.4	−152.7	−81.9
	ψ	171.7	169.9	169.1	116.0
Pro^2	ϕ	−60.4	−59.8	−59.1	−44.2
	ψ	122.5	105.8	107.7	116.3
Phe^3	ϕ	78.7	83.9	80.7	121.3
	ψ	9.0	4.8	12.7	−68.8
Ala^4	ϕ	−156.6	−146.8	−152.7	−79.0
	ψ	171.7	167.6	166.7	165.1
Pro^5	ϕ	−60.4	−55.9	−56.4	−72.3
	ψ	122.5	101.8	108.9	96.1
Phe^6	ϕ	78.7	85.1	77.8	132.8
	ψ	9.0	11.8	9.9	−76.9

[a]The conformations of the two molecules in the experimental crystal structure are identical.

energy conformation obtained by minimizing structures taken at 3-psec intervals along the dynamics trajectory (TABLE 1). This comparison shows that the lattice forces considerably affect the molecular conformation. For example, Ala^1 exists in an extended conformation in the crystal, whereas in the isolated state, it is in a C_7 (equatorial) conformation. Similarly, the D-Phe residues are in an α_L conformation in the crystal, but are C_7 in the isolated state.

Calculation of the intramolecular energy of the peptide reveals that it is higher for the peptide in the crystal environment (228.3 kcal/mol) than for the isolated molecule (222.8 kcal/mol). This indicates that the crystal lattice is providing an environment that favors an energetically strained conformation of the peptide, which is being stabilized (by at least 6 kcal/mol) by intermolecular interactions, primarily an extensive network of hydrogen bonds in which bridging water molecules link together adjacent peptide molecules. One might expect the same situation, namely, induced conformational strain, to occur in the case of a biologically active peptide bound to a receptor, with the receptor providing the environmental forces.

As an illustration of the effects of lattice forces on the dynamic properties of the peptide, FIGURE 8 compares the fluctuations of the ϕ and ψ torsion angles of the D-Phe[6]

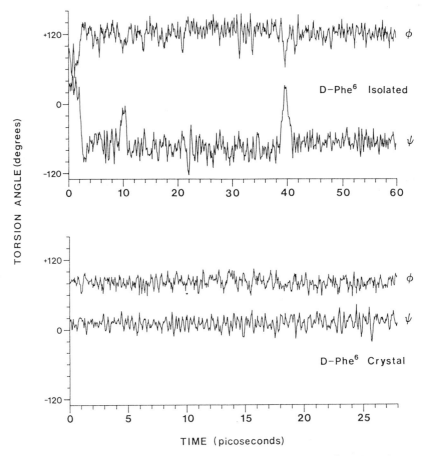

FIGURE 8. Fluctuations of the ϕ and ψ torsion angles of the D-Phe[6] residue of cyclo-(Ala-Pro-D-Phe)$_2$ over the course of dynamics simulations of an isolated peptide and of a peptide in the crystal environment. This figure shows the damping effect of the lattice forces on the fluctuations in the torsion angles.

residue over the course of the dynamics simulation for both the isolated and crystal systems. One can see immediately that the lattice forces act to damp out the fluctuations in the torsion angles. An interesting fluctuation can be seen at ≈ 40 psec in the case of the isolated system. In this simulation the peptide began with D-Phe[6] in the α_L state, which is observed in the crystal conformation. During the first 3 psec of the simulation this residue undergoes a conformational change to the C_7 region, around which it oscillates for the remaining 57 psec, with the exception of a brief return to the α_L state for ≈ 1–2 psec at 40 psec. This demonstrates the accessibility of the crystal conformation which is visited through dynamic fluctuations by the isolated molecule, and the ability of dynamics to take molecules into higher energy conformations.

Entropy

As we would expect, given the more restricted conformational freedom of the molecule in the crystal demonstrated here, calculation of the vibrational entropy of the peptide in the crystal and *in vacuo* shows that the isolated peptide is favored by ≈ 10 kcal/mol at 298K.

CYCLO-(GLY-PRO-GLY)$_2$

We have also performed simulations on the peptide crystal cyclo-(Gly-Pro-Gly)$_2$. Here we describe the results of the simulation in terms of the fit of calculated to experimental structure for the peptide and an analysis of the behavior of the water molecules.

Cyclo-(Gly-Pro-Gly)$_2$ crystallizes in the monoclinic space group P2$_1$.[43] The cell dimensions are a = 7.6194Å, b = 21.0709Å, c = 7.7018Å and β = 108.904°. There are two cyclic peptide and eight water molecules in the unit cell. The cyclic peptide conformation deviates from internal two-fold symmetry by forming one transannular hydrogen bond between the amide proton of Gly^1 and the carbonyl oxygen of Gly^4 (N-O distance = 2.91Å) in a type II β turn ($\phi_2, \psi_2 = -66, -36°$; $\phi_3, \psi_3 = -115, -7°$), while the other half of the molecule is in a type I β turn ($\phi_5, \psi_5 = -53, +126°$; $\phi_6, \psi_6 = +83, -3°$), without an internal hydrogen bond (distance between the carbonyl oxygen on Gly^1 and nitrogen on Gly^4 = 3.95Å) (FIG. 9). The neighboring peptide molecules are not directly linked by hydrogen bonds, but rather by bridging water molecules. The X-ray study indicated that the water molecules are fully ordered. Each hexapeptide molecule participates in 3 N-H $\cdots$ O hydrogen bonds and in 5 C=O $\cdots$ H hydrogen bonds to the water molecules. The remaining three hydrogen atoms in the water molecules are involved in hydrogen bonds to the other water molecules. All carbonyl oxygens of the peptide molecule participate in hydrogen bonds and the four water molecules form a linear array bridging one end of the peptide (C=O of Pro^5) to the other (NH of Gly^3) through a hydrogen-bonded network (FIG. 9).

We simulated one asymmetric unit of the crystal, i.e., one hexapeptide and the four water molecules. The energy of the system was minimized and the final coordinates from the minimization were used as the starting coordinates for a 20-psec molecular dynamics simulation.

Comparison of Experimental and Calculated Crystal Structures

One measure of the fit to experiment can be obtained by comparing the minimized, time-averaged dynamic and experimental structures. The torsion angles, which give

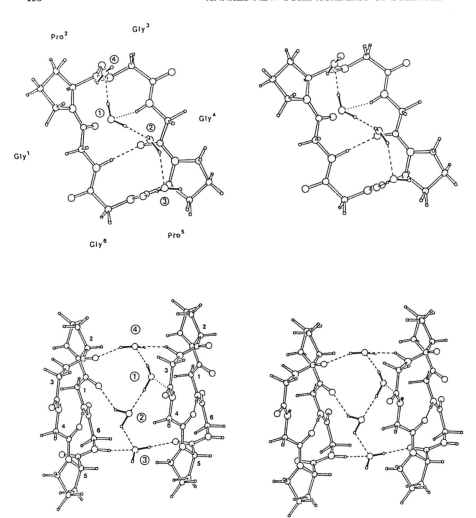

FIGURE 9. Stereo views which show the peptide conformation and the water positions in the crystal of cyclo-(Gly-Pro-Gly)$_2$ and part of the hydrogen bond network which connects these molecules together. Hydrogen bonds for which the H$\cdots$O distance is less than 2.2Å are indicated by a *dashed line* and hydrogen bonds where the distance is between 2.2 and 2.5Å are indicated by a *dotted line*. Hydrogen bonds are also formed between hydrogens on waters 3 and 4 and the C=O groups of the Gly^6 and Gly^3 residues, respectively, of peptides that are not shown in these views (see Table 3).

the overall conformation of the molecule, are compared in TABLE 2 for these three structures.

This table shows that there is a very good fit between the experimental and calculated torsion angles, the biggest difference being 16.1° for ψ of Gly^6. The minimized and time-averaged dynamics structures are very similar to one another, indicating that the average structure around which the system is fluctuating in the

TABLE 2. Torsion Angles of Experimental, Minimized, and Time-Averaged Dynamics Structures of the Crystal of Cyclo-(Gly-Pro-Gly)$_2$

Residue	Torsion Angle	Experimental	Minimized Crystal	Time-Averaged Dynamics
Gly^1	ϕ	-142	-128.6	-132.2
	ψ	-173	-179.2	-170.5
Pro^2	ϕ	-66	-60.6	-59.7
	ψ	-36	-42.8	-34.6
Gly^3	ϕ	-115	-104.9	-111.9
	ψ	-7	-12.4	-15.9
Gly^4	ϕ	-150	-134.7	-133.9
	ψ	178	-172.1	-171.4
Pro^5	ϕ	-53	-61.5	-61.0
	ψ	126	118.0	115.5
Gly^6	ϕ	83	91.3	93.4
	ψ	-3	-19.1	-16.0

dynamics is close to the minimized structure. Another test of the validity of the force field used is how well it can simulate the hydrogen bond network. In TABLE 3 we list the hydrogen bond distances for the experimental and the minimized crystal structures. As we can see, the differences are smaller than 0.12Å, except for the intramolecular transannular hydrogen bond, where the difference is 0.3Å. This increase in the length of the hydrogen bond is due to a movement of the Gly^1 amide hydrogen brought about by the changes in torsion angle around this residue (see TABLE 2).

Behavior of the Simulated Water

In the X-ray structure the water molecules are ordered and this is well reproduced in the molecular dynamics simulation, where water molecules oscillate about their equilibrium positions. During the dynamics simulation the hydrogen bond network is

TABLE 3. Hydrogen Bonds in Experimental and Minimized Crystal Structures of Cyclo-(Gly-Pro-Gly)$_2$

Donor	Acceptor	Experimental Distance (Å)	Minimized Distance (Å)
		Intramolecular	
Gly^1 NH	Gly^4 C=O	2.91	3.21
		Intermolecular	
Gly^3 NH	Water-4 O	3.12	3.08
Gly^4 NH	Water-1 O	3.16	3.06
Gly^6 NH	Water-3 O	2.99	3.11
Water-2 H2	Gly^1 C=O	2.81	2.78
Water-3 H1	Pro^5 C=O	2.79	2.78
Water-3 H2	Gly^6 C=O	2.76	2.75
Water-4 H2	Pro^2 C=O	2.82	2.72
Water-4 H1	Gly^3 C=O	2.85	2.76
Water-1 H1	Water-2 O	2.84	2.74
Water-1 H2	Water-4 O	2.84	2.70
Water-2 H1	Water-3 O	2.84	2.78

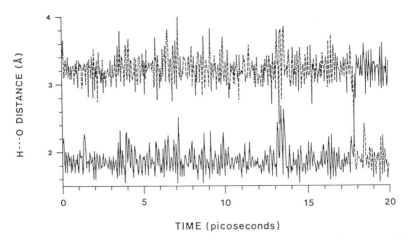

FIGURE 10. Histories of the hydrogen bond distances between the oxygen of water 3 and the hydrogens of water 2 during a 20-psec simulation of the crystal of cyclo-$(Gly-Pro-Gly)_2$. Note the interchange in the hydrogen atom positions which takes place at about 18 psec.

not disturbed. This means that the hydrogen-bonded network is retained, albeit constantly fluctuating. A rotational transition which occurs between 17–18 picoseconds of the dynamics gives insight into the dynamic nature of even an apparently ordered array. At that time the water molecule Wat^2 underwent rapid rotation in which the hydrogen atoms of Wat^2 interchange their positions in the hydrogen bond network (FIG. 10).

To investigate the energy of the water molecules and compare the minimized energies with time-averaged dynamic energies and with the energies of bulk water, we have calculated the intermolecular potential energies of each of the water molecules. TABLE 4 lists the energies for the minimized and time-averaged dynamics crystal structures.

All water molecules have a time-averaged potential energy greater than or equal to -10.1 kcal/mol, which is the average potential energy calculated for bulk water with the same potential field.[36] The absence of energies less negative than -10.1 kcal/mol is reasonable because there are no charges on the peptide molecule. The averaged potential energies of Wat^1, Wat^2 and Wat^4 are slightly less than the potential energy of bulk water due to the poorer hydrogen bonding that the waters are forced to adopt in the crystal.

TABLE 4. Minimized and Time-Averaged Intermolecular Potential Energies (kcal/mol) of the Water Molecules in the Crystal of Cyclo-$(Gly-Pro-Gly)_2$

Water	Minimized	Time-Averaged Dynamics
Wat-1	-10.1	-9.2
Wat-2	-11.2	-9.6
Wat-3	-11.8	-10.1
Wat-4	-9.5	-7.6

CONCLUSIONS

We have shown that, using a combination of molecular dynamics and energy minimization procedures, we can account for the structural properties of a peptide crystal and for the effects of the crystal forces on peptide conformation with a reasonable degree of accuracy. Further work remains to be done, however, including the use of constant-pressure dynamics simulations, tests of our ability to simulate thermal motion by the comparison of calculated with experimental temperature factors, and an extension of this study to a wider range of peptide crystals. The results emphasized that the conformation of the molecule is affected by its environment, so that one must be cautious when attempting to compare results of calculations on isolated systems with experimental crystal structures or when using crystal structures to rationalize biological activity.

REFERENCES

1. DESLAURIERS, R. & I. C. P. SMITH. 1980. *In* Biological Magnetic Resonance, Vol. 2. J. Reuben, Ed.: 243–344. Plenum Press. New York, NY.
2. KESSLER, H. 1982. Angew. Chem. Int. Ed. Engl. **21:** 512–523.
3. ROSE, G. D., L. M. GIERASCH & J. A. SMITH. 1985. Adv. Protein Chem. **37:** 1–109.
4. KARLE, I. L. 1981. *In* The Peptides, Analysis, Synthesis, Biology, Vol. 4. E. Gross and J. Meinhofer, Eds.: 1–54.
5. HAGLER, A. T. 1985. Theoretical Simulation of Conformation, Energetics, and Dynamics of Peptides. The Peptides, Vol. 7. J. Meienhofer, Ed.: 213–299.
6. HAGLER, A. T., D. J. OSGUTHORPE, P. DAUBER-OSGUTHORPE & J. C. HEMPEL. 1985. Science **227:** 1309.
7. HAGLER, A. T., P. S. STERN, R. SHARON, J. M. BECKER & F. NAIDER. 1979. J. Am. Chem. Soc. **101:** 6842–6852.
8. HAGLER, A. T., E. HULER & S. LIFSON. 1974. J. Am. Chem. Soc. **96:** 5319–5327.
9. HAGLER, A. T. & S. LIFSON. 1974. J. Am. Chem. Soc. **96:** 5327–5335.
10. HAGLER, A. T., S. LIFSON & P. DAUBER. 1979. J. Am. Chem. Soc. **101:** 5122–5130.
11. DAUBER, P. & A. T. HAGLER. 1980. Acc. Chem. Res. **13:** 105.
12. HALL, D. & N. PAVITT. 1984. J. Comp. Chem. **5:** 441–450.
13. HAGLER, A. T. & J. MOULT. 1978. Nature **272:** 222.
14. KUNTZ, I. D. Jr. & W. KAUZMANN. 1974. Adv. Protein Chem. **28:** 239–345.
15. HENDRICKSON, J. B. 1961. J. Am. Chem. Soc. **83:** 4537.
16. RAMACHANDRAN, G. N., C. RAMAKRISHNAN & V. SASISEKHARAN. 1963. J. Mol. Biol. **7:** 95.
17. SCHERAGA, H. A. 1968. Adv. Phys. Org. Chem. **6:** 103.
18. RAMACHANDRAN, G. N. & V. SASISEKHARAN. 1968. Adv. Protein Chem. **23:** 283.
19. LIQUORI, A. M. 1969. Q. Rev. Biophys. **2:** 65.
20. KARPLUS, M. & J. N. KUSHICK. 1981. Macromolecules **14:** 325.
21. DINOLA, A., H. J. C. BERENDSEN & O. EDHOLM. 1984. Macromolecules **17:** 2044.
22. TORRIE, G. M. & VALLEAU, J. P. 1977. J. Comp. Phys. **23:** 187–199.
23. VALLEAU, J. P. & G. M. TORRIE. 1977. *In* Statistical Mechanics, Part A. B. J. Berne, Ed.: 169. Plenum Press. New York, NY.
24. PATEY, G. N. & J. P. VALLEAU. 1975. J. Chem. Phys. **63:** 2334.
25. TORRIE, G. M. & J. P. VALLEAU. 1974. Chem. Phys. Lett. **28:** 578.
26. MEZEI, M., P. K. MEHROTRA & D. L. BEVERIDGE. 1985. J. Am. Chem. Soc. **107:** 2239–2245.
27. OKAZAKI, S., K. NAKANISHI & H. TOUHARA. 1979. J. Chem. Phys. **71:** 2421.
28. MRUZIK, M. R., F. R. ABRAHAM, D. E. SCHREIBER & G. M. POUND. 1976. J. Chem. Phys. **64:** 481–491.
29. MCCAMMON, J. A., B. R. GELIN & M. KARPLUS. 1977. Nature **267:** 585.

30. KARPLUS, M. & J. A. McCAMMON. 1981. CRC Crit. Rev. Biochem. **9:** 293.
31. VAN GUNSTEREN, W. F., H. J. C. BERENDSEN, J. HERMANS, W. G. J. HOL & J. P. M. POSTMA. 1983. Proc. Natl. Acad. Sci. USA **80:** 4315–4319.
32. HAGLER, A. T., J. MOULT & D. OSGUTHORPE. 1980. Biopolymers **19:** 395.
33. HAGLER, A. T., D. OSGUTHORPE & B. ROBSON. 1980. Science **208:** 599.
34. MOULT, J., A. YONATH, W. TRAUB, A. SMILANSKY, A. PODJARNY, D. RABINOVICH & A. SAYA. 1976. J. Mol. Biol. **100:** 179–195.
35. BROWN, J. N. & R. G. TELLER. 1976. J. Am. Chem. Soc. **98:** 7565.
36. BERENDSEN, H. J. C., J. P. M. POSTMA, W. F. VAN GUNSTEREN & J. HERMANS. 1981. *In* Intermolecular Forces. Proceedings of the Fourteenth Jerusalem Symposium on Quantum Chemistry and Biochemistry. B. Pullman, Ed.: 331–342. Reidel. Dordrecht, the Netherlands.
37. DAUBER-OSGUTHORPE, P., D. J. OSGUTHORPE, J. WOLFF & A. T. HAGLER. In preparation.
38. WOOD, W. W. 1968. *In* Physics of Simple Liquids. G. S. Rushbrooke, Ed.: 117. North-Holland. Amsterdam, the Netherlands.
39. LEVITT, M. 1983. J. Mol. Biol. **168:** 595–620.
40. NOVOTNY, J., R. BRUCCOLERI & M. KARPLUS. 1984. J. Mol. Biol. **177:**787–818.
41. HAGLER, A. T. & S. LIFSON. 1974. Acta Crystallogr. **B30:** 1336–1341.
42. EGGLESTON, D. S. Private communication.
43. KOSTANSEK, E. C., W. E. THIESSEN, D. SCOMBURG & W. N. LIPSCOMB. 1979. J. Am. Chem. Soc. **101:** 5811.

Progress in the Water Structure
of the Protein Crambin by X-Ray
Diffraction at 140 K[a]

MARTHA M. TEETER[b]

Department of Chemistry
Boston University
Boston, Massachusetts 02215

HÅKON HOPE

Department of Chemistry
University of California at Davis
Davis, California 95616

Diffraction data on crystals of crambin, a small, hydrophobic plant protein (MW = 4700), have been collected to 0.83 Å at 140 K without a capillary tube. It is important to collect X-ray diffraction data at the lowest temperature possible because large vibrational motion of molecules increases the uncertainty in atomic position. Also, protein data collection is particularly subject to inherent systematic errors because the capillary tube and the liquid of crystallization, which maintain the protein crystal hydration, can scatter anisotropically.

In order to minimize such errors, the method of H. Hope[1] for working with air-sensitive crystals has been adapted to proteins. Crystals of crambin were mounted without a capillary and rapidly transferred into a cold N_2 stream on the $P2_1$ diffractometer. The 26,970 observations made at the 2σ level to a minimum interplanar d spacing of 0.83 Å ($\sin \theta/\lambda = 0.6$) represent 81% of the accessible data at this resolution.

The starting model used was that refined at 0.945 Å without hydrogen atoms.[2] Konnert-Hendrickson restrained least-squares refinement proceeded (by stepwise additional of data with lower limit 1.5 to 1.0 Å) in 12 cycles to an R value of 14.0% (10.0 − 1.0 Å data). A difference map at this stage revealed additional waters as well as the anisotropic motion of the protein. Of the strongest 29 peaks, 40% (12) were assigned to additional waters, 31% (9), to protein motion and 28% (8) to alternate water positions. The largest peaks from anisotropic thermal vibration were associated with the sulfur atoms in crambin's three disulfide bonds. Ultimately, 15 new waters were located; 4 former water molecules were changed to alternates; 39 additional water alternates were located; and 2 new ethanols were modeled. The total amount of solvent at this stage is 72 waters with 60 alternates and 4 ethanols.

[a]This work was supported by Grant PCM 803224 from the National Science Foundation to M.M.T.
[b]Present address: Department of Chemistry, Boston College, Chestnut Hill, Massachusetts 02167.

163

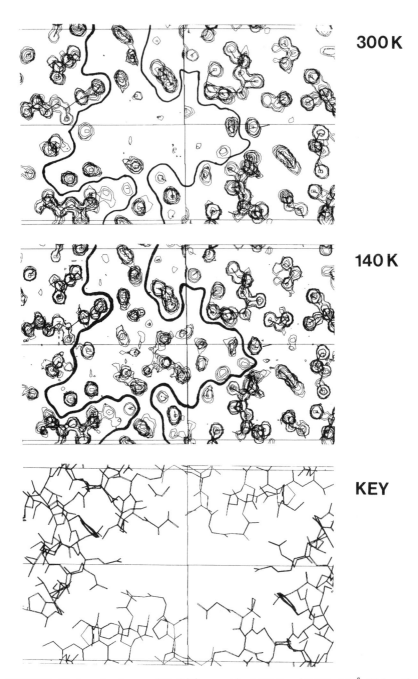

FIGURE 1. Electron density maps ($2F_o$-F_c) for crambin at 300 K and 140 K. A 5-Å-thick section of density is presented around the largest free area for water in the crystal. The *heavy lines* surrounding the central water channel designates the protein-water boundary. *Top:* 300 K structure; *middle:* 140 K structure; *bottom:* key indicating the bonds of the four symmetry-related protein molecules surrounding the water channel. Sections of the protein density are visible in the top two parts of the figure.

Refinement is currently proceeding with hydrogen atoms at calculated positions for the protein atoms. Anisotropic motion has been allowed for all atoms. The agreement factor for all of the data above 2σ is 13.3%.

The increased order in the water structure at 140 K is evident from FIGURE 1. The water molecules in the largest free area in the crystal, that region between adjacent molecules called the "hole,"[3] has weak electron density at 300 K. This is due to the mobility of the water molecules and their disorder at room temperature. Each water spends only a short time in one location and is vibrating strongly when it is localized. At 140 K, the electron density due to the water oxygens is higher. The peaks of density from oxygen atoms that are enclosed in the wavy black lines (the protein-water boundary) appear to be close packed across the "hole."

A comparison of the water positions in the two electron density maps reveals that, at the location of the weak peaks in the 300 K map, strong peaks of density are seen at 140 K. Regions in the "hole" farthest from the protein surface show new peaks of density at the lower temperature. The density at these positions is more diffuse, indicating partially occupied water sites.

REFERENCES

1. HOPE, H. 1985. New techniques for handling air-sensitive crystals. Presented at the annual meeting of the American Crystallographic Association (Stanford), PA3:24.
2. HENDRICKSON, W. A. & M. M. TEETER. Unpublished results.
3. TEETER, M. M. 1984. Water structure of a hydrophobic protein at atomic resolution: Pentagon rings of water molecules in crystals of crambin. Proc. Natl. Acad. Sci. USA **81:** 6014–6018.

Hydration of Nucleic Acid Crystals[a]

HELEN M. BERMAN

Institute for Cancer Research
Fox Chase Cancer Center
Philadelphia, Pennsylvania 19111

INTRODUCTION

All crystalline biological macromolecules are hydrated to varying degrees; some crystals contain as little as 20% water and others as much as 80%. It has been known since the first investigations of the DNA structure[1] that hydration affects its geometry. At high humidity DNA fibers are in the B conformation and at low humidity they assume the A conformation.[2] At the present time the mechanism for this conformational change is not understood, nor are the detailed interrelationships among the effects of sequence, humidity, and other environmental factors such as dielectric constant, temperature, and pressure. Because of the relatively recent availability of defined sequences of nucleic acids, it has become feasible to determine their crystal structures, thus enabling us to better understand the effects of these various features and phenomena. An excellent test for the level of our theoretical understanding of hydration is to be able to successfully simulate the water structure found in an experimentally determined nucleic acid crystal hydrate. The criteria for selecting a system whose water structure is appropriate for computer simulation are described.

DEFINITIONS

For the purposes of the discussion that follows some of the more important terms and concepts are defined. Nucleic acids are polymers consisting of repeating units of ribo or deoxyribo sugars, each of which are linked to one of five bases (FIG. 1). These sugar units are connected to one another via phosphodiester linkages. Although nucleic acids can be single or multistranded, the present discussion will be confined to those that form antiparallel, double-stranded, right-handed helical structures with the complementary bases hydrogen-bonded to one another via the standard Watson-Crick[1] hydrogen-bonding geometry (FIG. 2). When nucleic acids are in this general conformation, they form grooves named the major and the minor grooves. These grooves are defined with respect to the sides of the base pairs; the minor groove is on the side of the base pair with the O2 and N2 atoms; the major groove side has the N6, O6, and N4 atoms. A and B nucleic acids can be readily distinguished from one another by the fact that the former is a short, squat helix, with the bases pushed away from the helix axis, and has a very deep major groove (FIG. 3). B-DNA, on the other hand is a long, slim helix with the bases stacked along the helix axis and with a wide and relatively shallow

[a]This work was supported by Grants GM21589, CA06927, and RR05539 from the National Institutes of Health, and by an appropriation from the Commonwealth of Pennsylvania.

major groove (FIG. 4). Although in a qualitative sense the conformation angles of these two right-handed DNA forms are similar, the quantitative differences between them are sufficient to give rise to these very different macroscopic features. Moreover, the results of both fiber and single-crystal studies show that within each genus there are many polymorphic structures whose features are dependent on the many environmental factors described above.

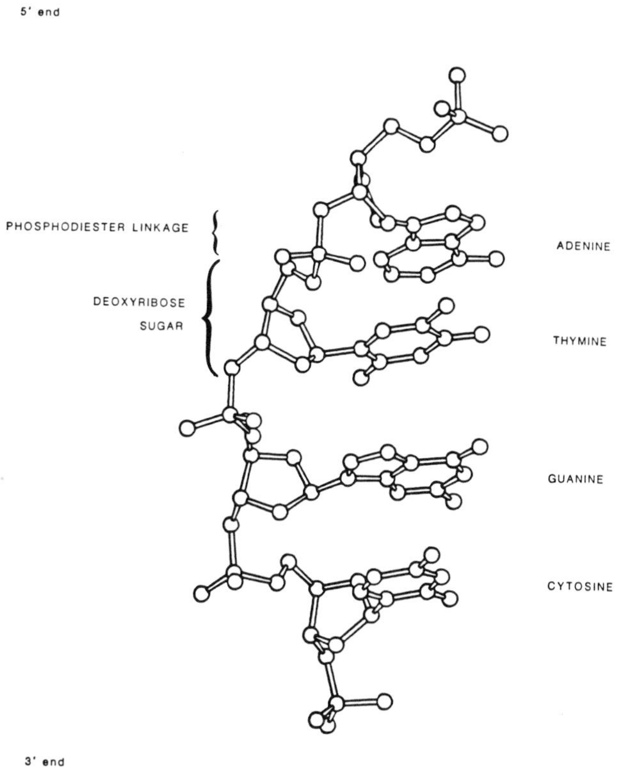

FIGURE 1. One strand of DNA showing the four bases, the phosphodiester linkage, and the deoxyribose sugar. In RNA there is a uracil in place of the thymine and a ribose in place of the deoxyribose sugar.

SELECTION OF A MODEL NUCLEIC ACID CRYSTALLINE HYDRATE

The earliest work on nucleic acids was done using fiber diffraction methods wherein it was possible to derive the average structure for any given sample.[3,4] The limitations of the method, however, made it impossible to determine the positions of the water molecules. In contrast, the methods of single-crystal structure determination enable one to derive a detailed view not only of the nucleic acid, but also of the ordered water in the crystal. There have been more than fifty crystal structures reported of defined-sequence nucleic acid fragments alone and complexed with other molecules such as those in antitumor drugs and antibiotics.[5,6] All of these crystals are highly hydrated and thus, in principle, any one of them would be interesting to model using theoretical methods. However, to properly simulate the water structure one must impose some rather rigid criteria on the experimental system under investigation.

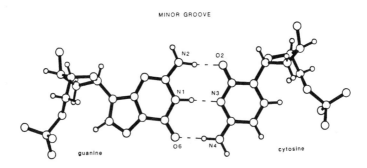

FIGURE 2. Watson-Crick type hydrogen bonding between adenine and thymine and guanine and cytosine.

Among the criteria are:

1. The structure should be well hydrated with a high percentage of ordered water;

2. The water should have a definable structure. Some of the water molecules should be beyond the first coordination shell;

3. The structure should be determined to reasonably high resolution to ensure that the positions of both the nucleic acid and the water molecules are well defined;

4. The agreement between observed data and calculated structure factors should be good;

FIGURE 3. (*Left*) A-RNA double helix showing deep major groove and shallow minor groove. (*Right*) stereo view of this same molecule. A-DNA is in this same general conformation.

5. The negative charges on the phosphate backbone should be balanced by known counterions.

With these criteria in mind, some of the more representative right-handed helical nucleic acid crystal structures are described. Reviews of single-stranded and left-handed crystal structures are left for future discussion and evaluation.

GpC and ApU

The first determination of the structure of the double helix at atomic resolution was reported in 1973,[7,8] some twenty years after the proposal by Watson and Crick of this

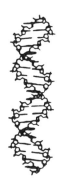

FIGURE 4. (*Left*) B-DNA double helix showing wide major groove. (*Right*) stereo view of the same molecule. RNA does not assume this conformation.

key geometry in molecular biology. Both GpC[9] and ApU[10] are self-complementary sequences and each form A-type RNA mini-helices. These structures are very well determined with agreement indices well below 10% and are better than 1-Å resolution. In GpC the negative charges on the phosphate backbone are balanced by sodium ions that bind to these phosphates. The mini-helices form base-stacked layers with a zig-zag water chain interconnecting minor groove base atoms (FIG. 5). All eighteen water molecules per double helix are in the first coordination shell; that is, each water is hydrogen-bonded either to the sodium ion or the nucleic acid fragment. The water molecules are arranged in a similar fashion in the calcium salt of this same sequence[11] and very differently in the ammonium salt.[12] There are fewer water molecules per dimer in Na ApU. As in the GpC structure the positions of the sodium counterions are well determined. In this case, however, one sodium ion is bound to a phosphate and the other to carbonyl oxygen atoms on the minor groove. All the water molecules are in the

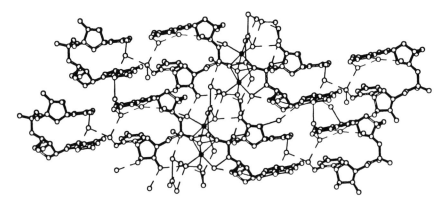

FIGURE 5. One view of the crystal structure of Na GpC. The sodium ions are represented by *solid circles* and hydrogen bonds by *dashed lines.* Note the octahedral coordination around the sodium ion.

first-coordination shell of hydration (FIG. 6). Thus, the solvent network is found predominately in the minor groove. The water molecules form a very elegant framework since they are all in the first-coordination shell, but they do not have the structural freedom to form clathrate-type arrangements. This has been attributed to the general lack of hydrophobicity of RNA sequences due mostly to the fact that uracil lacks the methyl group of the thymine that would be present in DNA.[10]

B-DNA

The structure of dCGCGAATTCGCG[13] showed for the first time a detailed view of a B-type helix. Three determinations were done, one at room temperature,[13] one at 7 degrees[14] with a high alcohol content, and one at 16K.[15] The high-alcohol crystal gave the best understanding of the hydration since in that form the water is more immobilized. The resolution of the determination is 2.3 Å and the agreement index

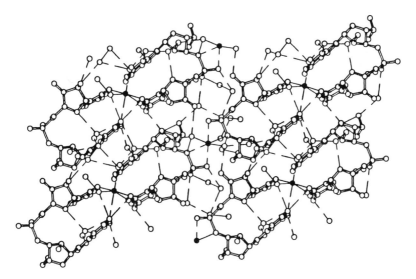

FIGURE 6. The crystal structure of Na ApU. The sodium ions are shown as *solid circles* and hydrogen bonds as *dashed lines*.

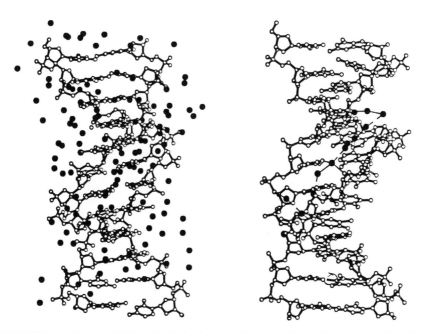

FIGURE 7. (*Left*) the B-DNA double helix showing all the ordered water in the crystal. (*Right*) the spine of hydration along the minor groove of B-DNA.

0.17. The structure consists of twelve base pairs arranged as a bent B-DNA helix, although the degree of bending is less than was found in the room-temperature form. About one-quarter of the solvent molecules were identified (FIG. 7, *left*). Thirteen of the more ordered water molecules arrange themselves along the minor groove, forming a zig-zag "spine of hydration" which consists of water molecules hydrogen-bonded to the N3 and O2 atoms of the A-T base pairs (FIG. 7, *right*). A second layer of water molecules completes the tetrahedral coordination around the waters. This spine is localized around the central A-T base pairs partly because of the potentially disruptive effect of the N2 hydrogen-bonding donors in the GC region and partly because of intermolecular packing constraints. The major groove is also hydrated, but because much of the water is disordered and not resolved, there is no recognizable pattern. There must be an abundance of positive ions to counter the negative phosphate backbone, but their locations are uncertain. Tentative assignments have been made for some Mg^{++} and spermine ions.

A-DNA

The structure of dCCGG[16] consists of a right-handed double helix with an A-type conformation. The resolution of the determination is 2.0 Å and the agreement is 0.16. There are 86 solvent molecules per asymmetric unit; the locations of the positive counterions is uncertain. In this structure the more ordered water molecules form a network across the major groove interconnecting the opposing phosphate backbones (FIG. 8). In another A-type structure, dGG $^{Br}UA^{Br}UACC$,[17] the ordered waters form pentagonal arrays in the major groove. In both structures the packing is such that the minor groove is blocked from attack by water molecules and instead base pairs from a neighboring molecule are packed against it. This packing pattern has been observed in other A-type structures.[18] It is possible that drying disrupts the potentially stabilizing

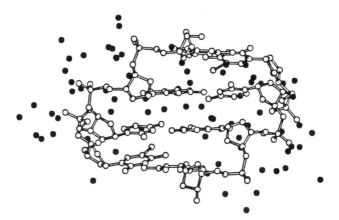

FIGURE 8. The water environment in dCCGG. Note that in this structure most of the water molecules are bound either to phosphate or major groove hetero atoms.

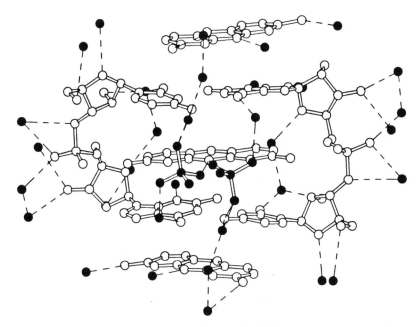

FIGURE 9. The 3:2 complex of CpG and proflavine. The *solid circles* represent water molecules and *dashed lines* show hydrogen bonds. The disordered sulfate ions are shown in the major groove.

minor groove spine of hydration of B-DNA to yield the type of very hydrophobic environment of the minor groove seen in A-DNA.

rCpG and dCpG Proflavine Complexes

The crystal structure of the 3:2 proflavine CpG complex was determined to 0.85-Å resolution with an agreement index of 0.10.[19] It consists of an intercalated duplex with a proflavine stacked above and below the base pairs. There is a disordered sulfate in the major groove of the double helix which balances the charges of the proflavine cations. This leaves a residual negative charge which must be balanced by either an ammonium ion or an additional charge on a proflavine. The location of this charge has never been resolved. There are twenty-four water molecules per duplex, all of which are hydrogen-bonded to the intercalated complex or to the sulfate ion. All but two water molecules are interconnected via the sulfate anions. There does not seem to be any particular preference for water binding in either the major or minor groove (FIG. 9). dCpG and proflavine form a 2:2 crystalline complex.[20] The resolution of that determination is 1.0 Å and the agreement index is 0.15. In this structure the negative charges on the phosphate backbone are balanced by the positively charged proflavine cations. The water molecules are bonded to the complex atoms and also to one another, forming a series of edge-linked pentagons in the major groove (FIG. 10).[21] This pattern resembles the semiclathrate structures of tertiary alkyl amines,[22] in which there are edge-linked polygon structures formed by the water molecules and the anions. The cationic guests form hydrogen bonds to the host cages. In the dCpG-proflavine crystal structure, one phosphate group participates in the water framework. The major groove side of the intercalated base pairs is positively charged by virtue of the protonated proflavines and is the side which "hydrogen bonds" with the "host" water molecules. The more hydrophobic minor groove has fewer water molecules and is arranged as a single polygon disk. This is consistent with the fact that the conformation of the nucleic acid fragment is more A-like in character and hence adopts the hydration pattern seen in other A-type structures.

D-CpG PROFLAVINE—A GOOD CANDIDATE FOR SIMULATION STUDIES

From the point of view of biological relevance the best candidates for simulation studies are the B-DNA and A-DNA oligomers. Examination of the ordered water shows that there are some interesting patterns in these structures. However, the relatively low percentage of the ordered water that is observed makes these structures less desirable for initial simulation work. In this sense, the ApU and GpC structures are ideal since they are clearly the best determined with the lowest R factors and highest resolution. The only possible drawback to using these structures is that they do

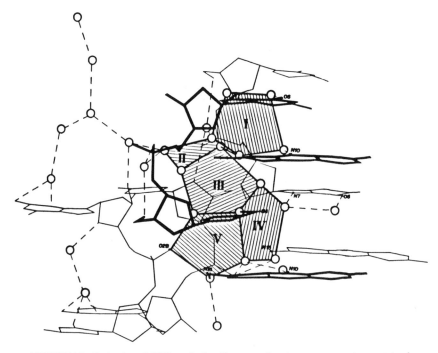

FIGURE 10. Hydration of dCG-proflavine. Pentagonal water arrangements are *striped*.

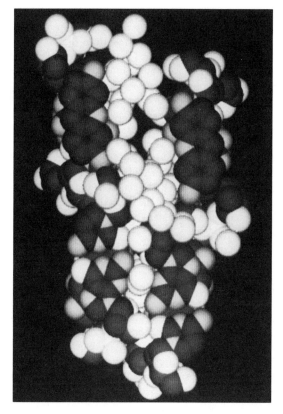

FIGURE 11. The crystal structure of dCpG-proflavine showing gaps containing disordered water.

not have second-shell water molecules and hence might not be as representative as more-hydrated samples. The rCpG-proflavine suffers as a potential candidate because the distribution of the charges is uncertain. At the present time, the dCpG-proflavine structure with its well-determined and interesting water structure is the best candidate for simulation studies; three such analyses have been reported.[23,24] The basic strategy in these studies was to fix the position of the solute atoms in the crystalline array and then try to computationally reproduce the water structure in the crystal. The results of these studies were somewhat promising, although they did not fully reproduce the water positions, as judged by calculating the distances between observed and calculated water positions or by checking the symmetry between asymmetric units. As pointed out by Beveridge,[24] one of the main problems is the potential functions for the solute atoms. The system also has some ambiguities experimentally. There were areas in which the waters are disordered, leading to gaps in the crystal (FIG. 11). Therefore, X-ray crystal structure determination was carried out at a lower temperature in an attempt to reduce the degree of disordering. It was discovered that the crystal could change forms in response to slight alterations of humidity within the capillary as well as by temperature changes. One crystal was obtained which, although very similar to the one described previously,[21] has some interesting differences. Since the temperature of the experiment was lower, it is not surprising that the temperature factors are all lower in this structure. All of the waters that are involved in the pentagonal network are present in this form. A few of them have moved by as much as 0.6 Å when compared to the previous dCpG-proflavine structure. Thus, some hydrogen bonds are broken while new ones are formed. The overall effect is that one pentagon is broken and replaced by a hexagon and a new pentagon is formed in a place that none existed before (FIG. 12). This crystal also has 2.5 more water molecules which are in the first-coordination shell and hydrogen-bonded to major groove hetero atoms, thus closing the previously existing gap in the crystal structure (FIG. 13). Of more importance is the conformational change of one of the sugars and in two other torsion angles. It is significant that the newly found water molecules are hydrogen-bonded to these regions, thus demonstrating that very small differences in the crystalline environment can perceptibly change the conformation of the molecular complex.[25]

SUMMARY

Can we make any generalizations from examination of the crystal structures in hand? The results of study of the very well-determined high-resolution structures indicate that the counterions have a very strong effect on organizing the water structure and that these counterions are bonded in a sequence-specific manner. Hence, the sodium ion bonds in the minor groove of ApU and only to the phosphate backbone in GpC. Not surprisingly then, the water network in ApU is predominately in its minor groove. Similarly, the negative sulfate counterion in the major groove of the 3:2 complex between proflavine and CpG has a significant influence on the water structure

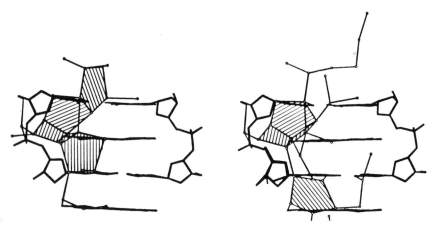

FIGURE 12. A comparison of the two dCpG-proflavine structures: room-temperature form (*left*) and low-temperature form (*right*).

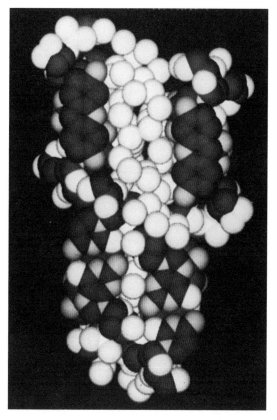

FIGURE 13. The lower-temperature form of dCpG-proflavine. Note the closing of the gaps.

in that crystal. The crystallization of two positive proflavine molecules with two negative nucleic acid chains obviates the need for inorganic ions and may provide additional insight about nucleic acid water structure. The presence of the charged aromatic hydrocarbon appears to provide the correct mixture of hydrophilicity and hydrophobicity that allows for both the gathering and ordering of water molecules around the nucleic acid molecule, not unlike what was previously observed in the semiclathrate structures. This same type of hydrophobic aggregation might pertain along the major groove side of structures containing the appropriate arrangement of methyl-containing thymine bases. Although it is very tempting at this point to make further rules and predictions, experience has shown that, especially in the case of nucleic acids, such prognostications would be premature. What is clearly needed are some more high-quality crystal structures of a variety of sequences under different and controlled conditions. Analyses of these may then put us in a position to successfully predict both the structure of water and its effects on nucleic acid conformation.

ACKNOWLEDGMENTS

I wish to thank David Beveridge, Stephan Ginell, Patricia Sawzik, and Stephen Neidle for their collaboration and their many stimulating discussions about this subject.

REFERENCES

1. WATSON, J. D. & T. H. C. CRICK. 1953. Nature 171: 737–738.
2. FRANKLIN, R. E. & R. G. GOSLING. 1953. Acta Crystallogr. 6: 673–677.
3. ARNOTT, S. 1970. Prog. Biophys. Mol. Biol. 21: 267–319.
4. LESLIE, A. G. W., S. ARNOTT, R. CHANDRASEKARAN & R. L. RATLIFF. 1980. J. Mol. Biol. 143: 49–72.
5. NEIDLE, S. & H. M. BERMAN. 1983. Prog. Biophys. Mol. Biol. 41: 43–66.
6. BERMAN, H. M. & P. YOUNG. 1981. Ann. Rev. Biophys. Bioeng. 10: 87–114.
7. DAY, R. O., N. C. SEEMAN, J. M. ROSENBERG & A. RICH. 1973. Proc. Natl. Acad. Sci. USA 70: 849–853.
8. ROSENBERG, J. M., N. C. SEEMAN, J. J. P. KIM, F. L. SUDDATH, H. B. NICHOLAS & A. RICH. 1973. Nature 243: 150–154.
9. ROSENBERG, J. M., N. C. SEEMAN, R. O. DAY & A. RICH. 1976. J. Mol. Biol. 104: 145–167.
10. SEEMAN, N. C., J. M. ROSENBERG, F. L. SUDDATH, J. J. P. KIM & A. RICH. 1976. J. Mol. Biol. 104: 109–144.
11. HINGERTY, B., E. SUBRAMANIAN, S. D. STELLMAN, T. SATO, S. B. BROYDE & R. LANGRIDGE. 1976. Acta Crystallogr. B32: 2998–3013.
12. AGGARWAL, A., S. A. ISLAM, R. KURODA, M. R. SANDERSON & S. NEIDLE. 1983. Acta Crystallogr. B39: 98–104.
13. WING, R., H. DREW, T. TAKANO, C. BROKA, S. TANAKA, K. ITAKURA & R. E. DICKERSON. 1980. Nature 287: 755–758.
14. KOPKA, M. L., A. V. FRATINI, H. R. DREW & R. E. DICKERSON. 1983. J. Mol. Biol. 163: 129–146.
15. DREW, H. R., S. SAMSON & R. E. DICKERSON. 1982. Proc. Natl. Acad. Sci. USA 79: 4040–4044.
16. DICKERSON, R. E., H. H. DREW, B. N. CONNER, R. M. WING, A. V. FRATINI & M. L. KOPKA. 1982. Science 216: 475–485.
17. KENNARD, O. 1984. Pure Appl. Chem. 56: 989–1004.
18. WANG, A. H.-J., S. FUJII, J. H. VAN BOOM & A. RICH. 1982. Proc. Natl. Acad. Sci. USA 79: 3968–3972.
19. BERMAN, H. M., W. STALLINGS, H. L. CARRELL, J. P. GLUSKER, S. NEIDLE, G. TAYLOR & A. ACHARI. 1979. Biopolymers 18: 2405–2429.
20. SHIEH, H.-S., H. M. BERMAN, M. DABROW & S. NEIDLE. 1980. Nucleic Acids Res. 8: 85–97.
21. NEIDLE, S., H. M. BERMAN & H.-S. SHIEH. 1980. Nature 288: 129–133.
22. JEFFREY, G. A. 1969. Acct. Chem. Res. 2: 344–352.
23. KIM, K. S. & E. CLEMENTI. 1985. J. Am. Chem. Soc. 107: 227–234.
24. MEZEI, M., D. L. BEVERIDGE, H. M. BERMAN, J. M. GOODFELLOW, J. L. FINNEY & S. NEIDLE. 1983. J. Biomol. Struct. Dyn. 1: 287–297.
25. BERMAN, H. M. & S. L. GINELL. 1986. J. Biomol. Struct. Dyn.: 131–136.

Monte Carlo Studies of Water in Crystal Hydrates

JULIA M. GOODFELLOW,[a,b] P. LYNNE HOWELL,[a,c] AND
FRANÇOISE VOVELLE[d,e]

[a]Department of Crystallography
Birkbeck College
London University
London WC1E 7HX, England

[d]Centre de Biophysique Moleculaire
45045 Orléans, France

INTRODUCTION

The computer simulation of solvent structures in crystal hydrates provides an obvious test of the predictive power of such methods to simulate solvent structures around molecules of biological interest. The main assumptions in these simulation methods include (*a*) the use of the "realistic" potential energy functions used to model each type of interaction and (*b*) the adequate equilibration and sampling of configurational space. Any comparison between two solvent structures requires a procedure by which the level of agreement can be assessed. If we wish to improve the reliability of simulated networks it is necessary to analyze the data such that the regions of disagreement can be found. However, initially it is necessary to assess the experimental data on solvent structures that are available from crystallography.

Experimental Structures

The best defined experimental solvent structures in crystal hydrates have been reviewed recently by Savage and Wlodawer.[1] The smallest structures, obtained from neutron data (so that deuterium/hydrogen positions can be determined), have led to a detailed appraisal of the water $\cdots$ solute hydrogen bond. A wide range of $O \cdots O$ distances have been found as well as a wide range for the $O \cdots O \cdots O$ angle. The hydrogen bond angle $O-H \cdots O$ tends to be nearly linear with the majority being greater than 150°. The seven medium-sized structures, at better than 1-Å resolution, begin to provide us with accurate models for networks of water molecules. These include rings of hydrogen-bonded water in α-cyclodextrin[2] and dCpG-proflavine.[3] These rings are often formed from pentagons and hexagons. These structures can also provide us with an insight into the complexity of the crystallographic disordered water

[b]Supported by Research Grants GR/D/14372 and GR/C/22417 from the Science and Engineering Research Council.
[c]Recipient of a Studentship from the Science and Engineering Research Council.
[e]Recipient of travel assistance from the Scientific Council at the French Embassy (London).

179

networks as seen in vitamin B_{12} coenzyme hydrate.[4] At this high level of resolution ($<1.0Å$) and with X-ray and neutron data, it is possible to model so-called disordered regions into several networks (of different occupancy) of major solvent sites and thence into "continuous" sites to fit elongated regions of electron density.[5]

An analysis of the 13 best protein crystal structures has also led to a detailed study of the water · · · main-chain hydrogen bond distance and angles.[6] However, the water-side-chain hydrogen bonds showed no preferred geometry, presumably because of an increased degree of disorder (higher temperature factors and lower occupancies) for side chains. Some protein structures have been studied with neutrons as well as X rays. Such studies show most agreement with ordered waters, but considerable disagreement in disordered regions. The best refined protein structures, those of crambin[7] and 2Zn insulin[8] show that about $\geq 70\%$ of the water can be modeled as ordered sites. These sites form rings which may be pentagonal as well as chains between polar atoms on the protein.

The oligonucleotide structures are now being refined at atomic resolution. The solvent structure of the "B"-DNA oligonucleotide dCGCGAATTCGCG[9,10] has shown a spine of hydration in the minor groove with each water linking polar atoms from one base to another on different chains and on an adjacent base pair. Ordered water molecules were seen around the phosphates group only when crystallization occurred at low temperature (16K) or at high MPD concentrations. So far, no counter-ions have been seen. Further refinement, of a different type, on this same structure[11] has led to model with hydrated minor groove but a relatively unhydrated major groove. The structures dGGBrUABrUACC[12] and dGGTATACC[13] have led to a detailed picture of the "A" form of DNA. In this structure, the relatively small major groove shows an extensive network of pentagonal solvent arrays while chains of water link phosphate groups along the backbone. A Z-DNA structure has also been refined with a view to understanding the water structure and shows a series of intra- and interstrand water bridges.[11]

In summary, the best refined high-resolution structures give us a detailed molecular model for the hydration of small biomolecule structures, some medium-sized structures, and a very few proteins and oligonucleotides. At this level, we can begin to see not only individual sites for solvent, but also networks of chains and loops of hydrogen-bonded water molecules surrounding biomolecules, as suggested by Saenger.[2]

Simulations

Computer simulations of complex aqueous systems are quite common even though the extension of these simulation techniques from the homogeneous liquid state to heterogeneous systems at interfaces is nontrivial.[14] The first data published on the simulation of solvent in a crystal hydrate were those of Hagler and Moult[15] on crystals of triclinic lysozyme. Unfortunately the experimental data were not at particular high resolution. The agreement between simulated and experimental data was greater than $1.0Å$ for a limited number of the best water positions. Therefore, investigators have studied much smaller crystal hydrates such as cyclic peptides,[16,17] amino acids,[18,19] and nucleotide hydrates.[20-23]

The aims of this study have been (a) to investigate the detailed nature of the model potential energy functions (using a system of a simple solute with one water molecule), (*b*) to assess quantitatively the results of full simulations of solvent in two well-defined crystal systems—B_{12} coenzyme and guanosine-5-phosphate—and (*c*) to analyze simulations of solutions of small solutes (amino acids and dinucleotides) in terms of water networks that can be compared with those in crystal hydrates, bulk water, clathrates, and ices.

METHODS

Water Models

Five water models have been used in the calculations presented in this paper and some of their characteristics are given in TABLE 1. These include two four-point charge models (namely the ST2 model of Rahman and Stillinger[24] and the EMPWI model of Vovelle and Ptak[25]), one three-point charge model (TIPS2 model of Jorgensen[26]), and two versions of the polarizable electropole model which employs an electropole expansion up to the quadrupole $\cdots$ quadrupole (PE(QQ)) or octupole $\cdots$ dipole

TABLE 1

Potential Function	Minimum Energy $(kcal \cdot mol^{-1})$	Distance (Å)	Acceptor Angle θ_A	Donor Angle θ_D
EMPWI	-7.5	2.75	130°	57°
TIPS2	-6.84	2.80	135°	47°
ST2	-6.20	2.85	130°	54°
PE(DO)	-4.5	2.90	150°	60°
PE(QQ)	-7.8	3.00	128°	57°

(PE(DO)) interactions.[27] The point charge models have Coulomb electrostatic interactions and inverse 6–12 nonbonded interaction coefficients. The polarizable electropole models use an electropole expansion for the electrostatic part together with a procedure that allows the dipole moment of each water molecule to increase due to polarization from the surrounding electrostatic fields. An inverse 6–9 nonbonded interaction is used in the PE(DO) model.

Solute Models

The solute atoms are represented by partial atomic charges calculated from quantum mechanics at either the CNDO/2 or SCF level. Large molecules, such as vitamin B_{12} coenzyme, are divided into overlapping fragments in order to calculate the charges. The nonbonded parameters have been taken from three sources—Momany *et al.*,[28] Lifson *et al.*,[29] and Weiner *et al.*[30] Water $\cdots$ solute potential energy functions are calculated using the usual combination rules. The coordinates of the solute atoms are taken from neutron or X-ray crystallographic studies and remain fixed throughout

TABLE 2

Model	Environment	NWAT[a]	NEQUIL[b]	NCONF[c]
Urea	*Vacuo*	1	—	—
Guanosine-5'-PO$_4$	Crystal	12	10^5	5×10^5
Vitamin B$_{12}$ coenzyme	Crystal	56	2×10^5	3×10^5
Glycine	Solution	109	10^6	5×10^5
Alanine	Solution	107	5×10^5	5×10^5
Valine	Solution	106	5×10^5	5×10^5
Leucine	Solution	106	5×10^5	5×10^5
Serine	Solution	105	5×10^5	5×10^5
Threonine	Solution	103	5×10^5	5×10^5
dCpG in B-form	Solution	118	5×10^5	5×10^5

[a]Number of water molecules.
[b]Number of configurations for equilibration.
[c]Number of configurations used to calculate properties of the system.

the energy minimization or Monte Carlo computer simulations. The properties of each system referred to in this paper are summarized in TABLE 2.

Energy Minimizations

The energy minimization method of Powell[31] was used to find the primary hydration sites around a small solute urea. Initially, the water molecule was placed at a point on a two-dimensional grid in the plane of the solute and the minimization carried out with respect to three rotational degrees of freedom. Approximate positions of local minima, found by this procedure, were then input to a full minimization routine with respect to three translational and three rotational degrees of freedom. The minimum

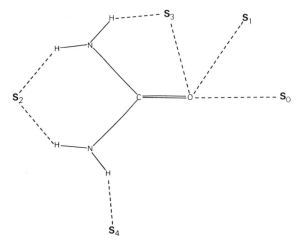

FIGURE 1. Schematic diagram of the urea molecule with possible hydration sites indicated as S$_0$, S$_1$, S$_2$, S$_3$ and S$_4$.

energy—primary hydration sites were found using four water models (ST2, TIPS2, EMPWI and PE[DO]), three sets of charges A, B and C obtained from CNDO/2 calculations, the literature[32] and SCF calculations adjusted to give the experimental dipole moment, respectively,[33] and three sets of nonbonded interaction parameters (I,[28] II[29] and III[30] obtained from the literature). Both the potential energy and the hydrogen bond geometry were calculated for each primary hydration site. FIGURE 1 is a schematic view of the possible (i.e., stereochemically acceptable) hydration sites around the carbonyl and amino groups of urea.

Computer Simulations

Monte Carlo simulations have been performed using the standard Metropolis algorithm at 20°C. All energy calculations are cut off at 7.5Å or half the box size (whichever is smaller). Periodic boundary conditions are employed. The number of

TABLE 3a

	Energy (kcal · mol^{-1})^b				
Model^a	S_0	S_1	S_2	S_3	S_4
ST2 IA	—	−6.03	(−2.31)	—	—
TIPS2 IA	−6.04	−6.43	(−4.21)	—	—
EMPWI A	—	—	−4.31	−8.55	—
PE(DO)A	—	−7.25	—	—	—
ST2 IIB	—	—	−12.01	—	−18.40

TABLE 3b

	Energy (kcal · mol^{-1})^b				
Model^a	S_0	S_1	S_2	S_3	S_4
ST2 IA		−6.03	(−2.31)		
ST2 IB		−5.89	−8.59		
ST2 IC		−4.93	−9.14		
EMPWI A			−4.31	−8.55	
EMPWI B			−10.39	−11.61	
EMPWI C			−8.33	−13.78	

TABLE 3c

	Hydrogen Bond Distance (Å)^b			
Model	S_0	S_1	S_2	S_3
ST2 IA		2.55	(3.52)	
ST2 IIA			3.08	2.63; 3.34
ST2 IIIA			3.06	2.63; 3.02
TIPS2 IB	2.65	—	(3.59)	2.65; 3.31
TIPS2 IIB	2.72	—	3.18	2.78; 3.05
TIPS2 IIIB	2.71	—	3.15	2.73; 2.96

^{a}The definition of the model used in each calculation is given in the METHODS section.
^{b}Sites S_0, S_1, S_2, S_3 and S_4 are defined schematically in FIGURE 1.

configurations sampled for equilibration and for the analysis are given in TABLE 2 for each system. In the crystal hydrate simulations, a unit cell of atoms is used as the box and all water molecules are moved independently of the crystal symmetry constraints.

RESULTS

Water · · · Solute Interactions in Vacuo

The results from the energy minimization of one water molecule around a small solute, urea, in vacuo are presented in TABLE 3. It can be seen that

(i) different water models can lead to different primary hydration sites (TABLE 3a). For example, the TIPS2 model leads to an out of plane. So site B from among the possible sites defined in FIGURE 1;

(ii) the use of different sets of partial atomic charges (A, B and C) leads to a different magnitude for the potential energy (TABLE 3b). For example, the EMPWI water model used with charge sets A, B and C leads to energies of -4.31, -10.39 and -8.33 kcal mol^{-1}, respectively, for the S_2 site; and

(iii) the use of different sets of nonbonded interaction coefficients leads to changes in the length of the water · · · solute hydrogen bond (TABLE 3c). For example, the

TABLE 4

	Average RMSD (Å)		AGF I (Å)		AGF II (Å)	
	Initial	Final	(A)^a	(B)^b	(A)^a	(B)^b
Sim 1	0.0	0.45	0.25	0.23	0.25	0.56
Sim 2	1.93	1.11	0.50	0.59	0.32	0.51
Sim 3	1.93	1.12	0.51	0.62	0.33	0.53
Sim 4	0.88	0.83	0.62	0.40	0.54	0.25

^{a}Solvent · · · solvent distances.
^{b}Solvent · · · solvent distances.

hydrogen bond length to the carbonyl oxygen from the S_3 site changed from 2.65Å to 2.78Å depending on the charge set (I, II or III) with the TIPS2 water model.

Initial Random Solvent Positions

Three Monte Carlo simulations of the solvent in guanosine-5'-PO$_4$ crystal hydrate were performed with different initial positions for the water molecule oxygen positions. These initial positions were chosen to be (i) the crystallographic positions (Sim 1), (ii) random positions calculated from a grid of points (Sim 2 and Sim 3), and (iii) random positions calculated by running the simulation at high temperature (Sim 4).

The initial and final root mean square deviations (RMSD) between the simulated and experimental water oxygen positions are given in TABLE 4. It can be seen that in Sim 1 the water molecules move away from the crystallographic positions by an

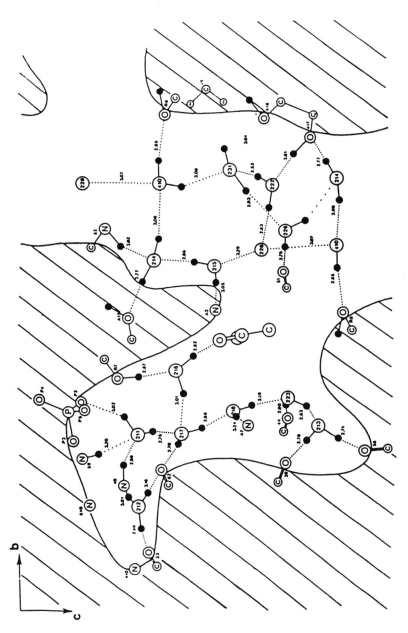

FIGURE 2. Diagram of the experimentally determined solvent network in B_{12} coenzyme crystal hydrate.

average of 0.45Å. In the random start simulations (Sim 2,3), the average deviation is reduced from 1.93Å to about 1.12Å (i.e., an average of 0.81Å), indicating movement towards the experimental positions. However, the final RMSD is greater for both Sim 3 and Sim 4 compared with Sim 1. Analysis of the water molecule environments (ageement factor I) during these simulations with initial random positions shows that (1) the solvent · · · solute distances are further from experimental values than the solvent · · · solvent distances and (2) the agreement between simulated asymmetric units (agreement factor II) is worse than that found in Sim 1.

B_{12} Coenzyme Crystal Hydrate

The major experimental solvent network in vitamin B_{12} coenzyme is shown in FIGURE 2 and can be described as an ordered pocket region and a disordered channel region to the left and right, respectively, of the central acetone molecule. Separate simulations were performed with this crystal system using five different water models while the solute model remained the same. The comparisons between each of the five simulated and the experimental solvent networks are summarized in TABLE 5. It can be seen that the lowest RMSD in water oxygen positions, the lowest agreement factor between simulated and experimental positions, and the lowest agreement factor between each simulated asymmetric unit are found with the EMPWI water model. The highest values (greatest disagreement) for these three parameters are found with the PE models. Use of the ST2 and TIPS2 models leads to intermediate values of these parameters.

Water Dimer

In order to investigate how best to improve the agreement between the simulated networks, using different models, and the experimental solvent networks, we need to look at the properties of the water model itself. This can be undertaken by considering properties of water dimer interaction drawn schematically in FIGURE 3. The energy surface can be represented by sections as a function of distance (R), donor angle (θ_D), and acceptor angle (θ_A). Such sections can be seen in FIGURES 4a, 4b and 4c for EMPWI TIPS2, ST2, PE(QQ), and PE(DO) models for water. Inspection of these

TABLE 5

Potential Functions	Average RMSD of Simulated to Experimental Waters Å	Agreement Factor between Experimental and Simulated Coordination (Å)	Agreement Factor between the Four Simulated Asymmetric Units (Å)
PE(DO)	1.26 ± 0.60	0.79	0.44
PE(QQ)	1.22 ± 0.73	0.65	0.35
ST2	1.02 ± 0.56	0.33	0.22
TIPS2	0.86 ± 0.55	0.37	0.30
EMPWI	0.75 ± 0.24	0.26	0.12

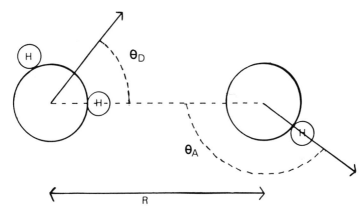

FIGURE 3. Schematic diagram of a water dimer defining the intermolecule distance (R), the donor angle (θ_D), and the acceptor angle (θ_A).

figures shows that the depth and position of the minima and the height of the repulsions whether lone pair · · · lone pair or hydrogen · · · hydrogen depend on the model. The EMPWI model always appears at one extreme (e.g., lowest minimum), whereas the PE models appear at the other end of the range (e.g., highest minimum). The ST2 and TIPS2 occur at intermediate positions.

Thus, the properties of the water model itself appear to correlate with the ability of the model to simulate solvent networks in crystal hydrates (previous section). In order to improve on the agreement between experimental and simulated solvent structure it will be necessary to modify the properties of the "best" model (i.e., EMPWI) in order to emphasize even further those properties that distinguish it from other water models, e.g., lower minima, higher repulsions.

Solution Studies

Analysis of the simulated structure of the hydrogen-bonded water networks around small solutes has been carried out for polar amino acids (serine and threonine) and for apolar amino acids (glycine, alanine, valine and leucine) in their zwitterionic form. The aim of this analysis was to look for patterns in the solvent structure, not just at individual hydration sites and to compare them qualitatively with those found experimentally. Therefore we considered all water molecules within 6.0Å of any solute atom and not just those hydrogen-bonded to polar solute atoms. Each of these complex networks of water molecules could be broken down into three patterns: namely, (i) self-bridging loops—those waters associated with any one polar solute atom (FIG. 5); (ii) polar bridging chains—chains of hydrogen-bonded water molecules linking different polar solute atoms (FIG. 6); and (iii) "apolar" networks—loops involving waters within 4.0Å of an apolar solute atom, but not hydrogen-bonded directly to the solute molecule (FIG. 7).

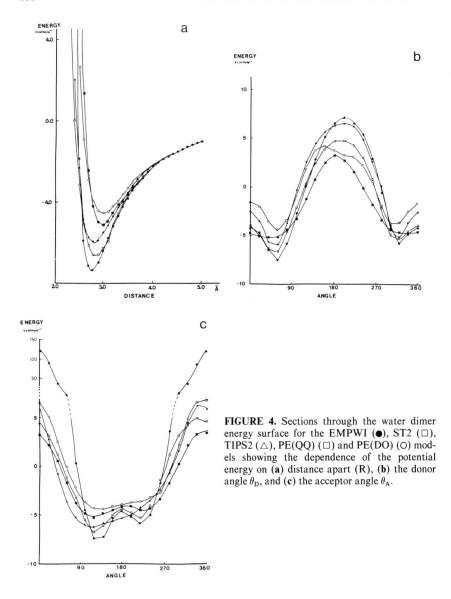

FIGURE 4. Sections through the water dimer energy surface for the EMPWI (●), ST2 (□), TIPS2 (△), PE(QQ) (□) and PE(DO) (O) models showing the dependence of the potential energy on (**a**) distance apart (R), (**b**) the donor angle θ_D, and (**c**) the acceptor angle θ_A.

Further simulations of solutions were carried out with the dinucleotide dCpG in the B-DNA form as solute and using the TIPS2 water model. FIGURE 8 shows a schematic representation of the self-bridging loops around the phosphate-sugar backbone of one chain. Similar loops are seen on the other chain of the duplex. Polar bridging chains are also seen linking polar atoms on the phosphate sugar backbone to other polar solute atoms on the same chain, the second chain, and to base atoms (FIG. 9). Similarly the networks are associated with the base atoms (FIG. 10).

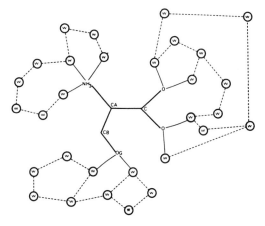

FIGURE 5. Schematic representation of self-bridging water networks around serine zwitterion in solution.

DISCUSSION

Crystallography of small, medium, and large biomolecules has led to a detailed knowledge of individual water molecule sites in terms of hydrogen bond geometries and also is beginning to give a picture of the hydrogen-bonded water molecule networks close to a solute molecule.[34] Computer simulation techniques offer us the opportunity to predict the energetics of the interactions of water molecules with a solute and to study disordered, more liquid-like, networks of solvent which are not seen crystallographically. However, it is necessary to investigate the accuracy of the potential energy functions which are used to model atomic interactions in any type of energy

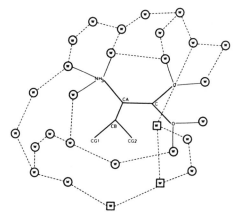

FIGURE 6. Schematic representation of the chains of water molecules linking polar solute atoms (polar bridging chains) of valine zwitterion in solution.

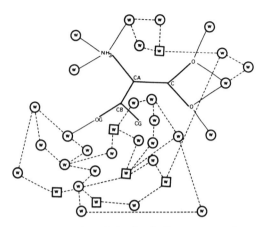

FIGURE 7. Schematic representation of the "apolar" water networks around the threonine zwitterion in solution.

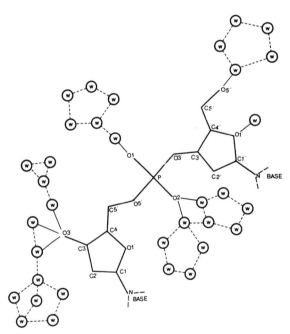

FIGURE 8. Schematic illustration of self-bridging water networks around the sugar-phosphate backbone of dCpG dinucleotide in solution.

calculation. In particular, it is possible to undertake simulations of solvent in crystal hydrates and to compare predicted and experimental networks of water molecules.

First, it is of interest to ascertain the sensitivity of the energy and geometry of water molecule sites around a solute to the potential energy functions used to describe the solute and the water molecules. Such calculations[35] have been performed with urea as solute and lead to considerable differences in both structure and energy of interactions, as shown in TABLES 3a, b and c. These differences are consistent with differences that can be seen if one compares the radial distribution functions for water around the atoms of urea from two independent molecular dynamics simulations.[36,37] Our calculations show that there are correlations between the position of the hydration site and

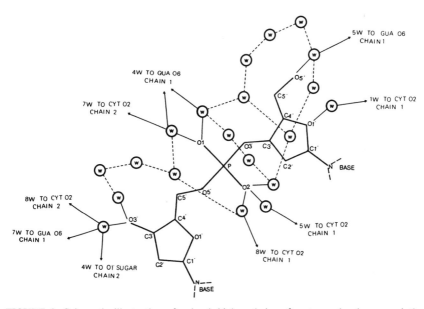

FIGURE 9. Schematic illustration of polar bridging chains of water molecules around the sugar-phosphate backbone of dCpG dinucleotide in solution.

their geometry and energy with, respectively, the water model, the solute nonbonded interactions coefficients, and the partial atomic charges. Thus, we can now ascertain which combination of models gives us, for example, low energies or long water · · · solute hydrogen bonds.

The simulations on guanosine-5′-phosphate crystal hydrate have highlighted the problems of convergence in simulations. Starting from random positions, the solvent molecules move towards the experimentally determined sites, but seem to become fixed in local minima with average root mean square deviations of around 1.0Å away from the experimental sites. Conversely, when the initial positions are chosen to be the crystallographically determined water molecule oxygen positions, the solvent molecules move away from these positions by about 0.5Å. Because this crystal hydrate has only 12 water molecule sites per unit cell, it may not be possible in a system with such small solvent channels for water molecules to move from random starting positions to the correct experimental positions.

A larger but still well-defined solvent network is found in vitamin B_{12} coenzyme crystal hydrate. Using this system, we have attempted to look at (a) the effects of different water models on the predicted solvent structure in simulations and (b) the

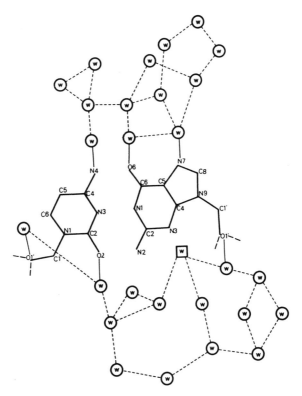

FIGURE 10. Schematic illustration of the hydration of a bone pair from a solution of dCpG dinucleotide.

properties of the water models themselves as indicated from sections through the dimer energy surface. It was seen clearly that the model which leads to the best fit with experimental data, in this crystal hydrate, had the most extreme properties (e.g., lowest energy minima), whereas the worst model had the opposite properties (e.g., highest minima). The implication of these results seems to be that to improve the agreement with the crystal hydrate data, one should exaggerate the properties of the "best" model. However, this leads to a water model that is not suitable for predicting properties of liquid water itself. Ways out of this impasse would include changing the functional form for the water model and looking at the effects of the solute model on the water · · · water interactions.

The results presented so far on crystal hydrates have been aimed at comparing the detailed geometry of the water molecule sites from simulations with those from experiments. The agreement found so far requires improvement if we wish to make detailed comments on the structure of solvent around biomolecules. However, it is still possible to use the present models to predict less quantitative properties of water molecule networks and to compare changes in these networks on the addition of a small solute to a bulk water system. With these aims, a series of simulations were undertaken with different small zwitterionic solutes. The analysis of the solvent networks was carried out in order to look at the overall structure around the solute. Two features of these networks include loops or rings of water molecules associated with each polar solute atom and chains of water molecules between different polar solute atoms. Water molecules close to apolar solute atoms were seen and these occurred in ring structures which were most frequently pentagonal, although 3-, 4-, 6-, and 7-membered rings were seen. A similar analysis can be carried out for bulk water in which the most frequent ring size, with this particular model, was five. These networks are qualitatively similar to those found in the best defined large biomolecule crystal hydrates.[2–5,7–13] Nonplanar pentagonal and hexagonal rings occur in ices, whereas planar rings occur in the clathrate hydrate structures. Thus the analysis of hydrogen-bonded water networks in terms of rings and chains of water molecules can be used to study and compare all aqueous systems from bulk water through aqueous solutions to crystal hydrates.

SUMMARY

Crystal hydrates provide an obvious test of the ability of computer simulations to predict water networks around biomolecules. Energy minimization studies on a simple water · · · solute (urea) system have delineated the idiosyncrasies of different models. More complex Monte Carlo simulations on vitamin B_{12} coenzyme crystal hydrate, using five different water models, show that modifications to the potential functions aimed at improving the agreement with the experimental solvent network lead to an unacceptable model for water itself. The implication of these results appears to be that the current functional forms for water models are not sufficient to predict the detailed geometry of these solvent networks. However, simulations of solvent around amino acids and dinucleotides in solution show that the hydrogen-bonded water networks can be broken down into different patterns or motifs associated with (i) each solute polar atom (self-bridging loops), (ii) chains between different solute atoms (polar-bridging chains), and (iii) ring networks associated with "apolar" water molecules. Such patterns are now being seen in the best refined crystal structure of sugars, oligonucleotides, and proteins.

REFERENCES

1. SAVAGE, H. & A. WLODAWER. 1985. *In* Methods in Enzymology; Biomembranes, Protons and Water Structure and Translocation. In press.

2. SAENGER, W. 1979. Nature **279**: 343–344.
3. NEIDLE, S., H. BERMAN & H. S. SHIEH. 1980. Nature **288**: 129–133.
4. SAVAGE, H. 1983. PhD thesis, University of London.
5. SAVAGE, H. 1985. Biophys. J. Submitted for publication.
6. BAKER, E. & R. HUBBARD. 1984. Prog. Biophys. **44**: 97–179.
7. TEETER, M. 1984. Proc. Natl. Acad. Sci. USA **81**: 6014–6018.
8. BAKER, E., G. DODSON, D. HODGKIN & R. HUBBARD. 1985. *In* Crystallography of Molecular Biology, NATO Summer School held at Bischenberg, Alsace, September 1985.
9. DREW, H. R. & R. E. DICKERSON. 1981. J. Mol. Biol. **151**: 535–556.
10. KOPKA, M. L., A. V. FRATINI, H. R. DREW & R. E. DICKERSON. 1983. J. Mol. Biol. **163**: 129–146.
11. WESTHOF, E., Th. PRANGÉ, B. CHENTIER, & D. MORAS. 1985. Biochimie. In press.
12. KENNARD, O. 1984. Pure Appl. Chem. **56**: 989–1004.
13. KENNARD, O. *et al* 1985. Private communication.
14. FINNEY, J. L., J. M. GOODFELLOW, P. L. HOWELL & F. VOVELLE. 1985. J. Biomol. Struct. Dynamics. In press.
15. HAGLER, A. T. & J. MOULT. 1978. Nature **272**: 222–226.
16. HAGLER, A. T. & J. MOULT. 1980. Biopolymers **19**: 395–418.
17. MADISON, V., D. J. OSGUTHORPE, P. DAUBER & A. T. HAGLER. 1983. Biopolymers **22**: 27–31.
18. GOODFELLOW, J. M., J. L. FINNEY & P. BARNES. 1982. Proc. Roy. Soc. London Ser. B **214**: 213–228.
19. GOODFELLOW, J. M. 1982. Proc. Natl. Acad. Sci. USA **79**: 4977.
20. GOODFELLOW, J. M. 1984. J. Theor. Biol. **107**: 261–274.
21. MEZEI, M., D. L. BEVERIDGE, H. M. BERMAN, J. M. GOODFELLOW, J. L. FINNEY & S. NEIDLE. 1983. J. Biolmol. Struct. Dynamics **1**: 287–
22. HOWELL, P. L. & J. M. GOODFELLOW. 1984. J. Phys. (Paris). **45**: C7-211–218.
23. KIM, K. S. & E. CLEMENTI. 1985. J. Am. Chem. Soc. **107**: 227–234.
24. STILLINGER, F. H. & A. RAHMAN. 1974. J. Chem. Phys. **60**: 1545–1557.
25. VOVELLE, F. & M. PTAK. 1979. Int. J. Pept. Protein Res. **13**: 435.
26. JORGENSEN, W. L. 1982. J. Chem. Phys. **77**: 4156–4163.
27. GELLATLY, B. J., J. QUINN, P. BARNES & J. L. FINNEY. 1983. Mol. Phys. **59**: 949–970.
28. MOMANY, F. A., L. M. CARRUTHERS, R. F. MCGUIRE & H. A. SCHERAGA. 1974. J. Phys. Chem. **78**: 1595–1620.
29. LIFSON, S., A. T. HAGLER & P. DAUBER. 1979. J. Am. Chem. Soc. **101**: 5111–5121.
30. WEINER, S. J., P. A. KOLLMAN, D. A. CASE, U. CHANDRA SINGH, C. GHIO, S. PROFETA, JR. & P. WEINER. 1984. J. Am. Chem. Soc. **106**: 765.
31. POWELL, M. J. D. 1964. Computer J. **7**: 155
32. ORITA, Y. & A. PULLMAN. 1977. Theoret. Chim. Acta **45**: 257
33. BAUM, J. O. Personal communication.
34. GOODFELLOW, J. M., P. L. HOWELL & R. ELLIOTT. *In* Crystallography of Molecular Biology. Proceedings of a Conference held at Bischenberg, September, 1985. NATO Advanced Study Institute. In press.
35. VOVELLE, F. & J. M. GOODFELLOW. 1985. J. Biomol. Struct. Dynamics. In press.
36. TANAKA, H., H. TOUHARA, K. NAKANISHI & N. WATANABE. 1984. J. Chem. Phys. **80**: 5170–5186.
37. KUHARSKI, R. & P. J. ROSSKY. 1984. J. Am. Chem. Soc. **106**: 5786–5793.

Computer Simulation of Nucleotide Crystal Hydrates and Solutions

P. LYNNE HOWELL AND JULIA M. GOODFELLOW

Department of Crystallography
Birkbeck College
London WC1E 7HX, England

Nucleic acids require solvent not only for their stability, but also to regulate transitions between different helical forms. We have used Metropolis Monte Carlo simulation techniques to study the molecular structure and energetics of water at nucleic acid interfaces in a (1) crystal hydrate guanosine 5' phosphate trihydrate, a system with 12 water molecules and 136 solute atoms per unit cell, and (2) in solution, a dinucleotide dCpG in a classical B fiber form with 118 water molecules and 124 solute atoms per unit cell.

In each of the simulations performed, the solute has been represented by two nonbonded Lennard-Jones[1] coefficients and by partial charges calculated using the semiempirical quantum mechanical method, CNDO/2. Two water models have been used, namely, the polarizable electropole model[2] for the crystal hydrate study and the three-point charge model of Jorgensen[3] for the solution study. For each simulation the equilibration of the system was monitored using the potential energy, and the analysis

TABLE 1. Average RMSD in Å[a]

	Initial	Trial	Solvent Positions
Simulation 1	0.0	0.45	Crystal
Simulation 2	1.93	1.11	Random
Simulation 3	0.88	0.83	Random

[a]Root mean square deviation between simulated and crystal water molecule positions.

was performed over the last 100K for the crystal hydrate and the last 300K configurations for the solution study.

RESULTS AND DISCUSSION

Crystal Hydrate

We have undertaken a series of simulations with different initial configurations for the solvent molecules. We find that the RMS (TABLE 1) deviation of simulated to experimental water oxygen positions is greater for those simulations using random initial positions for solvent compared with those using experimental crystal oxygen positions, so that the average RMS deviation in solvent positions shows that the water

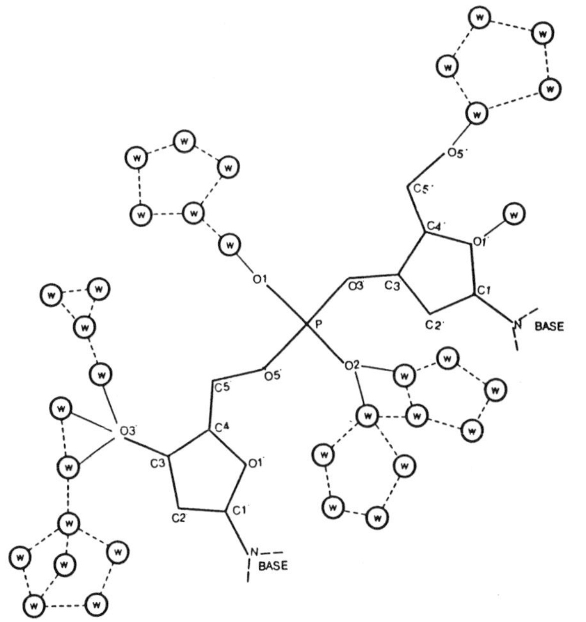

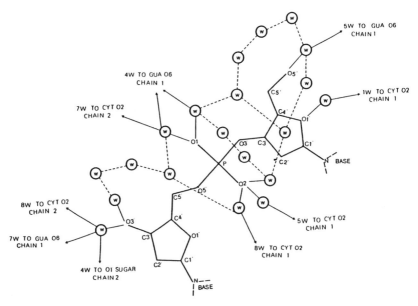

FIGURE 1a. Diagram showing the water networks associated with each polar or charged atom on the phosphate-sugar backbone (self-bridging loops).

FIGURE 1b. Diagram showing the water networks with between-polar groups on the phosphate-sugar backbone (polar-bridging chains).

molecules (1) move towards the experimental water structure when started from initial random positions, and (2) move away from the crystallographic positions when these are input as initial positions. Although the former movement is larger on average, it is not sufficient to get good agreement between experimental and simulated structures.

Dinucleotide Solutions

In our studies of a dinucleotide (dCpG) we found that the patterns of hydration could be divided into two different motifs associated with (1) each solute polar atom (self-bridging loops) (FIG. 1a) and (2) chains between different solute atoms (polar-bridging loops) (FIG. 1b). Further networks were associated with the hydration of the base pairs. These water networks are similar to those found in a series of simulations on amino acid zwitterions in solution. Moreover, well-refined crystal structures of sugars, nucleotides and proteins are exhibiting loops and chains of water molecules similar to those described in this study.

REFERENCES

1. LIFSON, S., A. T. HAGLER & P. DAUBER. 1979. J. Am. Chem. Soc. **53:** 4544–4556.
2. GELLATLEY, B. J., J. E. QUINN, P. BARNES & J. L. FINNEY. 1983. Mol. Phys. **59**(3): 949–956.
3. JORGENSEN, W. L. 1982. J. Chem. Phys. **77:** 4156.

Computer Simulations of Organic Reactions in Solution[a]

WILLIAM L. JORGENSEN,

JAYARAMAN CHANDRASEKHAR,

J. KATHLEEN BUCKNER, AND

JEFFRY D. MADURA

Department of Chemistry
Purdue University
West Lafayette, Indiana 47907

INTRODUCTION

The *a priori* calculation of energy surfaces for organic reactions is both important and challenging. The information that can be obtained on reaction mechanisms and the nature of intermediates and transition states is essential for the better understanding of reactivity. Much progress has been made for gas-phase reactions through the application of *ab initio* quantum mechanics. Such computations have now been used to study many processes including substitution reactions,[1-3] nucleophilic[4-8] and electrophilic additions,[9-11] reactions of carbenes,[12-15] and pericyclic reactions.[16-19] This work has nicely complemented related experimental investigations using modern gas-phase procedures such as ion-cyclotron resonance, high-pressure mass spectrometry, and the flowing afterglow technique.[20-31] The *ab initio* calculations are particularly valuable in providing structural results for stationary points on the reaction surfaces, which are difficult to characterize experimentally.

The theoretical study of organic reactions in solution is much more in the formative stage. The importance of solvation on reaction surfaces is evident in striking medium dependence of reaction rates, particularly for polar reactions. For example, nucleophilic substitutions may be 10^5 times faster in DMF than in aqueous solution and an additional 10^{15} faster in the gas phase.[20] Furthermore, for competing reactions, the preferred pathways in solution and the gas phase may be remarkably different. In the case of hydroxide ion reacting with methyl formate, the proton transfer and substitution that accompany addition in the gas phase are suppressed in protic solvents.[29,31] Of course, the immediate difficulty with theoretical treatment of condensed phase reactions is the realization that for molecular level insights the behavior of large collections of atoms must be modeled. Thermal averaging is also required so quantum mechanics is not viable. The more appropriate approach is to utilize the now standard techniques for simulating condensed phase systems at the molecular level, specifically, Monte Carlo statistical mechanics or molecular dynamics. However, the technical problems are still formidable due to system size, time scale, the need for complete

[a]This work was supported by grants from the National Science Foundation and the National Institutes of Health.

specification of the intermolecular interactions at all points on the reaction surface and the need to cover substantial regions of the reaction surface. Nevertheless, progress has been made with this methodology in our laboratory for calculating energy profiles of organic reactions in solution. Key results of initial investigations of a prototype S_N2 reaction in water[32] and DMF[33] and of an addition reaction in water[34] are summarized here. Besides determination of free energy profiles in solution, the computations provide detailed information on the origin of solvent-induced activation barriers.

METHODOLOGY AND COMPUTATIONAL DETAILS

A principal goal of the computations is to determine the change in the energy profile for a reaction upon transfer from the gas phase to solution. At this time, we are primarily calculating the minimum energy reaction path (MERP) in the gas phase and then determining the effects of solvation on the energetics at each point on that path. The actual quantity that is computed in solution is the relative free energy along the reaction path or "potential of mean force." To determine the optimal reaction path in solution would require the construction of multidimensional potentials of mean force. Since the present computations for a one-dimensional reaction coordinate require the equivalent of 6 months to 1 year of computer time on a VAX 11/780, extension to several dimensions is beyond the practical range of most computational resources. However, the gas-phase reaction paths considered here are anticipated to also be reasonable for the reactions in solution (*vide infra*).

The computational procedure involves three main steps. First, the lowest energy reaction path is determined for the gas phase using *ab initio* calculations. The energetic and geometric variations along the reaction path are fit to analytical functions of the reaction coordinate. Then, intermolecular potential functions are obtained to describe the interactions between the reacting system and a solvent molecule. These vary with the reaction coordinate and are represented through Coulomb and Lennard-Jones interactions between sites normally situated on the atoms. For aqueous solutions, the potential functions are based on numerous *ab initio* calculations for complexes of the substrate and a water molecule. The variation of the charges and Lennard-Jones parameters for the reacting system are also expressed as smooth functions of the reaction coordinate. The solvent-solvent interactions are described by the well-proven TIP4P model for water[35,36] and the OPLS potentials for DMF.[37] Finally, with analytical descriptions of the gas-phase reaction surface and of the intermolecular potential functions, Monte Carlo simulations are carried out to calculate the free energy profile for the reaction in solution. In fact, a series of simulations, ca. 6–10, are run with "importance sampling" to cover the full range for the reaction coordinate.[38] Each simulation is constrained to a region of the reaction coordinate or "window" by using fictitious biasing functions or "umbrella potentials." The influence of the biasing functions on computed properties can subsequently be removed. The probability of occurrence of each value of the reaction coordinate, $g_i(r_c)$, is determined for window i. The $g_i(r_c)$s for adjacent windows are spliced together at overlapping points to construct the total $g(r_c)$ which is related to the potential of mean force by $w(r_c) = -kT \ln g(r_c)$. The importance sampling with heavy biasing in high-energy regions is essential to obtain convergence of $w(r_c)$ for the full range of r_c in a reasonable amount of computer time. Besides the simulations with importance sampling, we also perform Monte Carlo calculations for fixed values of r_c corresponding to key points such as reactants, transition state and products. This permits thorough structural characterization of the solvation for the reacting system at these important points.

Although full details on the computational procedure for the present reactions may be found elsewhere,[32–34] a few critical items are reviewed here. The S_N2 reaction that has been modeled is the degenerate exchange of $Cl^- + CH_3Cl$. The gas-phase MERP and the solute-water potential functions were obtained using *ab initio* calculations with the 6-31G* basis set. The Monte Carlo simulations were then performed for the reacting system plus 250 water molecules in a periodic cell with dimensions of ca. $17.2 \times 17.2 \times 25.8$ Å. The same reaction was studied in liquid DMF using 180 DMF molecules in a tetragonal cell (ca. $25 \times 25 \times 38$ Å) with periodic boundary conditions.[10] The prototype addition reaction that has been modeled is for hydroxide ion plus formaldehyde in water.[34] In this case, the larger 6-31+G* basis set was required to establish the gas-phase MERP and substrate-water potential functions. The Monte Carlo simulations then included 269 water molecules in a periodic cell, ca. $16.5 \times 19.8 \times 24.8$ Å. In each case, the NPT ensemble has been employed at 25°C and 1 atm. The run for a window typically consisted of an equilibration phase for 10^6 configurations followed by averaging over an additional 2×10^6 to 4×10^6 configurations. The averaging for the fixed-solute simulations normally covered 3×10^6 configurations. Metropolis sampling has been used appropriately modified for the importance sampling and preferential sampling of solvent molecules nearest the solutes.

S_N2 REACTION: $CL^- + CH_3CL$

The computed energy profiles for the S_N2 reaction are shown in FIGURE 1. The reaction coordinate has been defined as the difference in the two C—Cl distances since this provides symmetry about the transition state. The MERP for the gas phase shows the double-well form with ion-dipole complexes (*1*) as minima and a symmetrical transition state (*2*). The nucleophile approaches the

backside of the C—Cl bond along the dipole axis. The 6-31G* well-depth for *1* (10.3 kcal/mol) and intrinsic barrier (13.9 kcal/mol) are consistent with experimental estimates of 8.6 and 10–13 kcal/mol, respectively.[20,32]

Before this work, the forms of the reaction profiles in solution were largely a matter of conjecture. From the classic studies of Ingold and his contemporaries, it is well known that such reactions follow bimolecular kinetics and have substantial activation energies in water, ca. 18–30 kcal/mol.[39,40] Furthermore, the free energies of activation are reduced by 5–8 kcal/mol upon transfer to DMF.[41,42] Unimodal reaction profiles with a single barrier and no intermediates are widely assumed in the literature. The implication is that the charge-localized nucleophile and product ion are so much better solvated than the ion-dipole complexes and transition state that the two ends of the gas-phase profile are pulled down to the point where the ion-dipole minima vanish and there is only a single barrier.

This conjecture is supported by our results for the free energy profile in water. An essentially flat surface is found up to the position of the ion-dipole minima. In this region, the increasing ion-dipole attraction is offset by some weakening of the nucleophile-water hydrogen bonds and by the loss of a water molecule on the nucleophile. The average $Cl^- \cdots H_2O$ hydrogen bond strength of 10–12 kcal/mol is

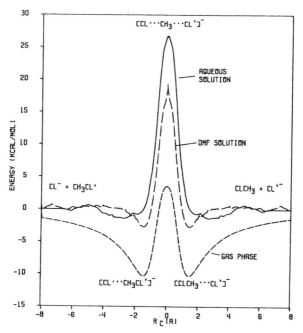

FIGURE 1. Calculated potential energy in the gas phase (*short dashes*) and the potential of mean force in DMF (*long dashes*) and in aqueous solution (*solid curve*) for the reaction of Cl^- with CH_3Cl. The reaction coordinate is the difference in the two C—Cl distances for the linear, backside attack.

matched by the $Cl^- \cdots CH_3Cl$ interaction of 10 kcal/mol.[32] Between the position of the ion-dipole complex and the transition state, there is the gas-phase barrier of 13 kcal/mol, which is augmented by a solvent-induced component of similar size to yield the net free energy barrier in water computed to be 26.3 ± 0.5 kcal/mol. This result is in exact agreement with the experimental value (26.6 kcal/mol).[42]

The origin of the solvent-induced barrier has been elucidated.[32] It can be attributed to the charge delocalization that occurs between the position of the ion-dipole complex and the transition state. At the ion-dipole complex, the charge on the nucleophile is not significantly reduced from the value for the bare ion. However, at the transition state both chlorines have a charge of $-0.77e$ in the potential functions. As a consequence, the optimal interaction energy for a single water molecule interacting with Cl^-, *1*, and *2* increases from -13.0 to -12.2 to -8.8 kcal/mol. Interestingly, we find that the number of strong water-solute interactions remains fairly constant at 7–8 along the reaction path. Thus, an average weakening of ca. 3 kcal/mol per hydrogen bond between *1* and *2* can more than account for the hydration-induced barrier.

Although the computed free energy profile in water is essentially unimodal, it seemed likely that there should be a solvent that could yield a nonconcerted profile, i.e., one with ion-dipole or solvent-separated intermediates. A solvent with diminished anion solvating ability should reduce the energy increase upon desolvation of the nucleophile. This could at least reintroduce the ion-dipole complexes as intermediates. Hydrocarbon solvents are good candidates, but unrealistic for solubility reasons. Consequently, we chose to try a dipolar aprotic solvent, namely DMF. As shown in FIGURE 1, this was adequate; the ion-dipole complexes are free energy minima in wells 2.7 kcal/mol deep with 3.3 kcal/mol barriers to dissociation. It should be noted that the ion-dipole complexes in the gas phase are in *free* energy wells only ca. 5 kcal/mol deep because of the substantial entropy loss upon their formation from the reactants.

Thus, our prediction for the mechanism of some S_N2 reactions in dipolar aprotic solvents is that they involve an initial complexation followed by the rate-determining step. Some support for this position can be found in NMR studies which have indicated the presence of complexes between chloride ion and benzyl chlorides in acetonitrile.[43–46] The lowering of the calculated free energy barrier in DMF (19.3 ± 0.5 kcal/mol) is also consistent with the experimental observations.[41,42] The smaller solvent-induced barrier is readily attributed to the uniformly weaker solute-solvent interactions in DMF. The increased charge delocalization on proceeding to the transition state still leads to diminished solvation, although the effect is less pronounced than in water.

SOLVENT EFFECT ON TRANSITION STATE STRUCTURE

A fundamental question concerns the effect of hydration on the structure of the S_N2 transition state. In the calculations described above, the structure of the transition state was not allowed to vary from the 6-31G* geometry for the gas phase. This approximation is consistent with Shaik's recent analysis.[47] He points out that the transition state *2* is relatively tight with modest distortion of the C—Cl bond lengths from equilibrium values for alkyl chlorides. Then, as with neutral molecules, solvation is anticipated to have little influence on bond lengths and bond angles.

Some quantification of these notions can be provided. First, in FIGURE 2, 6-31G* results are presented for the energies of symmetrically stretching the C—Cl bonds in *2*. Comparison is made with the stretching of the C—Cl bond in methyl chloride. Complete geometry optimization was performed at each fixed C—Cl distance. Distortion of *2* in the gas phase is easier than for methyl chloride; however, a stretch of 0.2 Å still requires 5 kcal/mol, and 15 kcal/mol is needed for 0.4 Å. It seems unlikely that the solvation energy could vary so rapidly, particularly since the *ab initio* results indicate the negative charge on chlorine only increases by 0.06e in going from a C—Cl bond length of 2.383 (*2*) to 2.70 Å. Parabolic fitting for the results in FIGURE 2 yields force constants of 1.70 and 3.44 mdyn $Å^{-1}$ for *2* and methyl chloride. These may be translated into frequencies of 684 and 746 cm^{-1}, respectively.[48] Experimental values for the force constant and frequency of the C—Cl stretch in methyl chloride are 3.50 mdyn Å^{-1} and 733 cm^{-1}.[49]

Another theoretical result worth noting in this context has been obtained by Morokuma.[50] He performed geometry optimizations for *2* and its dihydrated form with one water molecule on each chlorine. *Ab initio* calculations and the 3-21G basis set were used. The C—Cl bond length changed by only 0.002 Å upon addition of the two water molecules; it actually shortened from 2.395 to 2.393 Å.

Thus, the available evidence suggests that distortion of the S$_N$2 transition state by hydration should be slight. More definitive results could be obtained through fluid simulations for the transition state with variable C—Cl bond lengths.[51]

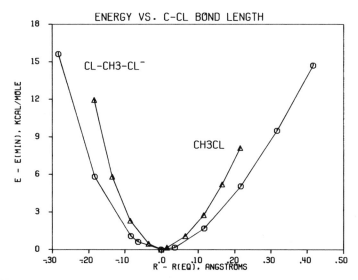

FIGURE 2. Dependence of the gas-phase energy on C—Cl bond length for the S$_N$2 transition state and methyl chloride. Results from *ab initio* 6-31G* calculations. Equilibrium C—Cl bond lengths are 2.383 Å for the transition state and 1.785 Å for methyl chloride.

ADDITION REACTION: $OH^- + H_2C=O$

For the addition reaction, the reaction coordinate was taken to be the carbonyl carbon-hydroxyl oxygen distance. The MERP for the gas phase was determined in C$_s$ symmetry with the 6-31+G* basis set. This includes *d*-orbitals and diffuse *s* and *p* functions on carbon and oxygen. Smaller basis sets were considered, but the diffuse functions were found to be essential for obtaining reasonable energetic results.[34] Similar observations have been made by others for describing the electronic structure of charge-localized anions with first-row elements.[52]

The energetic results for the lowest-energy C$_s$ pathway are given by the solid curve in the bottom part of FIGURE 3. Key optimized structures along the path are illustrated in FIGURE 4. The approach at large separation is coplanar with the hydroxide ion on the dipole axis of formaldehyde. An apparent ion-dipole minimum occurs at a C—O

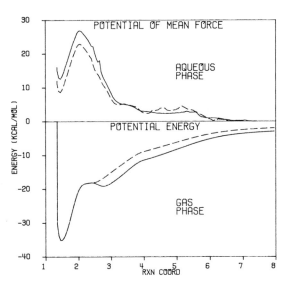

FIGURE 3. Calculated potential energy in the gas phase (*bottom*) and the potential of mean force in aqueous solution (*top*) for the addition of hydroxide ion to formaldehyde. The reaction coordinate is the carbon-hydroxyl oxygen distance in Å. Solid lines are for the initially coplanar trajectory, while the dashed lines are for the more tetrahedral approach.

separation of 2.748 Å (B in FIG. 4) with a binding energy of 19 kcal/mol. However, an activation energy of only 1.1 kcal/mol is needed to reach the transition state (C) at R_{CO} = 2.391 Å. The geometric changes between B and C in FIGURE 4 are substantial since the hydroxyl fragment lifts out of the plane to assume a more tetrahedral, final approach. The energy change is then rapid as covalent bonding sets in between the transition state and the tetrahedral product (D) with R_{CO} = 1.470 Å. The overall energy change for the reaction is calculated to be -35.2 kcal/mol. Zero-point and correlation corrections were shown to nearly cancel for the process.[34] Frequency calculations verified that the tetrahedral intermediate is a true minimum and that the transition state C has only one imaginary frequency. However, B is actually a saddle-point between two mirror-image minima with C_1 symmetry; these are essentially hydrogen-bonded forms with the hydroxide interacting with the formaldehyde hydrogens. The energy lowering from B is only about 0.1 kcal/mol, so the Monte Carlo calculations have been performed on the fully C_s pathway for convenience.

A second trajectory was also studied as represented by the dashed curves in FIGURE 3. In this case, the OCO angle was fixed at its value of 126.9° at the transition state (C) for all larger values of R_{CO}. Thus, this trajectory is more tetrahedral throughout and corresponds to traditional ideas for approach vectors in addition reactions.[53] The ion-dipole minimum vanishes for this trajectory, although there are still two distinct phases to the reaction separated near the transition state. The existence of the minimum is controversial. Other *ab initio* studies for addition reactions including hydride ion and OH⁻ with formaldehyde and OH⁻ with formamide have found the tetrahedral intermediate as the only minimum.[4-7] In contrast, Asubiojo and Brauman have interpreted gas-phase kinetics for additions to acyl halides in terms of a double-well energy surface with ion-dipole complexes as minima and the tetrahedral species as a transition state.[22] The discrepancy with the prior *ab initio* results may be attributed to the use of smaller basis sets in those studies. This causes the overall exothermicity to be overestimated, although it should be kept in mind that the present ion-dipole minimum is very shallow. The viability of such intermediates is anticipated to be enhanced by an increased dipole moment as with the acyl halides. At this time, a triple-well surface also seems possible for these substrates with the tetrahedral species still as an energy minimum.

Upon transfer to aqueous solution, the energy profiles undergo dramatic change. The upper curves in FIGURE 3 give the relative free energy for the two trajectories in water as obtained from the Monte Carlo simulations with importance sampling. In aqueous solution, the free energy rises only gradually from the reactants to $R_{CO} = 3$ Å. In this region, weakening solvation is being balanced by the increasing ion-dipole attraction. Then, between 2 and 3 Å the gas-phase energy is relatively constant. The desolvation continues and the free energy curves in water rise rapidly. The transition state is located at $R_{CO} = 2.05$ Å. It is 28 and 24 kcal/mol above the reactants for the two trajectories. From there, the gas-phase energy descends quickly and is mirrored in the potential of mean force. The tetrahedral intermediate is predicted to be 14 and 10 kcal/mol higher in free energy than the reactants in water. The discrepancy of 4 kcal/mol between the results from the two trajectories may reflect a difference in the free energies for the two geometries of the reactants or, more likely, it is a reasonable estimate of the absolute uncertainty in the calculations. However, the potentials of

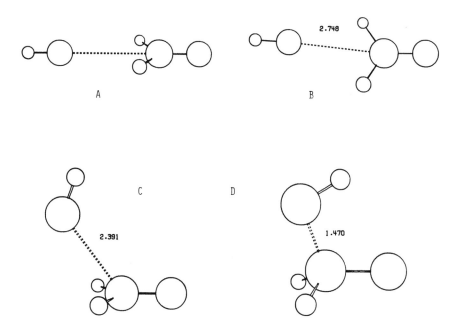

FIGURE 4. Optimized 6-31+G* structures on the reaction path for OH$^-$ + H$_2$C=O.

mean force beyond 2.4 Å were obtained independently and have the same qualitative shape. No evidence is found for intermediates other than the tetrahedral complex.

The energetic results are consistent with experimental data of Guthrie[54] and theoretical predictions of Weiner et al.[7] for alkaline hydrolyses of amides. For their systems, the activation energy for the addition step is about 22 kcal/mol and the tetrahedral intermediate is 9–18 kcal/mol above the reactants. Ester hydrolyses are typically more facile with activation energies of 15–20 kcal/mol.[54,55] The combined quantum and molecular mechanics approach of Weiner et al. also led to a predicted CO distance of about 2.0 Å for the aqueous transition state. Thus, the transition states for these endoergic processes are very product-like. For formaldehyde, the computed endoergicity significantly overestimates the available experimental data; the computed result is more in line with data for ketones, where formation of hydrates is less favorable.[56]

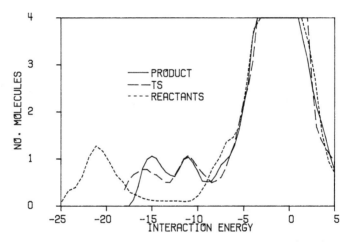

FIGURE 5. Computed solute-water energy pair distributions for the reactants $(OH^- + H_2C\!\!=\!\!O)$, aqueous transition state, and product. Units for the ordinate are molecules per kcal/mol.

From FIGURE 3 it is clear that the substantial barrier in solution is essentially entirely solvent-induced. The origin of the barrier has been extensively studied with the aid of the fixed solute simulations for the reactants, aqueous transition state, and tetrahedral intermediate.[34] Again, charge delocalization with concomitant weakening of hydrogen bonds is the dominant factor. The average number of strong water-solute interactions remains at 6–7 along the entire reaction path. However, the average hydrogen bond strength for hydroxide ion of about 21 kcal/mol declines to 12–13 kcal/mol for the transition state and product. This is illustrated in FIGURE 5, which gives the distributions of individual solute-water interaction energies. The bands at low energy are for the strong hydrogen bonds, while the spikes centered near 0 kcal/mol represent the weak interactions with the many remote water molecules. For the reactants, integration of the band from -25 to -11.4 kcal/mol reveals 6.0 water molecules solvating the hydroxide ion. For the transition state and product the low-energy band is shifted to significantly higher energy, -18 to -8.5 kcal/mol, and split because of distinction between interactions with the two oxygens of the solute. Integration of the split bands to -8.5 kcal/mol yields 7.0 and 6.4 water molecules for the transition state and product. Thus, the barrier for the reaction can be attributed primarily to reduction in hydrogen bond strengths rather than in numbers of hydrogen bonds. The same observation was made above for the solvent-induced component of the barrier for the S_N2 reaction.

The implied greater charge delocalization for the tetrahedral intermediate than for the hydroxide ion may seem surprising in view of notions based on Lewis structures. Nevertheless, the *ab initio* calculations show this to be the case and it is reflected in the parameterization of the intermolecular potential functions.[34] The charge on hydroxyl oxygen varies from $-1.30e$ for hydroxide ion to $-0.76e$ for the product, whereas the carbonyl oxygen begins with $-0.38e$ and ends at $-0.90e$. Thus, the formally negative oxygen of the tetrahedral intermediate is only 0.14e more electron-rich than the hydroxyl oxygen. The inadequacy of Lewis descriptions is also well illustrated by ions like H_3O^+ and NH_4^+, where the positive charge clearly resides primarily on the hydrogens.

CONCLUSION

These initial efforts have demonstrated the feasibility of performing computer simulations to study organic reactions in solution. The joint quantum and statistical mechanics approach is general and provides molecular-level details on the origin of activation barriers in solution. The results for prototype S_N2 and addition reactions have been shown to be consistent with experimental kinetic and thermodynamic data. An important general observation for these polar reactions has emerged; namely, it is more the change in hydrogen bond strengths accompanying charge delocalization along the reaction path than in numbers of hydrogen bonds that is the primary origin of the solvent-induced activation barriers. Furthermore, the findings for the S_N2 reaction suggest that a nonconcerted mechanism may be operative in dipolar aprotic solvents. Much additional, detailed information of this type is needed to obtain better understanding and control of organic reactivity.

SUMMARY

Quantum and statistical mechanics have been used to determine energy profiles for the S_N2 reaction of $Cl^- + CH_3Cl$ in the gas phase, in aqueous solution, and in liquid DMF. The energy profile in the gas phase has the characteristic double-well form featuring unsymmetrical ion-dipole complexes as minima and a symmetrical transition state. Hydration causes the reaction surface to become almost unimodal and increases the barrier significantly. The reaction profile in DMF is intermediate between those for the gas phase and aqueous solution. The ion-dipole complexes are still free energy minima in DMF. Thus, the reaction in DMF involves initial formation of the complex before the rate-determining step. The computed results are shown to be in good accord with experimental free energies of activation.

The same technique has been applied to the addition reaction of $OH^- + H_2C{=}O$ in the gas phase and aqueous solution. *Ab initio* $6\text{-}31+G^*$ calculations indicate that the reaction proceeds essentially without activation in the gas phase. Hydration introduces a substantial energy barrier. The transition state in water has been located at a C—O separation of roughly 2 Å. A key finding for both reactions is that the activation barriers induced by hydration result primarily from change in strengths rather than in numbers of solute-water hydrogen bonds along the reaction paths.

REFERENCES

1. BERTHIER, G., D.-J. DAVID & A. VEILLARD. 1969. Theoret. Chim. Acta **14:** 329.
2. WOLFE, S., D. J. MITCHELL & H. B. SCHLEGEL. 1981. J. Am. Chem. Soc. **103:** 7692, 7694.
3. URBAN, M., I. CERNUSAK & V. KELLO. 1984. Chem. Phys. Lett. **105:** 625.
4. BURGI, H. B., J. M. LEHN & G. WIPFF. 1974. J. Am. Chem. Soc. **96:** 1956.
5. WILLIAMS, I. H., G. M. MAGGIORA & R. L. SCHOWEN. 1980. J. Am. Chem. Soc. **102:** 7831.
6. ALAGONA, G., E. SCROCCO & J. TOMASI. 1975. J. Am. Chem. Soc. **97:** 6976.
7. WEINER, S. J., C. SINGH & P. A. KOLLMAN. 1985. J. Am. Chem. Soc. **107:** 2219.
8. HOUK, K. N., N. G. RONDAN, P. v. R. SCHLEYER, E. KAUFMANN & T. CLARK. 1985. J. Am. Chem. Soc. **107:** 2821.
9. NAGASE, S., N. K. RAY & K. MOROKUMA. 1980. J. Am. Chem. Soc. **102:** 4536.
10. CLARK, T., D. WILHELM & P. v. R. SCHLEYER. 1983. J. Chem. Soc. Chem. Commun.: 606.
11. HOUK, K. N., N. G. RONDAN, Y.-D. WU, J. T. METZ & M. N. PADDON-ROW. 1984. Tetrahedron **40:** 2257.
12. RONDAN, N. G., K. N. HOUK & R. A. MOSS. 1980. J. Am. Chem. Soc. **102:** 1770.
13. HOUK, K. N., N. G. RONDAN & J. MAREDA. 1984. J. Am. Chem. Soc. **106:** 4291.
14. SCHAEFER, H. F., III. 1979. Accts. Chem. Res. **12:** 288.
15. MOROKUMA, K., K. OHTA, N. KOGA, S. OBARA & E. R. DAVIDSON. 1984. Faraday Symp. Chem. Soc. **19:** 49.
16. TOWNSHEND, R. E., G. RAMUNNI, G. SEGAL, W. J. HEHRE & L. SALEM. 1976. J. Am. Chem. Soc. **98:** 2190.
17. HOUK, K. N. & K. YAMAGUCHI. 1984. *In* 1, 3-Dipolar Cycloaddition Chemistry. A. Padwa, Ed.: 407. John Wiley & Sons. New York, NY.
18. KOMORNICKI, A., J. D. GODDARD & H. F. SCHAEFER, III. 1980. J. Am. Chem. Soc. **102:** 1763.
19. RONDAN, N. G. & K. N. HOUK. 1984. Tetrahedron Lett. **25:** 2519.
20. OLMSTEAD, W. N. & J. I. BRAUMAN. 1977. J. Am. Chem. Soc. **99:** 4019.
21. PELLERITE, M. J. & J. I. BRAUMAN. 1980. J. Am. Chem. Soc. **102:** 5993.
22. ASUBIOJO, O. I. & J. I. BRAUMAN. 1979. J. Am. Chem. Soc. **101:** 3715.
23. FAIGLE, J. F. G., P. C. ISOLANI & J. M. RIVEROS. 1976. J. Am. Chem. Soc. **98:** 2049.
24. COMISAROW, M. 1977. Can. J. Chem. **55:** 171.
25. FUKUDA, E. K. & R. T. McIVER. 1979. J. Am. Chem. Soc. **101:** 2498.
26. BOHME, D. K., G. I. MACKAY & S. D. TANNER. 1980. J. Am. Chem. Soc. **102:** 407.
27. BARTMESS, J. E., R. L. HAYES & G. CALDWELL. 1981. J. Am. Chem. Soc. **103:** 1338.
28. McDONALD, R. N. & A. K. CHOWDHURY. 1982. J. Am. Chem. Soc. **104:** 901.
29. DePUY, C. H., E. W. DELLA, J. FILLEY, J. J. GRAWBOWSKI & V. M. BIERBAUM. 1983. J. Am. Chem. Soc. **105:** 2481.
30. CALDWELL, G., T. F. MAGNERA & P. KEBARLE. 1984. J. Am. Chem. Soc. **106:** 959.
31. JOHLMAN, C. L. & C. L. WILKINS. 1985. J. Am. Chem. Soc. **107:** 327.
32. CHANDRASEKHAR, J., S. F. SMITH & W. L. JORGENSEN. 1984. J. Am. Chem. Soc. **106:** 3049; 1985. J. Am. Chem. Soc. **107:** 154.
33. CHANDRASEKHAR, J. & W. L. JORGENSEN. 1985. J. Am. Chem. Soc. **107:** 2974.
34. MADURA, J. D. & W. L. JORGENSEN. 1986. J. Am. Chem. Soc. **108:** 2517.

35. JORGENSEN, W. L., J. CHANDRASEKHAR, J. D. MADURA, R. W. IMPEY & M. L. KLEIN. 1983. J. Chem. Phys. **79:** 926.
36. JORGENSEN, W. L. & J. D. MADURA. 1986. Mol. Phys. **56:** 1381.
37. JORGENSEN, W. L. & C. J. SWENSON. 1985. J. Am. Chem. Soc. **107:** 569.
38. PATEY, G. N. & J. P. VALLEAU. 1975. J. Chem. Phys. **63:** 2334.
39. INGOLD, C. K. 1969. Structure and Mechanism in Organic Chemistry, 2nd ed. Cornell University Press. Ithaca, NY.
40. BATHGATE, R. H. & E. A. MOELWYN-HUGHES. 1959. J. Chem. Soc.: 2642.
41. ALBERY, W. J. & M. M. KREEVOY. 1978. Adv. Phys. Org. Chem. **16:** 87.
42. MCLENNAN, D. J. 1978. Aust. J. Chem. **31:** 1897.
43. HAYAMI, J., N. TANAKA, N. HIHARA & A. KAJI. 1973. Tetrahedron Lett.: 385.
44. HAYAMI, J., T. KOYANAGI, N. HIHARA & A. KAJI. 1978. Bull. Chem. Soc. Jpn. **51:** 891.
45. HAYAMI, J., N. HIHARA, N. TANAKA & A. KAJI. 1979. Bull. Chem. Soc. Jpn. **52:** 831.
46. HAYAMI, J., T. KOYANAGI & A. KAJI. 1979. Bull. Chem. Soc. Jpn. **52:** 1441.
47. SHAIK, S. S. 1985. Prog. Phys. Org. Chem. **15:** 197.
48. COTTON, F. A. 1979. Chemical Applications of Group Theory. John Wiley & Sons. New York, NY.
49. JONES, E. W., R. J. L. POPPLEWELL & H. W. THOMPSON. 1965. Spectrochim. Acta **22:** 669.
50. MOROKUMA, K. 1982. J. Am. Chem. Soc. **104:** 3732.
51. JORGENSEN, W. L. & J. K. BUCKNER. In press.
52. See, for example: CLARK, T., J. CHANDRASEKHAR, G. W. SPITZNAGEL & P. V. R. SCHLEYER. 1983. J. Comp. Chem. **4:** 294.
53. BURGI, H. B. & J. D. DUNITZ. 1983. Acta Chem. Res. **16:** 153.
54. GUTHRIE, J. P. 1973. J. Am. Chem. Soc. **95:** 6999. 1974. *ibid.* **96:** 3608.
55. BENDER, M. L., R. D. GINGER & J. P. UNIK. 1958. J. Am. Chem. Soc. **80:** 1044.
56. GUTHRIE, J. P. 1978. J. Am. Chem. Soc. **100:** 5892.

Ionic Association in Water:
From Atoms to Enzymes[a]

J. ANDREW McCAMMON,[b] OMAR A. KARIM,
TERRY P. LYBRAND,[c] AND CHUNG F. WONG

Department of Chemistry
University of Houston
Houston, Texas 77004

INTRODUCTION

Chemistry and biochemistry are largely concerned with the association and transformation of molecules in water. Theoretical studies of such processes have in the past been hindered by a number of difficulties. Within an aqueous system, there are strong, directional, attractive forces among the water molecules, and often also between solute and solvent molecules, in addition to the excluded volume forces that have made even simple liquids a challenging subject.[1,2] For a model system comprising a few solute molecules and a few hundred water molecules, the competition among these interactions produces a complicated potential energy surface with many local minima. To calculate structural or thermodynamic properties, one must evaluate averages of certain quantities over a representative set of those configurations that have low enough energy to be thermally populated. To calculate kinetic properties, one must consider motions over energy barriers and, in the case of molecular association, motions corresponding to large displacements over the potential surface.

From the perspective of computer simulations, the difficulties that arise in any of the calculations mentioned above are largely associated with the time scales involved. In conventional molecular dynamics simulations, where one solves Newton's equations for the atoms in a model system, the accessible times on conventional computers have been too short for brute-force simulation of many systems. For example, a simulation to generate a fairly representative set of instantaneous hydration structures of a small univalent ion might involve about two hundred molecules and 20 psec of simulation; this would require about 60 hours of CPU time on a VAX 11/780; this is quite manageable. To study the hydration of a moderately large enzyme such as trypsin, however, a 20-psec simulation might require 3500 hrs on a VAX; this is cumbersome at best. For kinetic properties such as rate constants for barrier crossing or diffusional encounter, the situation can be worse by many orders of magnitude.

Happily, advances in the theory underlying computer simulations, in the algorithms used, and in computers themselves, have greatly expanded the range of

[a]This work was supported in part by the National Science Foundation and the Robert A. Welch Foundation.
[b]Camille and Henry Dreyfus Teacher-Scholar.
[c]Recipient of a Damon Runyon-Walter Winchell Cancer Fund Fellowship Award (DRG-888).

chemical phenomena that can be simulated in aqueous systems. In this paper, we will illustrate several of these advances by reference to three molecular dynamics studies of ionic-association phenomena in water. The studies deal with systems of increasing structural complexity. In the first study, we consider the thermodynamics and kinetics of the transition of a sodium chloride ion pair between the solvent separated and contact geometries. In the second study, the new "thermodynamic cycle–perturbation approach" is used to determine the relative affinity of chloride and bromide ions for a macrocyclic receptor in water. Finally, we describe the dynamic simulation of the enzyme trypsin in water, and the use of the thermodynamic cycle–perturbation approach to calculate the relative affinity of different cationic inhibitors for this enzyme. It may be worth noting here that another simulation method, Brownian dynamics, has recently been developed to determine the mechanisms and rate constants of diffusional bimolecular encounters.[3,4] Applications of this last method have shown that the enzyme superoxide dismutase achieves its high catalytic rate in part by electrostatic steering of substrate anions into its active sites.[5]

NA$^+$CL$^-$ INTERACTIONS IN WATER

To characterize the detailed role of ion-pair solvation in a relatively simple aqueous system, we carried out a series of molecular dynamics studies of a sodium chloride ion pair in water. These studies have focused on the thermodynamics and equilibrium structure of the system as functions of the ion-pair separation,[6,7] and on the dynamics of ionic dissociation or association.[8] In each study, Jorgensen's TIPS2 molecular model was used for water[9] and a compatible model was used for the ions.[10,11] These models consist of Lennard-Jones potentials centered on the oxygen and ion centers plus point charges embedded within the repulsive cores of the Lennard-Jones potentials. The models have been shown to provide a reasonable description of the hydration of the individual ions.[10,11] The system used in the ion-pair simulations consisted of 295 water molecules plus a single sodium chloride pair. These particles occupied a rectangular box with dimensions $18.62 \times 18.62 \times 25.46$ Å and were subject to periodic boundary conditions. To minimize image effects, the ions were placed along the long axis and near the center of the box. During the dynamic simulations, the temperature of the system fluctuated around an average value of 298K. Other details of the model, the dynamic equilibrations, and the simulations are fairly standard and are given in the references.[6–8]

Our first calculations were designed to determine the potential of mean force between the ions over a range of several Å starting from the ion-ion contact geometry.[6] An umbrella sampling procedure[12–15] was used to ensure uniform sampling of the system configurations over this range. In this procedure, one ion was fixed and the other was constrained to move in a sequence of overlapping small volumes along the long axis of the box in a sequence of simulations. Each simulation was of 10.5 psec duration and was preceded by a short equilibration period. After correcting for the effect of the constraining potentials, the potential of mean force was calculated; this result is displayed in FIGURE 1. For this model system, there are evidently stable states for both contact and solvent-separated ion pairs. The distance between the ions in the latter state is somewhat too small to accommodate a water molecule directly between the ions.

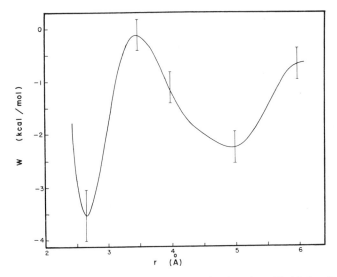

FIGURE 1. Potential of mean force W as a function of sodium ion–chloride ion distance r in water.

The equilibrium structures characteristic of four different ion-ion separations were determined by carrying out 20 psec simulations with the ions fixed at each separation.[7] In the contact state, the most prominent feature is the octahedral first shell around sodium; this comprises five water molecules plus the chloride at the sixth vertex. In the transition state, the "half-octahedron" of waters is tilted slightly, with one or two waters inclined toward the chloride ion. In the solvent-separated state, the sodium has an octahedral first shell of six waters; one face of the octahedron is tilted toward the chloride so that at least one water is in a suitable position to bridge the two ions.

The transitions of the ion pair between the contact and solvent-separated states have been explored in subsequent simulations.[8] Here, a representative set of transition states was generated by carrying out a long simulation with a constraining potential that held the ions near a separation of 3.4 Å. A number of such states were then propagated forward and backward in time. From these, a number of trajectories that successfully traversed the free energy barrier were selected for detailed analysis.

Two distinct mechanisms of ion-pair separation are apparent in the trajectories. The evolution of the ion-ion distance is shown for a representative trajectory of each type in FIGURE 2. The dominant ("direct") mechanism involves the direct transfer of a water molecule from the first hydration shell of the sodium ion to a bridging position off the ion-ion axis. The transition-state structure is one in which the sodium ion still has five water molecules in its first shell, but the "half-octahedron" of waters has tilted slightly to bring one of these molecules close to the chloride. Also, this nascent bridging water molecule reorients somewhat so that one of its hydrogens can interact favorably with the chloride while its oxygen continues to interact favorably with the sodium. The transition state rapidly falls apart (within about a picosecond) when the chloride and sodium ions separate slightly and a sixth water molecule moves from the second hydration shell to the first shell of sodium.

The secondary ("indirect") pathway is slightly more complicated. Here, a nascent bridging water molecule from the first hydration shell of chloride forms a hydrogen bond with one of the first-shell waters of sodium. Direct bonding of the chloride water to the sodium is prevented by the difference in sizes of the ions. This transition-state structure can persist for several picoseconds, during which an additional water molecule may enter the first shell of chloride. The transition state falls apart when the second-shell waters participate in a transient rearrangement of hydrogen bonds, allowing the first-shell water of chloride to reorient and form a direct interaction with sodium ion.

Comparing these two mechanisms, one sees that they have several important similarities. First, the structures of the contact and solvent-separated states are not correlated in any obvious way with the type of transition. Indeed, the type of transition that will occur changes several times during a 10-psec simulation of the transition state region. Second, the transitions generally involve the transfer of a first-shell water from one of the ions in the contact geometry, to a bridging position off the axis of the solvent-separated ions. Although the simple descriptions given above focused on the formation of bridges consisting of single water molecules, slightly more complex variations do occur within either class of mechanism (e.g., simultaneous formation of bridges by two waters from sodium in the direct mechanism). Finally, and perhaps of greatest importance, both mechanisms are essentially collective in character. In particular, the second hydration shell plays an important role in each mechanism. The fact that a number of molecules are intimately involved is what gives these transitions their diffusional character.

The diffusional character of the contact to solvent-separated transitions is clear not only from the frequent reversals of direction in the ion-pair motion (FIGURE 2), but also

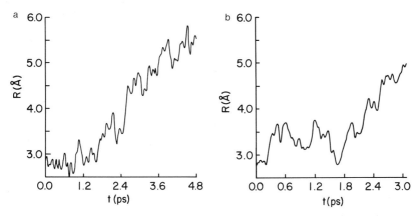

FIGURE 2. Sodium ion–chloride ion distance as a function of time in trajectories corresponding to the two different transition mechanisms discussed in the text. (**a**) The direct mechanism; (**b**) the indirect mechanism.

from the time required to complete the transitions. For noninteracting particles with the diffusion constants of separated sodium and chloride ions in water, the time required for relative diffusion through a distance of 2 Å is approximately 2 psec[8]; this is similar to the time required for the transitions.

HALIDE-RECEPTOR INTERACTIONS IN WATER

The umbrella sampling method described above is useful for evaluating the free energy of interaction of simple molecules and ions in water. One might therefore hope that the method could be used to study the interactions of more complicated molecules. In particular, the recognition and binding of small molecules (ligands) by receptor molecules (e.g. enzymes,[16] antibodies,[17] or chelating agents such as crown ethers[18]) are of great interest and importance in many branches of chemistry. Theoretical techniques that reliably predict the relative free energy of binding of different ligand-receptor pairs would be helpful in the interpretation of experimental studies and would have potential utility as a predictive tool in areas such as drug design.

The formation of a ligand-receptor complex is normally a complicated process. Both the ligand and receptor site must first be desolvated, and the ligand or receptor (or both) may undergo conformational changes as steric and Coulombic interactions are developed.[19] In principle, the umbrella sampling technique could be used to calculate the relative free energy of binding $\Delta\Delta A$ for two ligands at a common receptor

$$L + R \rightarrow L{:}R \qquad \Delta A_1 \qquad\qquad (1)$$

$$M + R \rightarrow M{:}R \qquad \Delta A_2 \qquad\qquad (2)$$

by computing ΔA_1 and ΔA_2 separately, then combining these results to yield $\Delta\Delta A = \Delta A_2 - \Delta A_1$. In practice, however, this approach will usually have severe limitations. Extremely long simulations may be required to model the desolvation processes and conformational changes, especially if entry of the ligand into a binding pocket creates a steric barrier to the escape of solvent molecules. Even without these complications, a sequence of long simulations is necessary to calculate ΔA for each reaction, and the final result $\Delta\Delta A$ would be computed as the difference of two large, rather imprecise numbers.

An alternate simulation technique, the thermodynamic cycle–perturbation method has recently been developed to circumvent many of the problems associated with the umbrella sampling method.[20] This technique applies perturbation-theory concepts[21–23] to a set of reactions forming a closed thermodynamic cycle. In this procedure, two hypothetical reactions are defined.[20]

$$L + R \rightarrow M + R \qquad \Delta A_3 \qquad\qquad (3)$$

$$L{:}R \rightarrow M{:}R \qquad \Delta A_4 \qquad\qquad (4)$$

Because Reactions 1–4 form a closed thermodynamic cycle and A is a thermodynamic state function, $\Delta\Delta A = \Delta A_2 - \Delta A_1 = \Delta A_4 - \Delta A_3$. Perturbation theory techniques can be used to compute the hypothetical free energy changes ΔA_3 and ΔA_4. (Other techniques can also be used to compute these changes.[24]) If V_L and V_M are potential-energy functions for the L/R/solvent and M/R/solvent systems, respectively, then a "mixed" potential-energy function for a hybrid system is

$$V_\lambda = \lambda V_M + (1 - \lambda)V_L \qquad (5)$$

Other types of mixed function (e.g., ones in which V_λ varies nonlinearly with λ) can also be defined. The free energy change is computed by carrying out a sequence of simulations corresponding to a sequence of increasing values λ_i of λ. From simulation i, the free energy is obtained for a certain range of λ about λ_i from

$$A(\lambda) - A(\lambda_i) = -\beta^{-1} \ln \langle \exp [-\beta(V_\lambda - V_{\lambda_i})] \rangle_{\lambda_i} \qquad (6)$$

where β^{-1} = kT (k = Boltzmann's constant, T = temperature in Kelvin) and $\langle \ \rangle_{\lambda_i}$ indicates an average for the simulation based on V_{λ_i}. If temperature, volume, and the number of particles are fixed in the simulation, A is the Helmholtz free energy.[21] If pressure is fixed instead of volume, A is the Gibbs free energy.[22,23] The increment $\lambda_{i+1} - \lambda_i$ is chosen to assure adequate statistics in the region of perturbation parameter λ where results $A(\lambda)$ from simulation i overlap those from simulation i + 1. Several simulations may be necessary to span the full perturbation from L to M. For example, a two-step simulation for reaction 3 might start with a simulation of L in the solvent environment. Results from this simulation could be used to compute the free energy difference between L and a hybrid (50%L/50%M) ligand in solution. A second simulation could then be performed on the hybrid ligand (50%L/50%M) in the solvent environment, and these results would be used to compute the free energy difference between the hybrid and M species in solution. The free energy differences are determined according to Equation 6. In this example, the potential energy V_L is computed for a large number of system configurations saved from the first simulation. Then, the "mixed" potential energy, V_λ, is computed for each configuration saved from the first simulation for discrete values of the perturbation parameter λ. When $\lambda = 0$, $V_\lambda = V_L$. When $\lambda = 0.5$, V_λ is the potential energy for the hybrid species in solution. Configurations saved from the second simulation are used to compute the free energy difference between the hybrid and M species in solution, where V_{λ_i} is now V_{hybrid} and V_λ is the potential energy for the M species in solution when $\lambda = 1$. The results of these two simulations are then combined to yield the free energy difference for L and M in solution.

In this study, the relative free energy of binding for Cl^- and Br^- to the macrotricyclic molecule SC24 (FIGURE 3) in water has been calculated.[25] The relevant thermodynamic cycle is

$$Cl^- \subset SC24 \xrightarrow{\Delta A_4} Br^- \subset SC24$$

$$\Delta A_1 \uparrow \qquad\qquad \uparrow \Delta A_2$$

$$Cl^- + SC24 \xrightarrow[\Delta A_3]{} Br^- + SC24 \qquad (7)$$

The symbol $\subset$ denotes an inclusion complex. The perturbation technique was used to compute the free energy changes ΔA_3 and ΔA_4. All simulations were performed using the canonical ensemble (i.e., constant temperature, volume, and number of particles)

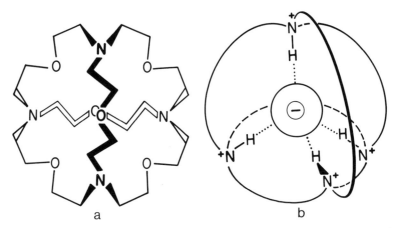

FIGURE 3. The SC24 receptor molecule. (**a**) Structural formula; (**b**) schematic illustration of the complex formed with a halide ion.

at 300 K. For the computation of ΔA_3, the system consisted of Cl^- and 214 water molecules in a cubic box of length 18.62 Å with periodic boundary conditions. The system for calculation of ΔA_4 contained SC24, Cl^-, and 191 water molecules in a cubic box as above. The initial structure for the $Cl^- \subset$ SC24 complex was taken from an x-ray crystallographic study,[26] and sufficient water molecules were included in each system to completely solvate the anion or anion/receptor complex and yield reasonable densities. The SPC potential functions were used for water.[27] Lennard-Jones parameters for Cl^- and SC24 were taken from the GROMOS molecular modeling program, and the partial charges for SC24 were from work by Kollman and Wipff (unpublished). Lennard-Jones parameters for Br^- were adjusted to reproduce experimental results for relative energies of $Cl^- - (H_2O)$ and $Br^- - (H_2O)$ complexes.[28] The Cl^- parameters are $r^* = 2.4954$ Å, $\epsilon = 0.107$ kcal/mol, and the corresponding Br^- parameters are $r^* = 2.5950$ Å, $\epsilon = 0.090$ kcal/mol. The molecular dynamics simulations were performed using AMBER[29] and were of the constant-temperature type.[30] The SHAKE procedure was used to constrain covalent bonds to equilibrium lengths.[31] All hydrogen masses were increased to 10 amu to allow a long dynamics time step, $\Delta t = 4$ femtoseconds, by slowing the librational motions of groups containing hydrogen.[32] This mass adjustment has no effect on equilibrium properties of a classical system, but does result in a more efficient sampling of configurations.

After extensive equilibration of each system, simulations were continued for 30 psec. Configurations were saved every 0.1 psec, and were used to calculate ΔA_3 and ΔA_4 as outlined above. Both one-step and two-step simulation procedures were used to compute ΔA_3 as discussed above, and both procedures gave the same results. Additional simulations were done for Br^-/H_2O and $Br^-/SC24/H_2O$ systems. In these calculations, Br^- was slowly changed to Cl^- to compute $-\Delta A_3$ and $-\Delta A_4$.

The free energies $A_i(\lambda) - A_i(0)$ for $i = 3$ and $i = 4$ are shown in FIGURE 4. The dashed curve ($i = 3$) displays the free energy change as Cl^- is slowly perturbed to Br^- in solution. The solid curve ($i = 4$) displays the corresponding free energy change for anions bound to SC24 in water. Our computed value for $\Delta A_3 = 3.35 \pm 0.15$ kcal/mol. This free energy change represents the relative free energy of hydration for Cl⁻ versus Br⁻, and our computed result is in excellent agreement with the tabulated experimental value ($\Delta\Delta G_{Hydr} = 3.3$ kcal/mol, where G is the Gibbs free energy).[33] Our computed $\Delta\Delta A$ values should be quite similar to $\Delta\Delta G$ results; for condensed-phase systems Helmholtz (A) and Gibbs (G) free energies will normally be nearly identical. Our computed value for $\Delta A_4 = 7.50 \pm 0.20$ kcal/mol, so that $\Delta\Delta A = 4.15 \pm 0.35$ kcal/mol from our calculations. This value is likewise in good agreement with experimental results, which suggest a value of approximately 4.3 kcal/mol.[34,35] The Br⁻ to Cl⁻ perturbation calculations also predict a value of $\Delta\Delta A = 4.15 \pm 0.35$ kcal/mol. The results suggest that selective binding of Cl⁻ to SC24 is due to the highly favorable interaction of Cl⁻ with the receptor, which more than compensates for the unfavorable free energy of desolvation of Cl⁻ versus Br⁻. The more favorable interaction of Cl⁻

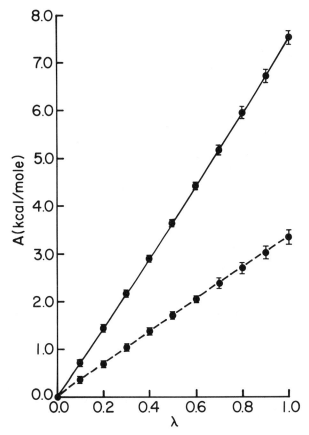

FIGURE 4. Free energy changes for transformation of Cl⁻ to Br⁻ in water ($\cdots\cdots$) and in the SC24 complex ($\cdot\!-\!\cdot\!-\!\cdot$). *Error bars* indicate the range of results obtained by separate analysis of 10-psec blocks of the full 30-psec simulation.

with the receptor relative to Br^- arises because the Br^- anion is slightly too large to be comfortably accommodated in the relatively rigid SC24 molecule.

TRYPSIN-BENZAMIDINE INTERACTIONS IN WATER

As was mentioned in the introduction, the most complicated simulation problems that can currently be handled require the use of new theoretical approaches, new algorithms, and new computers. All three of these have been applied in recent molecular dynamics studies of the enzyme trypsin in water. These studies have two general objectives. First, the simulations will be analyzed to add to our understanding of the dynamics of proteins in solution,[36] focusing particularly on the effects of protein-solvent interactions. Second, the simulations will serve as reference data for determining the relative free energies of binding various benzamidine inhibitors to native trypsin and to modified forms of this enzyme.

The benzamidine-trypsin system is a good candidate for initial studies of biomolecular interactions using the thermodynamic cycle–perturbation method. Benzamidine is one of the simplest inhibitors of trypsin. An X-ray structure of the enzyme-inhibitor complex is available.[37,38] In this, it is apparent that the benzene moiety of the inhibitor contacts several hydrophobic groups in the walls of the enzyme's specificity pocket, and that the positively charged amidinium moiety forms a salt bridge with the carboxyl group of an aspartic acid residue at the base of this pocket. Thermodynamic data are available for the binding of benzamidine with various substituent groups to trypsin.[39,40] Also, a variety of modified trypsins have been produced by genetic manipulations[41]; these and other modified forms of the enzyme are obvious candidates for model studies of ligand binding.

The first step of our work was to carry out a molecular dynamics simulation of benzamidine-inhibited trypsin in water. The system is a very large one, comprising 4,785 water molecules plus the inhibited enzyme in a box of dimensions 49.15 $\times$ 54.43 $\times$ 64.28 Å. Hydrogen atoms capable of participating in hydrogen bonds were included explicitly. The total number of atoms in the system was 16,384. Periodic boundary conditions were used; the box dimensions were chosen to be large enough that all solute atoms were separated from the closest atoms of image solutes by at least four layers of solvent. The enzyme was so oriented that the binding site was far removed from solute images. The dynamics calculations were carried out on a CYBER 205 supercomputer using a vectorized version of the GROMOS program kindly provided by Wilfred van Gunsteren. The SHAKE algorithm was used to constrain bond lengths.[31] Starting with the X-ray structure for the complex and a bulk solvent configuration, the system was relaxed and equilibrated at 300 K during a period of 16.3 psec. A subsequent 28.4-psec simulation was carried out with the system coupled to a constant temperature bath[30] at 300 K.

The general results that have been obtained from the simulation to date are in accord with previous simulation studies of proteins.[36] The rms deviation of the average structure from the X-ray structure of the complex is 2.2 Å for all atoms and 1.7 Å for the alpha carbons. The average rms fluctuations of the atom positions are 0.68, 0.52 and 0.59 Å for all atoms, the alpha and beta carbons, respectively. Corresponding fluctuations for a 16 psec simulation of ferrocytochrome c without explicit solvent surroundings[42] were 0.72, 0.57 and 0.65 Å. The anisotropy of the atomic motion and the tendency toward larger fluctuations near the surface are as observed in previous studies.

To calculate the relative free energy of binding of differently substituted benzamidines by the thermodynamic cycle–perturbation approach, it was necessary to carry out a reference simulation of benzamidine in water. This model comprised 212 water molecules plus the benzamidinium ion in a cubic box of edge length 18.6 Å. Using the same procedures as for the enzyme-inhibitor complex, the system was equilibrated for 36 psec and then simulated for 64 psec. The one modification of benzamidine for which we have results at present is the parafluoro compound. Designating this by M and the unsubstituted compound by L, the free energy changes obtained for reactions 3 and 4 (see the preceding section) are $\Delta A_3 = -0.8 \pm 0.1$ kcal/mol and $\Delta A_4 = 0.1 \pm 0.5$ kcal/mol. Thus, the predicted relative free energy of binding is $\Delta\Delta A = \Delta A_4 - \Delta A_3 = 0.9 \pm 0.5$ kcal/mol at 300 K. The experimental result,[39] approximately corrected from 288 K to 300 K, is $\Delta\Delta A = 0.5 \pm 0.3$ kcal/mol. The error bars in the predicted values correspond to the deviations in results obtained for three equal subintervals of each simulation from those for the full simulations. That calculations and experiment agree in indicating that the unsubstituted benzamidine binds slightly more strongly is encouraging. Also, the theoretical results indicate that, unlike the SC24-halide case, the difference in ligand-solvent interactions is more important than the difference in ligand-receptor interactions in determining binding specificity in the present case. Clearly, however, a wider range of enzyme-inhibitor pairs must be explored to validate and/or motivate further development of the theoretical methods. Such studies are currently in progress.

CONCLUDING REMARKS

The types of work described above can be extended in a variety of directions. For the more complicated systems such as the trypsin-benzamidine complex, it will be desirable to develop analyses of the solvent-solute interactions at the level of detail obtained for the simpler systems. For the enzyme-inhibitor system and other large systems, the computational requirements for relative free energy calculations can probably be greatly reduced by the application of stochastic boundary conditions that allow one to limit detailed modeling to the binding site regions.[43,44] Such calculations have clear potential in the area of drug design; corresponding calculations in which one considers modified receptors also have real potential for practical applications in chemistry and molecular biology.[36]

ACKNOWLEDGMENTS

We are grateful to Professors S. Allison, M. Berkowitz, H. Friedman, W. Jorgensen, P. Kollman, S. Northrup, P. Rossky, W. van Gunsteren, and G. Wipff for helpful discussions, and also to Professors Kollman, Rossky and van Gunsteren for providing important computer programs.

REFERENCES

1. HANSEN, J. P. & I. R. MCDONALD. 1976. Theory of Simple Liquids. Academic Press. New York, NY.
2. FRANKS, F. 1983. Water. Royal Society of Chemistry. London, England.
3. NORTHRUP, S. H., S. A. ALLISON & J. A. MCCAMMON. 1984. J. Chem. Phys. 80: 1517.
4. ALLISON, S. A., S. H. NORTHRUP & J. A. MCCAMMON. 1985. J. Chem. Phys. 83: 2894.
5. ALLISON, S. A. & J. A. MCCAMMON. 1985. J. Phys. Chem. 89: 1072.
6. BERKOWITZ, M., O. A. KARIM, J. A. MCCAMMON & P. J. ROSSKY. 1984. Chem. Phys. Lett. 105: 577.
7. BELCH, A. C., M. BERKOWITZ & J. A. MCCAMMON. 1986. J. Am. Chem. Soc. 108: 1755.
8. KARIM, O. A. & J. A. MCCAMMON. 1986. J. Am. Chem. Soc. 108: 1762.
9. JORGENSEN, W. L. 1982. J. Chem. Phys. 77: 4156.
10. CHANDRASEKHAR, J. & W. L. JORGENSEN. 1982. 77: 5080.
11. CHANDRASEKHAR, J., D. C. SPELLMEYER & W. L. JORGENSEN. 1983. J. Am. Chem. Soc. 106: 903.
12. PANGALI, C., M. RAO & B. J. BERNE. 1979. J. Chem. Phys. 71: 2975.
13. NORTHRUP, S. H., M. PEAR, C. Y. LEE, J. A. MCCAMMON & M. KARPLUS. 1982. Proc. Natl. Acad. Sci. USA 79: 4035.
14. CHANDRASEKHAR, J., S. F. SMITH & W. L. JORGENSEN. 1985. J. Am. Chem. Soc. 107: 154.
15. MEZEI, M., P. K. MEHROTRA & D. L. BEVERIDGE. 1985. J. Am. Chem. Soc. 107: 2239.
16. FERSHT, A. 1985. Enzyme Structure and Mechanism, 2nd ed. W.H. Freeman. New York, NY.
17. DAVIES, D. R. & H. METZGER. 1983. Annu. Rev. Immunol. 1: 87.
18. LEHN, J. M. 1985. Science 227: 849.
19. KOLLMAN, P. A. 1979. In Burger's Medicinal Chemistry, 4th ed. M. E. Wolff, Ed. Vol. 1: 313. John Wiley & Sons. New York, N.Y.
20. TEMBE, B. L. & J. A. MCCAMMON. 1984. Comput. Chem. 8: 281.
21. FRIEDMAN, H. L. 1985. A Course in Statistical Mechanics. Prentice Hall. Englewood Cliffs, NJ.
22. POSTMA, J. P. M., H. J. C. BERENDSEN & J. R. HAAK. 1982. Faraday Symp. Chem. Soc. 17: 56.
23. JORGENSEN, W. L. & C. RAVIMOHAN. 1985. J. Chem. Phys. 83: 3050.
24. BENNETT, C. H. 1976. J. Comp. Phys. 22: 245.
25. LYBRAND, T. L., J. A. MCCAMMON & G. WIPFF. 1986. Proc. Natl. Acad. Sci. USA 83: 833.
26. METZ, B., J. M. ROSALKY & R. WEISS. 1976. J. Chem. Soc. Chem. Commun.: 533.
27. BERENDSEN, H. J. C., J. P. M. POSTMA, W. F. VAN GUNSTEREN & J. HERMANS. 1981. In Intermolecular Forces. B. Pullman, Ed.: 331. Reidel. Holland.
28. ARSHADI, M., R. YAMDAGNI & P. KEBARLE. 1970. J. Phys. Chem. 74: 1475.
29. SINGH, U. C. & P. A. KOLLMAN. 1984. J. Comput. Chem. 5: 129.
30. BERENDSEN, H. J. C., J. P. M. POSTMA, W. F. VAN GUNSTEREN, A. DINOLA & J. R. HAAK. 1984. J. Chem. Phys. 81: 3684.
31. RYCKAERT, J. P., G. CICCOTTI & H. J. C. BERENDSEN. 1977. J. Comput. Phys. 23: 327.
32. WOOD, D. W. 1979. In Water: A Comprehensive Treatise. F. Franks, Ed. Vol. 6. Plenum. New York, NY.
33. FRIEDMAN, H. L. & C. V. KRISHNAN. 1972. In Water: A Comprehensive Treatise. F. Franks, Ed. Vol. 3: 1. Plenum. New York, NY.
34. GRAF, E. & J. M. LEHN. 1976. J. Am. Chem. Soc. 98: 6403.
35. KAUFMAN, E. 1979. Études physico-chimiques de cryptates de cations et d'anions. Thesis. Université Louis Pasteur, Strasbourg, France.
36. MCCAMMON, J. A. & S. C. HARVEY. 1987. Dynamics of Proteins and Nucleic Acids. Cambridge University Press. London.
37. BODE, W. & P. SCHWAGER. 1975. J. Mol. Biol. 98: 693.
38. BERNSTEIN, F. C., T. F. KOETZLE, G. J. B. WILLIAMS, E. F. MEYER, M. D. BRICE, J. R. ROGERS, O. KENNARD, T. SHIMANOUCHI & M. TASUMI. 1977. J. Mol. Biol. 112: 535.
39. MARES-GUIA, M., D. L. NELSON & E. ROGANA. 1977. J. Am. Chem. Soc. 99: 2331.

40. NARAY-SZABO, G. 1984. J. Am. Chem. Soc. **106:** 4584.
41. CRAIK, C. S., C. LARGMAN, T. FLETCHER, S. ROCZNIAK, P. J. BARR, R. FLETTERICK & W. J. RUTTER. 1985. Science **228:** 291.
42. NORTHRUP, S. H., M. R. PEAR, J. D. MORGAN, J. A. McCAMMON & M. KARPLUS. 1981. J. Mol. Biol. **153:** 1087.
43. BERKOWITZ, M. & J. A. McCAMMON. 1982. Chem. Phys. Lett. **90:** 215.
44. BROOKS, C. L., A. BRUNGER & M. KARPLUS. 1985. Biopolymers **24:** 843.

Dynamic Simulations of Oxygen Binding to Myoglobin

DAVID A. CASE[a]

Department of Chemistry
University of California
Davis, California 95616

J. ANDREW McCAMMON[b]

Department of Chemistry
University of Houston
Houston, Texas 77004

INTRODUCTION

An important goal in the theoretical study of biochemical mechanisms is the characterization of activated processes, in which motion from one stable configuration to another is limited by the rate of motion through a "transition state" of relatively high energy. For barriers much larger than thermal energies, such crossings occur infrequently, and are difficult to observe in equilibrium simulations. This is true even though the transitions themselves are often instrinsically rapid and can be followed on a picosecond time scale. Classical mechanical simulation methods in which the probability of finding reactive events is greatly enhanced have been known for many years,[1,2] and recent advances in computers have made it practical to apply these to kinetic problems in condensed phases.[3,4] In the field of protein dynamics, the first applications were to the study of tyrosine ring rotations in the pancreatic trypsin inhibitor.[5-7] Here we report similar simulations of the entrance and exit of dioxygen to the heme pocket region of myoglobin.

The basic simulation procedure consists of two parts. In the first, an equilibrium distribution of molecular configurations is determined as a function of some reaction coordinate connecting reactants to products. "Umbrella" sampling techniques enhance the statistical sampling of the transition state region, through separate but overlapping simulations that are carried out in the presence of extra potential terms that constrain the system to "windows" along the reaction path. The logarithm of this probability distribution gives a free energy profile for motion along the reaction coordinate, and from this information the transition state theory (TST) estimate of the rate constant can be determined. In a second step, corrections to TST are determined by following trajectories whose initial conditions are on a dividing surface located at the top of the free energy barrier; those trajectories that cross the dividing surface more than once lower the transition state theory estimate. Within classical mechanics, this

---

[a]Alfred P. Sloan Foundation Fellow.
[b]Supported by the National Science Foundation and recipient of a Dreyfus Teacher-Scholar Award.

procedure is formally exact, and is independent of the choice of reaction coordinate or dividing surface (provided that all reactive trajectories cross the dividing surface). In practice, it can be extremely difficult to carry out the dynamic equilibrations required by this scheme, so that a judicious choice of reaction coordinate is important.

The physical system we study here, the binding of dioxygen to myoglobin, has been extensively studied for many years. The most detailed information about rates and intermediates in this process is obtained by following the course of rebinding of ligand after photodissociation by a short laser pulse. Both low-temperature and room temperature experiments of this sort have been carried out,[8–11] and the results can generally be characterized by the following kinetic scheme:

$$MbL \equiv [Mb + L] \equiv Mb + L \qquad (1)$$

in which $L = O_2$ or CO, and the intermediate species has the ligand trapped inside the protein but not bonded to the iron atom. Additional intermediates are required to describe the rebinding kinetics at low temperature, but it is not clear whether these represent additional stable configurations along a binding pathway or are the result of parallel kinetic processes. At room temperature, the observed forward rate constant for the second step (above) is 5×10^6 sec^{-1} for carbon monoxide and 13×10^6 sec^{-1} for dioxygen.[10] It is this escape process that we model in the present calculations.

Although the kinetic rebinding studies do not identify the physical configurations involved in the various intermediates, an examination of the crystal structure of myoglobin[12,13] makes it plausible that the principal intermediate has the ligand not bound to the iron atom but still near to it in a heme "pocket" region, and that egress to the solvent is most likely *via* a path approximately parallel to the plane of the heme group and which passes between the side chains of valine E11 and histidine E7. A contour map of the potential that would be felt by a ligand in the presence of the rigid protein (in its X-ray conformation) is shown in FIGURE 1.[14,15] The plane of this figure is parallel to the heme plane and displaced from it by 3 Å in the distal direction; the iron atom is at the origin in this coordinate system. (See Refs. 14 and 15 for details of the calculation.) The low-potential region in the center corresponds to the heme pocket, and the solvent region is to the upper left. As the figure indicates, in the absence of protein relaxation, a ligand would experience a barrier of more than 100 kcal/mol upon entering or leaving the pocket. As shown by Case and Karplus,[14] there are plausible motions of the side chains of Val E11 and His E7 that will allow access with much smaller barriers. Our purpose here is to gain a more complete characterization of the way in which this "gate-opening" motion controls the kinetics of ligand binding.

COMPUTATIONAL DETAILS

We have modeled the protein using the AMBER program package,[16] with the empirical energy potentials described elsewhere.[17] We use the "larger" united atom van der Waals parameters, dividing the 1–4 interactions by a factor of 8.0, as described in Ref. 17, and we neglect nonbonded interactions beyond 8 Å. The oxygen atoms in the O_2 ligand are assigned as atom type "O" and have no partial charges. The O-O bond is given a stretching force constant of 848 kcal/mol-Å^2 and an equilibrium length of 1.25Å, values appropriate for the free ligand. No bond is included between the iron

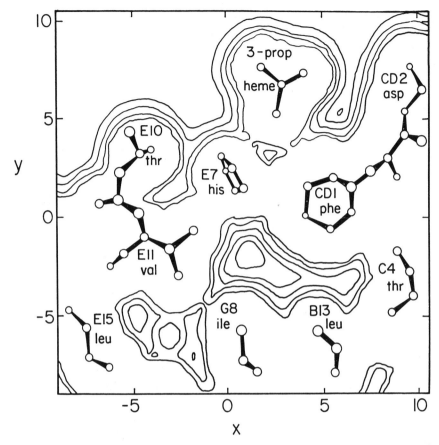

FIGURE 1. Contour map of energy seen by a ligand in a plane through the X-ray structure of myoglobin. In this coordinate system, the heme group is in the xy plane, with the iron atom at the origin. Contours are at 90, 30, 10, 0 and -5 kcal/mol relative to a ligand at infinity. See Refs. 14 and 15 for details of the calculation.

atom and the ligand, so that the simulation corresponds to the state following photodissociation; rebinding to the iron atom is also not allowed in these simulations. In order to reduce the required computation time and to eliminate possible slowly relaxing energy fluctuations outside the region of interest, only residues with atoms within 9 Å of the center of the heme pocket are included in the analysis; the remainder are kept fixed in their X-ray positions. TABLE 1 lists the residues we include, with a total of 275 atoms.

The initial dynamic equilibration of the molecule was carried out in a manner similar to that used in earlier studies: the X-ray coordinates were first subjected to 500 steps of conjugate gradient energy minimization to eliminate most of the components of the force. This was followed by constant-temperature dynamic simulations (of 1 psec each) at 50, 100, 150, 200, 250 and 300 K. This in turn was followed by 8 psec of constant total energy dynamic simulation, during which time the average kinetic energy corresponded to a temperature of 306 K. Throughout this procedure, the ligand remained trapped in the heme pocket region shown in FIGURE 1.

Our reaction coordinate is based upon the model of Case and Karplus,[14] and on earlier test calculations. Its value (δ) is defined as the perpendicular distance from the center of mass of the dioxygen ligand to the plane defined by $C\beta$ of Val E11, $C\gamma$ of His E11, and the nitrogen of pyrrole 1 of the heme group. The orientation is such that negative values are "inside" the protein, and positive values are toward the outside. As we show below, the bottleneck (i.e., the maximum in the free energy profile) occurs at about $\delta^+ = -0.5$ Å, slightly to the inside of the plane containing the three atoms listed above. The equations needed to calculate δ and its derivatives with respect to cartesian coordinates are given in the APPENDIX.

The umbrella sampling was performed by carrying out a series of dynamic calculations with constraining potentials of the form $U(\delta) = K(\delta - \delta_0)^2$, where $K = 10$ kcal/mol-Å^2 and $\delta_0 = -2.5$, -1.75, -1.25, -0.75 and -0.25 Å; an additional simulation was carried out in the absence of constraining potential. Each simulation was begun from the endpoint of the previous one, and was equilibrated for 2 psec, during which time the velocities were reassigned to maintain an average temperature of 305 K. This was followed by 18 psec of dynamic simulation, from which the equilibrium distribution in the presence of the umbrella potential, $\rho^*(\delta)$, and the average potential energy $\langle V(\delta) \rangle$ were calculated, using bins 0.1 Å wide. From ρ^* the potential of mean force in the absence of the umbrella potential can be determined as[1–3]:

$$W(\delta) = -k_B T \ln [\rho^*(\delta)] - U(\delta) + C_i \qquad (2)$$

where C_i is a different constant for each "window." Since the distributions for the various windows overlap, the potentials of mean force can be joined to form a continuous function, and this procedure determines the constants C_i. The true equilibrium distribution is obtained from this continuous function as $\rho(\delta) = \exp [-W(\delta)/k_B T]$.

From the equilibrium distribution, the transition state theory estimate for the rate constant is[1–7]:

$$k_{TST} = (1/2) \langle |d\delta/dt| \rangle_{eq} [\rho(\delta^+) \{ \int \rho(\delta) \, d\delta \}^{-1}] \qquad (3)$$

Here the time derivative of δ is evaluated from an equilibrium distribution at the top of the barrier, and the integral is over the initial state—in this case over the region where the ligand is in the heme pocket. All of the factors that appear in this expression are

TABLE 1. List of Residues Included in the Simulation

Leu	B10	Leu	B13	Phe	B14	Thr	C4
Lys	C7	Phe	CD1	Arg	CD3	Phe	CD4
His	E7	Gly	E8	Val	E9	Thr	E10
Val	E11	Leu	E12	Ala	E14	Leu	F4
Ser	F7	His	F8	His	FG2	Ile	FG4
Leu	G5	Ile	G8	Phe	H15	Heme +	O$_2$

time-average or equilibrium properties, and it may be shown that k_{TST} represents a rigorous upper bound to the true (classical mechanical) rate of crossing the barrier.

Dynamical features of actual reactive trajectories are needed to determine corrections to the transition state theory rate, and these corrections define the "transmission coefficient" κ. As we noted above, these corrections fundamentally arise when trajectories cross the dividing surface (defined by $\delta = \delta^+$) more than once, i.e., when the dynamics is not "straight through" in the transition state region. We use the following expression for κ[5–7,18]:

$$\kappa(t) = D\langle (d\delta/dt)^+ \, H(\delta(t) - \delta^+) \rangle \qquad (4)$$

where D is a normalization constant chosen so that $\kappa(0) = 1$ and H is a step function that is equal to 1 if its argument is positive, and is zero otherwise. Thus, $\kappa(t)$ is a time-correlation function that measures the net flux into the final-state region for trajectories that start at the barrier top. The flux time-correlation decays rapidly to a plateau value that is taken to be the actual transmission coefficient. In the present case (as we show below,) there are few trajectories that recross the surface, so that our transmission coefficient is nearly unity.

RESULTS AND DISCUSSION

Transition State Results

FIGURE 2 shows the potential of mean force and the average potential energy as a function of the reaction coordinate δ. The barrier to exit from the heme pocket is *not* dominated by potential energy—in this model the gate atoms can move out of the path of ligand with no net overall increase in potential energy. This result differs from that of an earlier study[14,15] in which energy minimization calculations starting from the bottleneck region suggested a potential barrier of 10–15 kcal/mol. The two models are not the same—we now use a diatomic ligand rather than a sphere, and large parts of the protein are now kept fixed. However, with the new model, calculations comparable to those of the original study[14,15] (i.e., the use of energy minimization to move the ligand in several steps towards the bottleneck region) again yields a barrier of about 15 kcal/mol. It is only when the ligand is allowed to equilibrate dynamically in the bottleneck region that the potential barrier disappears. This equilibration procedure greatly increases the range of configurations sampled by the system, and allows it to escape local minima in which energy minimization calculations may become trapped. To support this picture, we have performed additional energy minimization ("quenching") from various points along the dynamical equilibration in both the bottleneck region and in the heme pocket. These results (which will be described in a subsequent publication) also show that the total potential energy of the system does not have a maximum in the region between residues E7 and E11.

There *is* a free energy barrier in this region, however, presumably arising in part from the loss of translational and rotational entropy of the ligand as it is forced to squeeze through a narrow opening. The potential of mean force is about 6.5 kcal/mol higher at $\delta = \delta^+ \equiv -0.5$ than in the heme pocket region, which is centered around $\delta = -3.2$ A. As Equation 3 indicates, the rate constant depends not only upon the change in potential of mean force, but also upon the "width" of the initial state through the configuration integral in the denominator. This is in accord with the intuitive picture of an effusive exit: the larger the heme pocket, the longer it will take on the average for the ligand to find the escape path.

We are now in a position to estimate the rate constant for escape from the pocket in the transition state theory approximation. The velocity along the reaction coordinate can be determined from a numerical simulation in which the system is constrained to stay near the dividing surface with a strong ($K = 50$ kcal/mol-Å^2) window potential. The average absolute velocity, $\langle |d\delta/dt| \rangle$, evaluated at the crossing points is 5.0×10^4 cm/sec. (This is close to the value predicted from a one-dimensional effusive distribution, for which $\langle |d\delta/dt| \rangle = (2k_B T/\pi\mu)^{1/2}$; for $\mu = 16$ amu [the mass of the

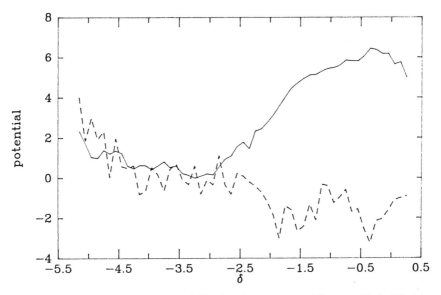

FIGURE 2. Potential of mean force (*solid line*) and average potential energy (*dashed line*) in kcal/mol, as a function of the reaction coordinate δ, in angstroms.

ligand and the "gate atoms" are about the same] this gives 3.1×10^4 cm/sec.) The equilibrium density at the top of the barrier is $\rho(\delta^+) = \exp(-6.5/0.6) = 2.0 \times 10^{-5}$, and the integral in the denominator of Eq. 3 is 1.2×10^{-8} cm. (The last two values depend upon the choice of zero for the potential of mean force, but their ratio is independent of this choice.) This gives $k_{TST} = 4.1 \times 10^7$ sec^{-1}, which may be compared to the value of 1.3×10^7 sec^{-1} inferred by Henry *et al.* from studies of geminate recombination following flash photolysis.[10,11] Although the closeness of agreement with experiment may be the result of some cancellation of errors, it does indicate that the physical picture embodied in these calculations—that the ligand leaves between residues E7 and E11 with little potential barrier—is consistent with the observed results.

Analysis of Trajectories.

Corrections to transition state theory may be estimated by analyzing trajectories generated by forwards and backwards integration from points chosen at the top of the potential of mean force. We have analyzed 24 trajectories whose starting points were taken from a simulation in the presence of a window potential with $K = 50$ kcal/mol-$Å^2$ and $\delta_0 = -0.6$ Å. Starting points for the trajectories were chosen about 0.7 psec apart in order to minimize correlations between the positions or velocities; the trajectories thus sample a time period of about 18 psec. Each trajectory was run for 4 psec; in all but one case this was sufficient to allow it to escape the protein or to become "thermalized" in the heme pocket region. Four sample trajectories are shown in FIGURE 3. Here the contours for the X-ray conformation (as in FIGURE 1) are shown

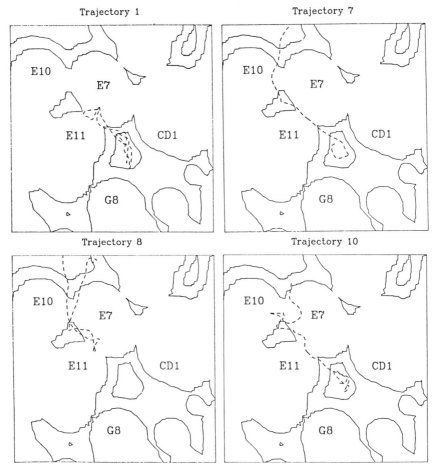

FIGURE 3. Sample trajectories (*dotted lines*). Contours are drawn as in FIGURE 1, with values at 90 and 0 kcal/mol.

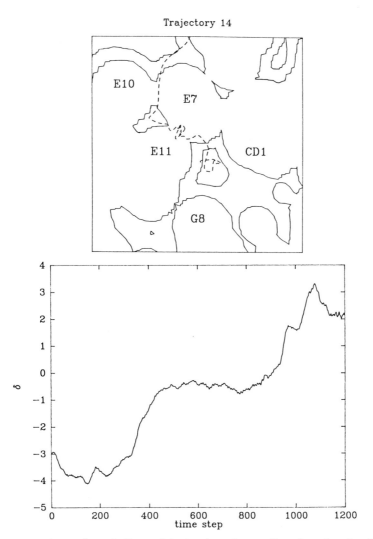

FIGURE 4. Trajectory #14 as in FIGURE 3 (*top*) and reaction coordinate δ as a function of time (*bottom*). Each time step corresponds to 5×10^{-15} sec.

merely to guide the eye—unlike our earlier calculations,[14,15] all 275 atoms in the active region of the simulation are allowed to move as determined by Newton's equations of motion. It is still useful to plot just the motion of the center-of-mass of the ligand, however, since its motion is much larger than that of the atoms of the protein.

As FIGURE 3 indicates, we encounter both nonreactive trajectories (e.g., 1 and 8) as well as reactive trajectories (7 and 10). It is also possible for reactive trajectories to cross the dividing surface defined by $\delta = \delta^+$ more than once, although few of this type have been encountered so far. In fact, the only nonreactive trajectories encountered in this sample were those shown in FIGURE 3. All of the rest, (except #14, discussed below) were reactive trajectories which had "straight-through" dynamics in the transition state region. Although it is clear that a larger sample is needed to obtain good statistics, this result makes it plausible that the transmission coefficient for this reaction is near unity, and that transition state theory may be used to estimate the rate constant.

Results of trajectory #14 are shown in FIGURE 4. This clearly shows diffusive motion near the top of the barrier, with the trajectory crossing the dividing surface at $\delta = -0.6$ seven times, as shown at the bottom of the figure. If our initial sampling of phase space points on the dividing surface is accurate, however, trajectories like that shown in FIGURE 4 must be rare ones.

A "smooth" trajectory is further analyzed in FIGURE 5. Here we plot the side-chain dihedral angles for residues E7 and E11 as a function of time. The ligand passes through the transition state region between time steps 150 and 200 during this trajectory. For valine E11, the χ_1 angle (defined as N-Cα-Cβ-Cγ1) increases from its average value of about 170° to 195° just before the ligand passes the bottleneck region. This behavior is found for the other trajectories we have examined as well, and appears to be important in allowing the ligand to escape. During this time period, χ_1 for histidine E7 is also changing, going from a "low" value near 180° to a value around 210°. Case and Karplus[14] argued earlier that exactly such an increase in this angle was likely to be involved in the gate-opening process, and this interpretation is supported by the trajectories analyzed here. Further analysis of the nature of the reactive (and nonreactive) trajectories found here will be presented elsewhere.

Comparison with Other Results

Although the rate-constant calculations presented here are of interest in their own right, and provide a new picture of the process of ligand binding in myoglobin, it is also of interest to compare these results to those for the rotation of tyrosine side chains in the pancreatic trypsin inhibitor, since the latter represents the only other dynamic process in proteins that has been studied by these methods. The nature and length of the simulation, as well as the statistical methods used, are similar in the two calculations. Two striking differences appear in the results, however, that may reflect real differences in the underlying physical phenomena being modeled. First, for the tyrosine rotation, the potential of mean force, W, and the average potential energy, $\langle V \rangle$, were nearly identical, suggesting that entropy differences are small between the initial and transition state regions. This makes sense since there is no obvious way in which the state of disorder of the protein should change substantially upon such a process. For oxygen migration, on the other hand, one may well expect both the ligand and the gate atoms to be less mobile at the transition state than when the ligand is in the heme pocket. For example, if we adopt a model in which the O_2 molecule rotates freely in the heme pocket but not at all at the transition state, the loss of entropy is about 10 cal/mol-K, corresponding to a gain in free energy of 3 kcal/mol upon going to the transition state. The corresponding result for total loss of translational freedom has been estimated to be a loss of 16 cal/mol-K in entropy, or a gain of 5 kcal/mol in free energy.[11] Restrictions in the motion of the side chains E7 and E11 may also contribute to the free energy barrier. We plan further analyses of the trajectories along these lines,

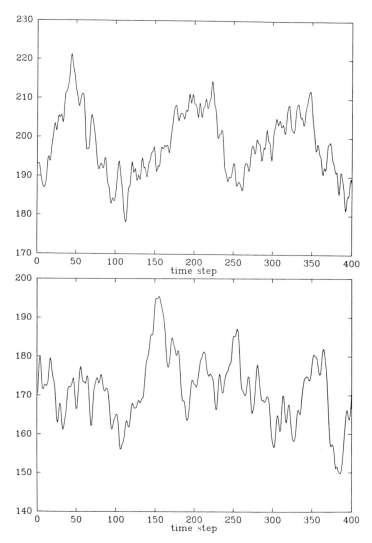

FIGURE 5. Values for χ_1 for His E7 (*top*) and Val E11 (*bottom*) as a function of time for trajectory #7. Each time step is 5×10^{-15} sec.

bearing in mind that there is no rigorous way to divide entropy into "pieces," and that heuristic attempts to do so must be viewed with caution.

The second principal difference between the present results and those for the tyrosine ring rotation lies in the dynamics at the transition state region. For the ring rotation, there was a substantial diffusive character, with the result that the transmission coefficient κ had a value of about 0.2, rather than the value near unity found here. In physical terms, a rotation of the ring experiences (on the average) many impulsive collisions, and many of the trajectories that reach the top of the barrier return to the original state rather than proceeding on towards the final state "products." For O_2 in myoglobin, on the other hand, collisions serve mainly to open a pathway along which the ligand moves in a much more ballistic fashion. This difference may be due to the smaller size, and consequently smaller collisional cross section, of the oxygen molecule compared to the tyrosine ring.

It should be emphasized that the results presented here are based upon a relatively small number of trajectories, that only a fraction of the protein was allowed to participate in the simulation, that no inclusion of solvent was attempted (including water molecules that may enter the heme pocket), and that only a single path out of the heme pocket was explored. For these reasons, the results should be considered as preliminary ones, and we have plans to extend the scope of the simulations, as well as to study the way the barrier changes as the temperature is lowered. Nevertheless, the results do point towards a new way of thinking about the migration of small ligands inside proteins. The methods described here should be applicable to the study of other ligands entering the heme pocket, as well as to the study of ligands such as xenon, which bind at other sites inside the protein.[20,21]

SUMMARY

We report dynamic simulations of the process by which a dioxygen molecule enters or leaves the heme pocket region of myoglobin along a path between the distal histidine (E7) and valine (E11). Our reaction coordinate measures the distance of the ligand from a "dividing plane" defined by three protein atoms. The equilibrium probability distribution as a function of this coordinate is determined by a series of molecular-dynamic simulations with overlapping "umbrella" constraining potentials; the resulting potential of mean force has a barrier of about 7 kcal/mol for exit from the heme pocket. A comparison of this free energy profile with the corresponding potential energy profile suggests that entropy effects dominate the kinetic barrier. Reactive trajectories are generated from dynamic simulations beginning at the top of the potential of mean force; only a small fraction of these recross the dividing surface, indicating that transition state theory may be a good approximation for this process.

ACKNOWLEDGMENT

We thank Dzung Nguyen for many useful discussions and for assistance in preparing the figures.

REFERENCES

1. KECK, J. C. 1962. Discuss. Faraday Soc. **33:** 173.
2. ANDERSON, J. B. 1973. J. Chem. Phys. **58:** 4684.
3. BENNETT, C. H. 1975. Diffusion in Solids. J. J. Burton & A. S. Nowick, Eds.: 73. Academic Press. San Francisco, CA.
4. PANGALI, C., M. RAO & B. BERNE. 1979. J. Chem. Phys. **71,** 2975.
5. NORTHRUP, S. H., M. R. PEAR, C.-Y. LEE, J. A. MCCAMMON & M. KARPLUS. 1982. Proc. Natl. Acad. Sci. USA **79:** 4035.

6. McCammon, J. A., C.-Y. Lee, & S. H. Northrup. 1983. J. Am. Chem. Soc. **105:** 2232.
7. McCammon, J. A. 1984. Rep. Prog. Phys. **47:** 1.
8. Austin, R. H., K. W. Beeson, L. Eisenstein, H. Frauenfelder & I. C. Gunsalus. 1975. Biochemistry **14:** 5355.
9. Doster, W., D. Beece, S. F. Bowne, E. E. DiIorio, L. Eisenstein, H. Frauenfelder, L. Reinisch, E. Shyamsunder, K. H. Winterhalter & K. T. Yue. 1982. Biochemistry **21:** 4831.
10. Henry, E. R., J. H. Sommer, J. Hofrichter & W. A. Eaton. 1983. J. Mol. Biol. **166:** 443.
11. See also Findsen, E. W., T. W. Scott, M. R. Chance, J. M. Friedman & M. R. Ondrias. 1985. J. Am. Chem. Soc. **107:** 3355.
12. Hartmann, H., F. Parak, W. Steigemann, G. A. Petsko, D. R. Ponzi & H. Frauenfelder. 1982. Proc. Natl. Acad. Sci. USA **79:** 4967.
13. Takano, T. 1977. J. Mol. Biol. **110:** 537, 569.
14. Case, D. A. & M. Karplus. 1979. J. Mol. Biol. **132:** 343.
15. Case, D. A. 1982. Hemoglobin and Oxygen Binding. C. Ho, Ed.: 371. Elsevier. New York, NY.
16. Weiner, P. K. & P. A. Kollman. 1981. J. Comp. Chem. **2:** 287.
17. Weiner, S. J., P. A. Kollman, D. A. Case, U. C. Singh, C. Ghio, G. Alagona, S. Profeta & P. Weiner. 1984. J. Am. Chem. Soc. **106:** 765.
18. Chandler, D. 1978. J. Chem. Phys. **68:** 2959.
19. Frauenfelder, H. & P. G. Wolynes. 1985. Science **229:** 337.
20. Tilton, R. F., I. D. Kuntz, Jr. & G. A. Petsko. 1984. Biochemistry **23:** 2849.
21. Tilton, R. F., Jr., S. J. Weiner, U. C. Singh, I. D. Kuntz, Jr., P. A. Kollman, N. Max & D. A. Case. 1986. J. Mol. Biol. In press.
22. McConnell, A. J. 1957. Applications of Tensor Analysis.: 52–56. Dover. New York, NY.

Appendix

We give here an outline of the way in which the reaction coordinate is calculated. The equation for a plane passing through points (x_1, y_1, z_1), (x_2, y_2, z_2), (x_3, y_3, z_3) is[22]:

$$\begin{vmatrix} x - x_1 & x - x_2 & x - x_3 \\ y - y_1 & y - y_2 & y - y_3 \\ z - z_1 & z - z_2 & z - z_3 \end{vmatrix} = 0$$

This can be expanded to give the equation of the plane in standard form:

$$a_x x + a_y y + a_z z = b$$

The vector (a_x, a_y, a_z) is then normal to the plane. The perpendicular distance from a fourth point (x_4, y_4, z_4) to this plane is:

$$\delta = (b - a_x x_4 - a_y y_4 - a_z z_4)/a,$$

where $a = (a_x^2 + a_y^2 + a_z^2)^{1/2}$. Derivatives of δ with respect to the Cartesian coordinates may then be determined by straightforward, but tedious, algebra.

Modeling Complex Molecular Interactions Involving Proteins and DNA[a]

PETER A. KOLLMAN, SCOTT WEINER, GEORGE SEIBEL,
TERRY LYBRAND, U. CHANDRA SINGH,
JAMES CALDWELL, AND SHASHIDHAR N. RAO

Department of Pharmaceutical Chemistry
University of California, San Francisco
San Francisco, California 94143

INTRODUCTION

Applying molecular simulation techniques to molecules of biological interest is very exciting, because such molecules are often essential elements in life processes. But extracting useful information from such simulations is also tremendously challenging, because the systems are so inherently complex.

What does one want to extract out of such simulations? Our research efforts have focused on attempting to calculate and understand the structures and interaction energies of complexes involving proteins, DNA, or ionophores and appropriate ligands. The fundamental difficulty in such simulations lies in the fact that the systems have so many local minima. There are two challenges: first, to find all the relevant local minima, and, second, to accurately evaluate their energy. In the case of proteins, solving such problems would be equivalent to solving the protein-folding problem. However, the number of possible local minima are uncountably large, so a less brute-force approach must be taken. In our own calculations on protein ligand complexes, we phrase the question more narrowly. Given that we begin the simulation at or near the structure as determined by high resolution protein crystallography, can we derive sensible answers that are consistent with available experiments and suggest new ones? This narrows the "local minimum" search considerably, but still requires an accurate energy evaluation. Thus, one of the goals of this paper is to describe our efforts to improve the analytical energy functions for use in molecular mechanical and dynamical simulations of complex molecules. A second goal is to show, in the case of our simulations of DNA and DNA-drug complexes, how one can interleave theory with nuclear magnetic resonance (NMR) NOE measurements in a way that increases our understanding of the structure and dynamics of the system. A final goal is to show how such methods can give useful insight into the mechanism, stereochemistry, and energetics of enzyme-catalyzed reactions, as well as the effect of single amino acid substitutions on the energetics.

[a]This work was supported by Grants CA-25644 and GM-29072 (to P.A.K.) from the National Institutes of Health and Grant CHE-85-10066 from the National Science Foundation.

FORCE FIELD DEVELOPMENT

We began our molecular mechanics calculations on nucleic acids and proteins with a force field[1,2] that began with that developed by Gelin and Karplus[3] for peptides and modified it to allow the study of nucleic acids and the inclusion of hydrogen-bonded hydrogens explicitly. This proved a reasonable place to start and useful results were found using it, but its defects became clearer with more experience. We made an effort to develop an improved force field. For our purposes, the nonbonded (electrostatic plus van der Waals) interactions were the most important, so we made a considerable effort to improve these: first, we generalized the approach suggested by Momamy[4] and Cox and Williams[5] to determine partial charges on the various atoms of proteins and nucleic acids using quantum mechanical calculations on relevant fragments and fitting the quantum mechanically calculated electrostatic potential to the partial-charge model[6] (*not* using Mulliken populations). Second, it was clear that we needed to make the dispersion terms less attractive and the exchange repulsion terms more repulsive, in order to better "retain" observed X-ray structures. Finally, we sought a general approach to determine torsional, bond-bending and bond-stretching energies that was less arbitrary and used experimental structural and vibrational frequency data as a guide. Our united atom model (with CH, CH_2 and CH_3 groups treated as extended atoms) was reasonably successful in reproducing relative conformation energies and structures of small-model systems and gave much less protein compaction upon energy refinement than the previous model.[7] The fact that it was an improvement over previous force fields was independently validated by Hall and Pavitt.[8] However, it was clear to us already (and so noted) that compromises had to be made in its development that were less than satisfactory. First, we found that, using the larger and less deep van der Waals functions for the extended atoms CH, CH_2, and CH_3 led to too large repulsions for *gauche* and *cis* conformations of butane unless the van der Waals interactions for atoms separated by three bonds (1–4 interactions) were scaled down by a factor of two. This compromise gave a reasonable fit to furanose-ring puckering and butane conformation while allowing the use of larger van der Waals radii. Nonetheless, as we noted,[7] such a choice of van der Waals radii leads to an overestimate of the density of liquid dimethyl ether by 10% in Monte Carlo simulations. Using the much larger united-atom van der Waals radius suggested by Jorgensen[9] or those found by van Gunsteren[10] to give a good description of a small crystal and scaling 1–4 van der Waals interactions down by a factor of eight led to a good description of the butane conformational profile, but a poorer description of the sugar-pucker properties of deoxyadenosine, where the van der Waals repulsion between C5' and C2', which are separated by 4 bonds, is too large and leads to a minimum energy "C2'endo" sugar pucker with phase nearer 130–140° than 150–160°. In the case of proteins, such a model seems satisfactory, given the molecular dynamics simulation carried out by Tilton *et al.*[11] on myoglobin with Xe atoms in its hydrophobic cavities. In that simulation, the radius of gyration of the molecule fluctuated around the experimental value during the 100 psec of the run.

An all-atom version of the force field has recently been accepted for publication.[12] It gives superior ethidium intercalation energies[13] to the united-atom model albeit with similar structures and does not appear to have as severe a "compaction" problem as the united-atom model with the smaller van der Waals radii. Nonetheless, the radii for O,

N and carbonyl C are the same as in the united-atom model, and, based on the Monte Carlo simulations of Jorgensen,[14] this model still may give too dense amide liquids. We are currently exploring this question by Monte Carlo simulations on N-methyl acetamide using both united- and all-atom parameters.

Jorgensen and coworkers[15] have nearly completed a set of nonbonded parameters for proteins, using a 12-6-1 potential and calibrating the parameters to fit liquid densities and energies. It is reasonable to combine these with the internal (bond length, angle, dihedral, improper torsions) parameters of standard protein force fields, and, by suitable calibration of the V_1, V_2, and V_3 Fourier torsion coefficients, have a simple force field that avoids "compaction" and fits local conformational energies well. Nonetheless, it is not clear at this point that such a model is generally superior to one with smaller van der Waals radii, since the large van der Waals radii needed to reproduce liquid densities may lead to too little inherent conformational flexibility in the interior of macromolecules. Nonetheless, it will be very useful to have both types of force fields to use in comparative protein calculations. Another concern is that the potential developed by Hagler et $al.$[16] to reproduce crystal lattice properties of amides does only an adequate job in reproducing the liquid properties of amides,[14] underestimating the density of formamide by 5% and the enthalpy of vaporization of formamide by 13%. Are proteins more "solid" or more "liquid-like," and, thus, which is the more appropriate potential to use?

An implicit element in the above potentials for water and proteins is the incorporation of many-body effects only in an average way or not at all. Thus, the charge distributions in TIPS[17] and SPC[18] water have dipole moments significantly higher than an isolated water molecule. This is a very reasonable approximation, but it will fail most severely for systems including highly charged ions. We have begun[19] the development of a set of potentials that aim to give a good representation of water in gas, liquid and solid phases as well as in ion-water interactions. This potential, which is a modification of the RWK2 potential but includes polarization effects, has already been shown to reproduce the second virial coefficient and water-dimer energy and structure for gas-phase properties, and the lattice energy and density of ice I and VII for solid-phase properties. It has not yet been tested on liquid properties. The use of this model and explicit exchange-repulsion/charge-transfer three-body nonadditive terms based on ion $-(H_2O)_2$ quantum mechanical calculations has led to $Na^+(H_2O)_n$ (n = $1 \rightarrow 6$), $K^+(H_2O)_n$ (n = $1 \rightarrow 6$), and $Cl^-(H_2O)_n$ (n = $1 \rightarrow 4$) hydration enthalpies in essential agreement with Kebarle's gas-phase experiments.[20] Molecular mechanical calculations on liquid models for Na^+, K^+, Cl^- and Mg^{+2} were also encouraging[19] and suggest that both the structures and the ionic-solution enthalpies can be reproduced with such a potential. The above study is representative of our approach to potential function development—the fitting of empirical data as a prerequisite, using the results of ab $initio$ quantum mechanical calculations as an essential guide.

MOLECULAR MECHANICAL AND MOLECULAR DYNAMICS STUDIES OF DNA, DNA-NETROPSIN AND DNA-ACTINOMYCIN D COMPLEXES

We have carried out a number of simulations of DNA and DNA-drug complexes using molecular mechanics and dynamics methods. The molecular dynamics studies on DNA recently completed in our lab include a study of d(CGCGA).d(TCGCG) with and without counterions[21] and with[22] and without[21] explicit solvation by 830 water molecules, a study of mismatched analogs of the above sequence d(CGCGA) · d(TCACG) and d(CGAGA) · d(TCGCG),[23] studies of the dynamics, helix-repeat and sugar-pucker profiles of the sequences dA$_{10}$ · dT$_{10}$, d(ATATATATAT)$_2$ and d(GCGCGCGCGC)$_2$,[24] studies of triostin A-DNA bisintercalation[25] and studies of DNA loop structures.[26] But here we focus on three studies, which are, in our opinion, the clearest cases of the interleaving of theory and experiment that is the cornerstone of our approach to computer modeling.

The first study, which appeared in print earlier this year,[27] suggests that the following experimental phenomena can be understood in a simple and clear fashion from the results of molecular mechanics optimizations on d(ATATAT)$_2$, d(TATA-TA)$_2$, d(CGCGCG)$_2$, d(GCGCGC)$_2$ and dA$_6$ · dT$_6$: the molecule d(ATAT)$_2$ crystallizes[28] with an "alternating B" conformation, sugar puckers A(C3'endo), T(C2'endo); poly d(A-T) · poly d(A-T) has a doublet ^{31}P spectrum,[29] poly d(G-C) · poly d(G-C) a singlet[30]; Raman studies[31] suggest only C2'endo sugars for calf thymus DNA and GC polymers, but two peaks of approximately equal intensity in poly d(A-T), poly d(A-T) and in poly dA · poly dT at low temperature and moderate salt, with the "C3'endo" peak getting weaker at higher T and lower salt.

Our molecular mechanical studies show[27] that a pure C2'endo B DNA structure is preferred over C3'endo-containing variants by two effects: longer phosphate-phosphate distances and the C2'endo pucker is ~0.6 kcal/mol intrinsically more stable for a deoxyribose sugar. However, in the case of T-containing nucleic acids, there is an especially favorable nonbonded interaction involving the thymine C5, C6 and C5-methyl and the DNA backbone which occurs only when the sugar at its 5' end is C3'endo. This helps explain at the same time why in poly d(A-T) · poly (A-T), the sugar puckering is most favorable for A(C3'endo), T(C2'endo) (with uniform puckering only slightly less stable), consistent with the X-ray result for d(ATAT)$_2$; and in poly dA · poly dT, the sugar puckering in the most favorable conformation is A(C2'endo), T(C3'endo), in contradiction to fiber-diffraction analysis of Arnott *et al.*,[32] with the uniform pucker model only ~2 kcal/mol less stable for dA$_6$ · dT$_6$.

It should be noted that our results can be considered a genuine prediction for poly dA · poly dT, since the Raman spectra suggest equal amounts of C2'endo and C3'endo pucker, but not an answer to the question as to which base's sugar corresponds to which pucker. Recent NMR studies by Sarma *et al.*[33] question the "heteronomous" nature of poly dA · poly dT, but we suggest they are not definitive and that smaller fragments of dA$_n$ · dT$_n$ must be studied to determine a definitive conformational profile for it.

We have carried out molecular mechanical studies of netropsin interacting with different sequences[34] of DNA: dA$_6$ · dT$_6$, d(ATATAT)$_2$, d(TATATA)$_2$, d(CGCGCG)$_2$ and d(CGCGAATTCGCG)$_2$. In the latter sequence we used the 4 drug-DNA NOE's determined by Patel[35] as a guide for molecular mechanical model building, followed by subsequent energy refinement. Upon submission of that paper, both referees noted that we should compare our results to the X-ray structure of netropsin: d(CGCGAATTCGCG)$_2$ determined by Kopka *et al.*[36] and we did so. The results were most gratifying: even though we had used an Arnott model[37] rather than the Drew *et al.* structure[38] for the DNA portion, upon rms superposition of the phosphates of the two structures, our calculated netropsin position in the groove, was in very good agreement with that reported by Kopka *et al.* With the exception of the amidinium fragment at one end of netropsin, which in the calculations swung around and formed hydrogen bonds with the phosphate oxygens but in the X-ray structure stayed parallel to the minor groove, the rms difference between the X-ray and calculated netropsin atoms was 0.6 Å. Most significantly, the drug was in correct register along the minor groove and many of the specific H bonds were identical in the two structures. This result suggests that a few key NMR data (not many may be required) may enable the theoretician to construct a very reasonable model for a drug-DNA complex.

We have used computer graphics model building as well as molecular mechanics[39] and molecular dynamics calculations on the actinomycin D-d(ATCGAT)$_2$ complex, for which 300 NOE distances have been determined by NMR[40] and for which the X-ray structure is being determined by Berman and co-workers.[41]

The structures for the actinomycin D complex with various DNA sequences d(GCXYGC)$_2$, XY = AT, TA, GC, CG and d(ATGCAT)$_2$ were determined by model building, using the suggested structure of Sobell, the X-ray structure of d(GC)$_2$:actinomycin, and other non-NMR solution experiments as qualitative guides. In particular, it is clear that actinomycin D interacts with DNA by intercalation and the X-ray structures and the shape of the drug strongly suggest that the drug sits in the minor groove. Using the experimental structure of actinomycin itself and the many van der Waals contacts the drug can make in the minor groove of DNA, there is surprisingly little ambiguity in fitting the drug with DNA in such a way to explain the GC preference of the drug, since only with this sequence can the drug form both an amide N-H . . . N H bond with the guanine N3 and an amide C = O. . .HNH H bond with the guanine 2-NH$_2$ group. For the other XY = AT, GC, TA, at most one minor groove H bond can be formed, both during model building and subsequent energy refinement. The sequence-dependent calculated DNA interaction energies are consistent with experimental inferences.

We have used both the energy-minimized structure and an 80 psec molecular dynamics trajectory starting with the energy-minimized structure to compare with the results of 2D NMR studies on the complex d(ATGCAT)$_2$ · Act D carried out by Brown *et al.*[42] We emphasize that no information from the NMR study was included in the molecular mechanical/dynamical studies and thus the comparison is a genuine test of the ability of the model building/molecular mechanics/dynamics approach using indirect experimental inferences to predict the structure of a drug-DNA complex. The all-atom molecular mechanical model[12] was used in the study in order to facilitate the comparison with the proton-proton distances inferred from the NOE measurements. A very simple model can be used to derive H. . .H distances from the NOE intensities observed by Brown *et al.*[40] and this involves using the ratio of the NOE intensity between the cytosine H5 and H6 protons and the 6th power dependence of NOE on distance:

$$R(X. . .Y) = R(H5. . .H6) \left[\frac{NOE(H5. . .H6)}{NOE(X. . .Y)} \right]^{1/6}$$

where $R(X. . .Y)$ is the unknown distance between protons X and Y, $NOE(X. . .Y)$ is the observed NOE between X and Y, $R(H5. . .H6)$ is 2.40 Å, the "observed" reference distance between cytosine H5 and H6 protons and the NOE(H5. . .H6) the observed NOE intensity between the cytosine H5 and H6 protons.

Of the 300 NOE distances suggested by the Brown study, for all but 26 the molecular mechanically calculated and NMR "determined" distances were within 1.5 Å of each other. For the remaining 26, where the difference was more than 1.5 Å, all the NMR-"determined" distances were shorter than found in the calculations. An 80-psec molecular dynamics trajectory reduced the experimental discrepancy for some distances but increased it for others. A detailed analysis of many of the largest discrepancies, many of which involved the terminal A-T base pairs, suggested either extensive conformational equilibria (e.g., single strand ⇌ double strand equilibria) in the structure which the theory doesn't sample or spin diffusion in the experimental study, which would lead to distances having an NOE much more intense than their average separation would suggest. Nonetheless, there is enough congruence between the distances inferred from the calculations and experiments to suggest that the calculated structure is qualitatively correct. A further test of this suggestion will be a comparison between the calculated and experimental distances in the X-ray structure now being determined by Berman and coworkers.[42]

SERINE PROTEASE ENZYME CATALYSIS—COMPUTER GRAPHIC, MOLECULAR MECHANICAL AND QUANTUM MECHANICS STUDIES OF ENZYME SUBSTRATE INTERACTIONS

We turn our attention to four studies which have been carried out in our lab over the last few years, whose focus has been to understand the nature of serine protease binding and catalysis. Our initial study used computer graphics and molecular mechanics to model-build and energy-refine noncovalent (Michaelis) and covalent (tetrahedral intermediate) complexes of D and L N-acetyl tryptophan and α-chymotrypsin,[43] using a rather low-resolution X-ray structure of the latter as a starting geometry for the active site. As we discuss elsewhere, this was a good place to start our studies because the question to answer was narrowly focused, i.e., could our calculations allow us to say why both stereoisomers bound to the active site with similar affinity, but only the L was a good substrate of the enzyme, being hydrolyzed at least 10^5 faster than the D isomer? In addition, there is no solvation-energy difference between stereoisomers and there is likely to be little difference in solvation of the enzyme-substrate complex; thus, the relative molecular-mechanical energies are likely to reflect the *relative* true energies (enthalpies and free energies) of interaction of these isomers. These expectations were realized in our study, and, as we have detailed more fully elsewhere, our calculated relative energies for the D and L non-covalent and covalent complexes were consistent with experimental inferences, gave clear mechanistic insight into why the L was the much better substrate of the enzyme, and made the prediction that the lowest energy conformer for the D isomer would be qualitatively different than that of the corresponding D N-acetyl tryptophan (carboxylic acid). Thus, the study was successful, but limited to comparing the relative energies of stereoisomers at selected points along a putative path for catalysis. To make the approach more general, both solvation effects and the explicit consideration of bond-making and bond-breaking energies had to be included. This led to our next study, the hydrolysis of formamide by hydroxide ion in the gas phase and in aqueous solution.[44]

We used *ab initio* gradient-optimization methods to evaluate the gas-phase hydrolysis of formamide by OH^-, with the second step in the process (following OH^- attack on the formamide $C = O$ to form a tetrahedral adduct) involving a single water catalysis to transfer the proton from the OH^- to the amine (formerly amide) N lone pair. In the gas phase, tetrahedral-adduct formation involves no barrier, with the adduct 26 kcal/mol more stable than the reactants, and the single H_2O proton shuttle involves surmounting a barrier of ~10–15 kcal/mol to reach the products, which are calculated ($\Delta E = -49$ kcal/mol) and found experimentally ($\Delta H = -46$ kcal/mol) to be much more stable than the reactants.

As a simple "first order" method to estimate the aqueous-solvation energy along this hydrolysis path and thus to estimate the reaction energetics for solution-phase amide hydrolysis, we embedded the structures taken from the gas-phase optimization into a box of 216 TIPS3P H_2O molecules and energy-refined the water molecules, keeping the solute positions fixed. Such a calculation led to a solution-phase pathway to hydrolysis dramatically different from that found in the gas phase, with a "solvent induced" barrier of 22 kcal/mol leading to a tetrahedral intermediate 13 kcal/mol higher in energy than reactants and a second barrier of about 22 kcal/mol higher than reactants involving proton shuttling to form the products, which are calculated ($\Delta E = -10$ kcal/mol) and found ($\Delta H = -6$ kcal/mol) somewhat more stable than reactants.

The dramatic difference between the gas-phase and solution-phase reaction profiles comes mainly from the fact that OH^- is a so much less stable ion than $HCOO^-$ or the delocalized tetrahedral intermediate that the overall reaction energy is ~50 kcal/mol downhill: this fact dominates the reaction profile. In solution, the localized anion, OH^-, is hydrated more effectively than $HCOO^-$ or the tetrahedral intermediate anion, leading to both an approximate energy balance between reactants and products ($\Delta H = -10$ kcal/mol) and a kinetic barrier to forming the tetrahedral intermediate.

The study of formamide/OH^- reaction provided us with two important bits of information to carry into our study of amide hydrolysis catalyzed by trypsin. First, we used the *ab initio* optimized geometries of the simpler formamide/OH^- reaction to build molecular mechanical models for the various structures along the serine protease pathway.[45] Secondly, gas-phase and aqueous-solution base-catalyzed amide hydrolysis provide appropriate reference points for the enzyme-catalyzed reaction.

Our model for the enzyme-catalyzed reaction consisted of bovine trypsin,[45] 100 X-ray diffraction located H_2O molecules, an additional 100 H_2O molecules added to the active site region, and a specific tripeptide substrate of the enzyme. After model-building the substrate into a position appropriate for catalysis (KM structure), we used molecular mechanical methods to force Ser 195 → His 57 proton transfer and subsequent attack of Ser 195 O_γ on the substrate carboxyl carbon. The resulting tetrahedral intermediate had the His 57 HN_ϵ in excellent position to transfer to the substrate leaving group amine nitrogen, which was an amide nitrogen before covalent attack. This proton transfer was forced by molecular mechanical methods and the substrate C-N bond lengthened. After bond cleavage, a water molecule had, during optimization, ended up in excellent position to function as the general base in the acyl enzyme, and water → His 57 and OH^- attack on the acyl bond of the substrate followed. Proton transfer from His to the Ser 195 O_γ then regenerated the native enzyme. At selected points along this pathway single point *ab initio* calculations with a 4–31 G basis set were carried out on the parts of the system where covalent bonds were formed/broken (e.g., imidazole/methanol/formamide for the first part of the reaction and imidazole/methyl formate/H_2O in the second part of the reaction). The total energy for the reaction was the sum of the quantum mechanical energies (which increased during the reaction) and molecular mechanical energies (which decreased during the reaction), leading to a reaction profile with a low activation barrier.[45] Based on our results, it appears that a key feature which distinguishes enzyme catalysis from the solution reaction is the fact that in the enzyme reaction, the OR^- does not have a sheath of water that needs to be stripped off prior to attack on the sciscile $C = O$. As

TABLE 1. Calculated Energy Differences for K_m and TET1 Structure and Experimental Kinetic Data for Trypsin Mutants

	Structure			
	$K_m{}^a$ $\Delta\Delta E_{calc}{}^c$	$K_m{}^d$	TET1[b] $\Delta\Delta E_{calc}{}^e$	$k_{cat}{}^f$
Native	0	144	0	1444
A 216[g]	10.6	3904	7.0	1017
A 226[h]	9.2	3665	9.7	13
A 216/226[i]	11.2	331	16.6	1
ASN 102[j]	2.9	—[k]	27.3	—[k]

[a]Michaelis complex structure (see Ref. 45).
[b]Model for tetrahedral intermediate for acyl enzyme formation (see Ref. 45).
[c]Difference in molecular-mechanical energy for KM (native) structure and that of mutant for residues within 10 Å of active site or mutation site.
[d]Experimental K_m (see Ref. 46).
[e]Difference in molecular-mechanical energy for model for tetrahedral intermediate in native and that of mutant for residues within 10 Å of active site or mutation site.
[f]Experimental k_{cat} measured in Ref. 46.
[g]Mutant with Gly 216 changed to Ala.
[h]Mutant with Gly 226 changed to Ala.
[i]Mutant with both Gly 216 and 226 changed to Ala.
[j]Mutant with Asp 102 changed to Asn.
[k]No data available.

soon as the proton is transferred from the Ser O_γ to the His 57, it is at an ideal (3 Å) separation from the $C = O$ and can attack with little further barrier due to protein "strain." Thus, the enzyme-catalyzed reaction is not a gas phase one, but has elements that are between the gas-phase and solution-phase OH^-. . .formamide reactions.

The results of the molecular mechanical optimizations on the various snapshots along the catalytic pathway of trypsin let us turn our attention to the site-specific mutagenesis studies carried out by Craik *et al.*[46] We can use our model for the Michaelis complex and the tetrahedral intermediate to carry out analogous molecular mechanical optimizations on models for the Michaelis complex and tetrahedral intermediate for various mutants: Gly 216 → Ala, Gly 226 → Ala, Gly 216, 226 both → Ala and Asp 102 → Asn 102 (TABLE 1). In these studies we make the assumption that the structures for the "quantum mechanical" part of the system can be taken as essentially identical (this is reasonable, because our constraints force this to occur) and that the relative molecular-mechanical energies for the Michaelis complex only including those residues within 10 Å of the serine O_γ or the C_α of Asp 102, Gly (Ala 216) or Gly (Ala 226) can be related to the relative K_m experimental values and the relative molecular mechanical energies for the tetrahedral intermediate can be related to changes in k_{cat}/K_m. A final assumption is that our calculated results on bovine trypsin are relevant to the experimental results on rat trypsin, but given the nearly complete conservation of the active-site region,[46] this seems reasonable. Thus, it is interesting that the $\Delta\Delta E_{calc}$ for the double Ala mutant Michaelis complex (K_m) is comparable to that for the single mutants, but $\Delta\Delta E_{calc}$ for TET1 is much higher. This suggests that the experimental results can be rationalized without invoking nonproductive binding. It is encouraging that the relative $\Delta\Delta E_{calc}$ for the two single Ala mutants are correctly ordered, in that the transition state (TET1) is correctly calculated to be less strained for Ala 216 than Ala 226. Finally, our calculations predict that the Asn 102 mutant will bind substrate about as well as native, but will be a very poor catalyst.

More precise agreement between calculations and experiments would not be expected given the simplicity of the model. However, two recent developments encourage us that more accurate calculations can be carried out. The first, from our laboratory, involves the capability of carrying out combined *ab initio* quantum/molecular mechanical optimizations, combining the energy gradients of each to search for minima and transition states on the surface.[47] Using this approach, we will no longer have to assume that the structures in the trypsin-catalyzed reaction are similar to those of the formamide/OH^- reaction. A second exciting development, from the groups of Berendsen/van Gunsteren,[48] McCammon,[49] and Jorgensen,[50] is the combined use of molecular dynamics or Monte Carlo methods with statistical mechanical perturbation theory to calculate free energy *differences* for closely related systems. We are currently carrying out molecular dynamics calculations on the isolated enzyme, the Michaelis complex and the tetrahedral intermediate in trypsin, in which we change the amino acids during the simulation, e.g., grow a Gly → Ala in residue 216 or 226, in order to more precisely calculate the free energy difference for Michaelis complex and tetrahedral-intermediate formation. We see this approach as an enormously exciting development that should open a whole new vista of applications of molecular dynamical simulations.

SUMMARY

We have presented a perspective of progress in three areas of simulations of complex molecules: (a) the development of force fields for molecular simulation; (b) the application of computer graphics, molecular mechanics and molecular dynamics in simulations of DNA and DNA-drug complexes and (c) the application of computer graphics, molecular mechanics and quantum mechanics in studies of enzyme substrate interactions. It is our perspective that improvements are being made in force fields, and these will allow a more accurate simulation of *structures and energies* of complex molecules. In the area of DNA molecular mechanics and dynamics, it is clear that the use of computer graphics model building combined with NMR NOE data is a potentially very powerful tool in accurately determining structures of drug-DNA complexes using molecular mechanics and dynamics. Finally, we are in a position to reasonably simulate structures and (qualitatively) energies for complete reaction pathways of enzymes using a combination of computer graphics, molecular mechanics and quantum mechanics. More accurate energies and pathways are sure to follow, using the combined molecular mechanics/quantum mechanics optimization developed by Singh and the free energy perturbation methods pioneered in Groningen and Houston.

ACKNOWLEDGMENTS

We are grateful to the UCSF Computer Graphics Lab (supported by Grant RR-1081) for essential facilities used in this research.

REFERENCES

1. KOLLMAN, P., P. WEINER & A. DEARING. 1981. Biopolymers **20:** 2583–2621.
2. BLANEY, J., P. WEINER, A. DEARING, P. A. KOLLMAN, E. C. JORGENSEN, S. OATLEY, J. BURRIDGE & C. BLAKE. 1982. J. Am. Chem. Soc. **104:** 6424–6434.
3. GELIN, B. & M. KARPLUS. 1979. Biochemistry **18:** 1256–1268.
4. MOMAMY, F. 1978. J. Phys. Chem. **82:** 592–601.
5. COX, S. R. & D. E. WILLIAMS. 1981. J. Comp. Chem. **2:** 304–323.
6. SINGH, U. C. & P. A. KOLLMAN. 1984. J. Comp. Chem. **5:** 129–145.
7. WEINER, S. J., P. A. KOLLMAN, D. A. CASE, U. C. SINGH, C. GHIO, G. ALAGONA, S. PROFETA & P. WEINER. 1984. J. Am. Chem. Soc. **106:** 765–784.
8. HALL, D. & N. PAVITT. 1984. J. Comp. Chem. **5:** 441–450.
9. JORGENSEN, W. 1981. J. Am. Chem. Soc. **103:** 335–340.
10. VAN GUNSTEREN, W. F. & H. J. C. BERENDSEN. 1985. Molecular Dynamics and Protein Structure. J. Hermans, Ed.: 5. Polycrystal Press. Western Springs, IL.
11. TILTON, R., U. C. SINGH & P. A. KOLLMAN. Molecular dynamics simulations on myoglobin and 5 Xe atoms located in the hydrophobic cavities of the molecule. Unpublished observations.
12. WEINER, S. J., P. A. KOLLMAN, D. T. NGUYEN & D. A. CASE, 1986. J. Comp. Chem. **7:** 230–252.
13. LYBRAND, T. & P. A. KOLLMAN. 1985. Biopolymers. **24:** 1863–1879.
14. JORGENSEN, W. L. & C. J. SWENSON. 1985. J. Am. Chem. Soc. **107:** 569–578.
15. JORGENSEN, W. Personal communication.
16. HAGLER, A., E. EULER & S. LIFSON. 1974. J. Am. Chem. Soc. **96:** 5319–5327.
17. JORGENSEN, W. L., J. CHANDRESEKHAR, J. D. MADURA, R. W. IMPEY & M. L. KLEIN. 1983. J. Chem. Phys. **79:** 926–935.
18. BERENDSEN, H. J. C., J. P. M. POSTMA, W. F. VAN GUNSTEREN & J. HERMANS. 1981. *In* Intermolecular Forces. B. Pullman, Ed.: 331–342. Reidel. Dordrecht, Holland.
19. LYBRAND, T. & P. A. KOLLMAN. 1985. J. Chem. Phys. **83:** 2923–2933.
20. DZIDIC, I. & P. KEBARLE. 1970. J. Phys. Chem. **74:** 1466–1474; ARSHADI, M., R. YAMDAGNI & P. KABARLE. 1970. J. Phys. Chem. **74:** 1475–1482.
21. SINGH, U. C., S. J. WEINER & P. A. KOLLMAN. 1985. Proc. Natl. Acad. Sci. USA **82:** 755–759.
22. SEIBEL, G. L., U. C. SINGH & P. A. KOLLMAN. 1985. Proc. Natl. Acad. Sci. USA. **82:** 6537–6540.
23. SINGH, U. C. & P. A. KOLLMAN. A comparative molecular dynamics analysis of d(CGCGA) · d(TCGCG) and mismatched AC and GA base pair analogies d(CGCGA) · d(TCACG) and d(dCGAGA) · d(TCGCG). Unpublished results.

24. SINGH, U. C., S. R. RAO & P. A. KOLLMAN. Unpublished material.
25. SINGH, U. C. & N. PATTIBIRAMAN. Unpublished material.
26. HAASNOOT, C., U. C. SINGH, P. BASH & P. KOLLMAN. Unpublished material.
27. RAO, S. & P. A. KOLLMAN. 1985. J. Am. Chem. Soc. 107: 1611–1617.
28. VISWAMITRA, M. A., O. KENNARD, P. G. JONES, G. M. SHELDRICK, L. A. SALIBURY, L. FALVELLO & Z. SHAKKED. 1978. Nature 273: 687–688.
29. PATEL, D. J., S. A. KOZLOWSKI, J. W. SUGGS & S. D. COX. 1981. Proc. Natl. Acad. Sci. USA 78: 4063–4067.
30. COHEN, J. S., J. S. WOOTEN & C. L. CHATTERJEE. 1981. Biochemistry 20: 3049–3055.
31. THOMAS, G. A. & W. L. PETICOLAS. 1983. J. Am. Chem. Soc. 105: 993–996.
32. ARNOTT, S., R. CHANDRASEKHARAN, A. BANERJEE, R. HE & J. K. WALKER. 1983. J. Biomol. Struct. Dyn. 1: 437–452.
33. SARMA, M. H., G. GUPTA & R. H. SARMA. 1985. J. Biomol. Struct. Dyn. 2: 1057–1084.
34. CALDWELL, J. & P. KOLLMAN. 1986. A molecular mechanical study of netropsin-DNA interaction. Biopolymers. 25: in press.
35. PATEL, D. J. 1982. Proc. Natl. Acad. Sci. USA 79: 6424–6428.
36. KOPKA, M., C. YOON, D. GOODSELL, P. PJURA & R. DICKERSON. 1985. Proc. Natl. Acad. Sci. USA 82: 1376–1380.
37. ARNOTT, S., P. CAMPBELL-SMITH & P. CHANDRESEKHARAN. 1976. CRC Handbook of Biochemistry, Vol. 2: 411–422. CRC. Boca Raton, FL.
38. DREW, H., R. WING, T. TAKANO, C. BROKA, S. TANAKA, K. ITAKURA & R. DICKERSON. 1981. Proc. Natl. Acad. Sci. USA 78: 2179–2183.
39. LYBRAND, T., S. BROWN, R. SHAFER & P. KOLLMAN. 1985. Computer modeling of actinomycin D. Interactions with double helical DNA. J. Mol. Biol. In press.
40. CREIGHTON, S., T. LYBRAND, U. SINGH, T. ANDREA & P. KOLLMAN. Unpublished material.
41. TAKUSAGAWA, F., B. GOLDSTEIN, S. YOUNGSTER, R. A. JONES & H. M. BERMAN. 1984. J. Biol. Chem. 259: 4714–4715.
42. BROWN, S. C., K. MULLIN, C. LEVENSON & R. H. SHAFER. 1984. Biochemistry 23: 403–408.
43. WIPFF, G., A. DEARING, P. WEINER, J. BLANEY & P. KOLLMAN. 1983. J. Am. Chem. Soc. 105: 997–1005.
44. WEINER, S. J., U. C. SINGH & P. A. KOLLMAN. 1985. J. Am. Chem. Soc. 107: 2219–2229.
45. WEINER, S. J., G. L. SEIBEL & P. A. KOLLMAN. 1986. Proc. Natl. Acad. Sci. USA 83: 649–653.
46. CRAIK, C. S., C. LARGMAN, T. FLETCHER, S. ROCZNIAK, P. J. BARR, R. FLETTERICK & W. J. RUTTER. 1985. Science 228: 291–297.
47. SINGH, U. C. & P. A. KOLLMAN. 1985. A combined *ab initio* and molecular mechanical method for carrying out simulations on complex molecular systems: Applications to the $CH_3Cl + Cl^-$ exchange reaction and gas phase protonation of polyethers. J. Comp. Chem. In press.
48. POSTMA, J. P. M., H. J. C. BERENDSEN & J. R. HAAK. 1982. Faraday Symp. Chem. Soc. 17: 55–67.
49. LYBRAND, T., J. A. McCAMMON & G. WIPFF. 1986. Proc. Natl. Acad. Sci. USA 83: 833–836.
50. JORGENSEN, W. 1985. Chem. Phys. 83: 3050–3054.

Salt Effects on Enzyme-Substrate Interactions by Monte Carlo Simulation[a]

R. J. BACQUET AND J. A. McCAMMON

Department of Chemistry
University of Houston
Houston, Texas 77004

It has been proposed that the long-range electrostatic field produced by an enzyme's charge distribution can be important in channeling diffusing substrate to an active site. In the case of superoxide dismutase (SOD) this steering effect competes with an overall repulsion which is due to the net negative charge on both substrate and enyzme. These opposing effects are both reduced by salt ions, which screen the electrostatic interaction of the reactants. The reaction rate of O_2^- with SOD is found experimentally[1] to decrease with increasing salt concentration, implying that impairment of the steering field is the dominant factor.

Theoretical calculation of the potential near an enzyme is complicated by the presence of small mobile salt ions which adopt an equilibrium distribution in response to the enzyme's innate field. The effect of this ionic atmosphere has been modeled only by application of Debye-Hückel screening to the enzyme-substrate interaction, which is of limited validity.[2] More sophisticated approaches, such as nonlinear Poisson-Boltzmann theory, integral equation methods and computer simulation, generally founder on the topographic and electrostatic complexity of an enzyme such as SOD. However, the salt distribution responds to the macromolecular potential over a large region of space and is not expected to be very sensitive to fine details of the enzyme's shape and charge distribution. Thus a very simple SOD model can capture the essential features of the potential in the region exterior to the protein, and the techniques mentioned above become feasible.

In the present work we apply Metropolis Monte Carlo (MC) computer simulation to a model system in which both SOD and the small ions are spheres with "soft" repulsive cores. The enzyme contains a few charges designed to closely reproduce the monopole, dipole, and quadrupole moments of the complete charge distribution. The SOD model also possesses a plane and axis of symmetry that are nearly present in the detailed structure and that facilitate the MC work. The solvent is a dielectric continuum with $\epsilon = 78.54$ everywhere including, for simplicity, the enzyme interior. MC simulation of this system is straightforward and gives the correct answer for the given model. The more analytic approaches require approximations beyond those implicit in the model, and their accuracy is best judged by comparison with a simulation.

The smoothed MC results in FIGURES 1 and 2 provide a striking picture of salt effects on the potential field of SOD. Approaches to the enzyme along an axis passing

[a]This work was supported in part by grants from the Robert A. Welch Foundation, the National Science Foundation, and the National Institutes of Health.

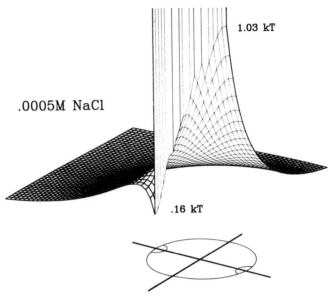

FIGURE 1. Potential energy of O_2^- for positions on a quadrant of a plane passing through the two active sites of SOD (the only unique region due to symmetry of the model). Values are given for enzyme-substrate contact (30.0 Å) at two points, the active site and midway between active sites. Mesh lines are 3.0 Å apart and are displayed out to 135.0 Å.

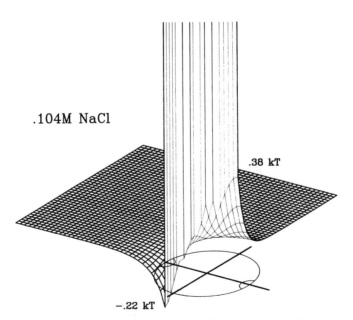

FIGURE 2. Same as FIGURE 1, but at higher salt concentration.

through the two active sites are greatly favored, but to a lesser degree, as the concentration is raised. The overall repulsive nature of the potential surface declines with increasing salt concentration. Which of these effects dominates the reaction rate is not clear from an examination of the potential fields. However, a recently developed Brownian dynamics trajectory method[3] allows determination of the rate constant from arbitrarily complex potential fields. Such calculations are under way and can be compared directly with experiment. We are currently incorporating into the MC studies a low value for the dielectric constant of the protein interior and an explicit treatment of the shape of the boundary separating the two dielectric regions.

REFERENCES

1. CUDD, A. & I. FRIDOVICH. 1982. J. Biol. Chem. **257:** 11443.
2. MATTHEW, J. B. 1985. Annu. Rev. Biophys. Biophys. Chem. **14:** 387, and references therein.
3. NORTHRUP, S. H., S. A. ALLISON & J. A. McCAMMON. 1984. J. Chem. Phys. **80:** 1517.

Energy Minimization Calculations on the Alkali Metal Cation Complexes of Valinomycin

JOSEPH N. KUSHICK AND REMO A. MASUT

Department of Chemistry
Amherst College
Amherst, Massachusetts 01002

We have carried out a series of molecular-mechanics calculations on the alkali ion complexes of valinomycin.[1,2] For the ions Na^+, K^+, Rb^+, and Cs^+ we have found three-fold rotationally symmetric conformations as the lowest-energy structures, whereas for Li^+ a markedly asymmetric configuration is preferred. The relative free energies of the complexes show that Li^+ is by far the poorest binding partner in solution, followed by Na^+, which is in turn far poorer than any of the three larger ions. The binding selectivity derives from the slower variation of the complexation free energy with ionic size than the ionic solvation free energy, so that the ionophore is unable to compete with the solvent for the smaller ions. Our calculated strain energies suggest that valinomycin's failure to form complexes with the smaller ions in solution is due partially to the rigidity of the ionophore structure, which prevents the central cavity from contracting to accommodate them. Certain geometric criteria indicate that K^+ provides the best fit to the binding site, although there is some inconsistency between the energetic and geometric criteria of binding ability.

REFERENCES

1. MASUT, R. A. & J. N. KUSHICK. 1984. J. Comput. Chem. **5:** 336–342.
2. MASUT, R. A. & J. N. KUSHICK. 1985. J. Comput. Chem. **6:** 148–155.

A Molecular Mechanics Study (AMBER) of the Displacement of Thyroxine from the Binding Pocket of Prealbumin by PCBs and PCB Analogues

T. DARDEN,[a] J. McKINNEY,[a] A. MAYNARD,[a]
S. OATLEY,[b] AND L. PEDERSEN[c,d]

[a]Laboratory of Molecular Biophysics
National Institute of Environmental Health Sciences
Research Triangle Park, North Carolina 27709

[b]Department of Chemistry
University of California, San Diego
La Jolla, California 92093

[c]Department of Chemistry
University of North Carolina
Chapel Hill, North Carolina 27514

It has recently been shown that polychlorinated biphenyls (PCBs) will quantitatively displace ^{125}I-thyroxine from its complex with prealbumin,[1] a major thyroxine transport protein. From this competition binding data it has been possible to determine equilibrium constants for PCB binding using nonlinear least-squares analysis and assuming a single binding site model. Given X-ray structures[2] for thyroxine-bound and unbound prealbumin, binding energies were found for a series of PCBs, polychlorinated dibenzodioxins, and dibenzofurans using the energy minimization program, AMBER.[3-5] Force-field parameters generated for the PCBs and analogues were based on quantum-mechanical calculations. Ortho-substituted PCBs are nonplanar, whereas nonortho-substituted PCBs become planar easily. The prealbumin residues which define the thyroxine pocket[2] were included in the calculations. The binding energy was defined as ΔE_g = E (complex) - E (protein) - E (substrate); the relative free energy of binding in solution was approximated as $\Delta(\Delta G_s = \Delta(\Delta E_g) - \Delta(\Delta G^{g->s})$). The free energies of solvation for the substrates, $\Delta G^{g->s}$, were found using MODEL 1.3,[e] which relates solvation free energies to the accessible polar surface area (APSA) of a substrate by $\Delta G^{g->s} = -.075 \times$ (APSA). This relation correlated well with experimental estimates.[6] Corrections for solvation leads to estimates for the differential free energies of complex formation. These are compared in TABLE 1 with the experimental binding ratios, which are related as ln [K(PCB) / K (reference)].

The theory correctly separates substrates into strong, intermediate, and weak binders. On the basis of the calculations 2,3,7,8-tetrachloro-dibenzo-p-dioxin and

---

[d]To whom correspondence may be addressed.
[e]Given to us by C. Still.

TABLE 1. Relative Binding Constants and Differential Free Energies for PCBs and Analogues

Compound	$\Delta(\Delta G_s)$	K (PCB)/K (Ref. 1)
1 2-hydroxy-3,5-dichlorobiphenyl	0.0	1.0
2 4-hydroxy-3,5-dichlorobiphenyl	−19.2	45.0
3 4'-hydroxy-2,3,4,5-tetrachlorobiphenyl	−5.0	24.3
4 4-hydroxy-3,5,4'-trichlorobiphenyl	−4.9	27.5
5 4,4'-dihydroxy-3,3',5,5'-tetrachlorobiphenyl	−25.2	45.5
6 3,3',4,4',5,5'-hexachlorobiphenyl	−10.1	27.7
7 2,2',6,6'-tetrachlorobiphenyl	6.4	a
8 4'-hydroxy-2,4,6-trichlorobiphenyl	2.3	2.9
9 biphenyl	8.6	a
10 3,3',5,5'-tetrachlorodiphenoquinone	2.2	2.9
11 2,4,6-triiodophenol	−6.9	33.8
12 2,3,7,8-tetrachlorodibenzo-p-dioxin	−10.9	—
13 octachlorodibenzo-p-dioxin	20.8	—
14 2,3,7,8-tetrachlorodibenzofuran	−26.8	—

NOTE: ΔG_s (kcal/mol) and K (PCB)/K (reference) were referenced to results for 2-hydroxy-3,5-dichlorobiphenyl.
[a] No competitive binding detected relative to thyroxine.

2,3,7,8-tetrachloro-dibenzofuran are predicted to be strong binders, 3,3',5,5'-tetra-chloro-diphenoquinone a weak binder, and octachlorodibenzo-p-dioxin is predicted to not bind competitively. The theoretical model developed for prealbumin interactions may be of use in estimating the toxicity properties of PCBs and related halogenated aromatic hydrocarbons of environmental importance.

REFERENCES

1. RICKENBACKER, U., J. D. MCKINNEY, S. J. OATLEY & C. C. F. BLAKE. 1986. A structurally specific binding of halogenated biphenyls to thyroxine transport protein. J. Med. Chem. **29**: 641–648.
2. OATLEY, S. J., J. M. BURRIDGE, & C. C. F. BLAKE. 1982. Hormone Antagonists. M. K. Agarwal, Ed.,: 705–715. Walter de Gruyter. Berlin.
3. BLANEY, J. M., P. K. WEINER, A. DEARING, P. A. KOLLMAN, E. C. JORGENSEN, S. J. OATLEY, J. M. BURRIDGE, & C. C. F. BLAKE. 1982. Molecular mechanics simulation of protein-ligand interactions: Binding of thyroid hormone analogues to prealbumin. J. Am. Chem. Soc. **104**: 6424–6434.
4. WEINER, P. K., & P. A. KOLLMAN. 1981. AMBER: Assisted model building with energy refinement. A general program for modeling molecules and their interactions. J. Comp. Chem. **2**(3): 287–303.
5. WEINER, S. J., P. A. KOLLMAN, D. A. CASE, U. C. SINGH, C. GHIO, G. ALAGONA, S. PROFETA & P. WEINER. 1984. A new force field for molecular mechanical simulation of nucleic acids and proteins. J. Am. Chem. Soc. **106**: 765–784.
6. WOLFENDEN, R., L. ANDERSON, P. M. CULLIS & C. C. B. SOUTHGATE. 1981. Affinities of amino acid side chains for solvent water. Biochemistry. **20**: 849–855.

Theoretical Models of Spermine/DNA Interactions[a]

BURT G. FEUERSTEIN,[b] NAGARAJAN PATTABIRAMAN,[d]
AND LAURENCE J. MARTON[b,c]

[b]*Brain Tumor Research Center*
Department of Neurological Surgery
[c]*Department of Laboratory Medicine*
School of Medicine
[d]*Department of Pharmaceutical Chemistry*
School of Pharmacy
University of California at San Francisco
San Francisco, California 94143

In order to study the interaction of polyamines and DNA we carried out conformational energy calculations on spermine and molecular-mechanics calculations on the spermine/DNA complex. From calculating spermine's relative conformational energy, we find that its central diaminobutane moiety is maintained in *trans* configuration. By searching for the distance defined by the central diaminobutane as a fixed distance between proton acceptors in B DNA, we found the N7 positions of purines in the major groove of alternating purine/pyrimidine sites to be likely sites of spermine binding. We were encouraged in this selection by the data of Drew and Dickerson[1] showing binding of spermine to alternating purine/pyrimidine sequences across the major groove in B DNA. We decided to compare this model to the model of Liquori *et al.*,[2] in which spermine bridges the minor groove. We constructed d(G-C)$_5$, d(G-C)$_5$ and d(A-T)$_5$, d(A-T)$_5$ from the coordinates of Arnott *et al.* for B DNA,[3] and docked spermine to DNA with the MIDAS program.[4] Molecular-mechanics calculations were carried out on the polymers both complexed and uncomplexed with spermine by means of the AMBER program.[5,6]

TABLE 1 presents total energies, intramolecular energies, and interaction energies of spermine and the two oligomers after our molecular-mechanics calculations in both complexed and uncomplexed structures. The large decrease in total energy brought about by causing spermine to interact with DNA can be seen by comparing the results of calculations with complexed and uncomplexed models. The comparison with the model of Liquori[2] in TABLE 1 shows it to be significantly less stable than the major groove model—the total energy of the minor groove model is more than 70 kcal/mol less stable than its major groove counterpart. The major groove models thus appear to be far more likely than this minor groove model. The stabilization of the major groove model is achieved by maximizing interactions between proton acceptors in DNA and proton donors in spermine. Each amine in spermine interacts with two proton acceptors in DNA. These interactions are made possible by bending the major groove over spermine, embracing it, and consequently opening the minor groove (FIG. 1). The bend

[a]This work was supported in part by Grant CA 13525 from the National Institutes of Health.

TABLE 1. Calculated Energies for Oligomers Alone and Spermine-Oligomer Complexes[a]

	Spermine-d(G-C)$_5$ · d(G-C)$_5$	d(G-C)$_5$ · d(G-C)$_5$	d(A-T)$_5$ · d(A-T)$_5$	Spermine-d(A-T)$_5$ · d(A-T)$_5$	Liquori Spermine-d(A-T)$_5$ · d(A-T)$_5$
Total energy	−818.7	−1170.0	−994.8	−1340.6	−1266.2
Intramolecular energy of DNA	−818.7	−737.0	−994.8	−922.5	−971.3
Interaction energy of spermine	—	−516.7	—	−494.4	−374.2
Intramolecular energy of spermine	—	+83.6	—	+76.3	+74.3

[a]Energies are in kcal/mol.

takes place over a space of several residues, with little disruption of base stacking. Electrostatic potential of the pocket embracing spermine becomes more electronegative because of the closer approaches of phosphate groups, allowing a better electrostatic fit of the electropositive spermine. To accommodate the bend, sugar puckering is altered from C3' endo to C2' endo. Further examination of the interaction reveals several different patterns of spermine binding to both bases and sugar/phosphate groups.

We conclude that spermine prefers certain positions over others in these oligomers. Its interaction with DNA also provides a large potential force for the modification of DNA conformation.

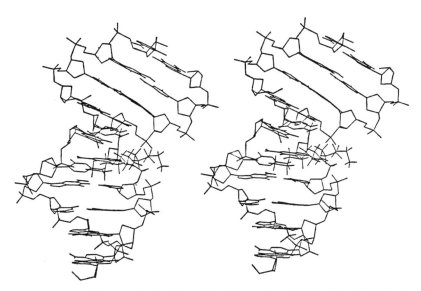

FIGURE 1. Stereodiagram showing interaction of spermine with $d(G-C)_5 \cdot d(G-C)_5$ after energy minimization. Note the major groove bending over spermine, becoming smaller, while the minor groove opens widely.

ACKNOWLEDGMENTS

We wish to thank U. Chandra Singh and Peter Kollman for helpful discussions, and Robert Langridge and the UCSF Computer Graphics Laboratory for the use of MIDAS and the computing facilities (NIH Grant RR 1081).

REFERENCES

1. DREW, H. R. & R. E. DICKERSON. 1981. J. Mol. Biol. **151:** 535–556.
2. LIQUORI, A. M., L. CONSTANTINO, V. CRESCENZI, V. ELIA, E. GIGLIO, R. PULITI, S. M. DeSANTI-SAVINO & V. VITIGLIANO. 1967. J. Mol. Biol. **24:** 113–122.

3. ARNOTT, S. & D. W. L. HUKINS. 1972. Biochem. Biophys. Res. Commun. **47:** 1504–1509.
4. LANGRIDGE, R., T. FERRIN, I. KUNTZ & M. L. CONNOLLY. 1981. Science **211:** 661–666.
5. SINGH, U. C. & P. A. KOLLMAN. 1984. J. Comp. Chem. **5:** 129–145.
6. WEINER, S. J., P. A. KOLLMAN, D. CASE, U. C. SINGH, C. GHIO, G. ALAGONA & P. K. WEINER. 1984. J. Am. Chem. Soc. **106:** 765–784.

Molecular Dynamics: Applications to Proteins

MARTIN KARPLUS

Department of Chemistry
Harvard University
Cambridge, Massachusetts 02138

A paper published in 1977 on the simulation of a small protein, the bovine pancreatic trypsin inhibition, described the first molecular dynamics study of macromolecules of biological interest.[1] Although the trypsin inhibitor is rather uninteresting from a dynamic viewpoint (its function is to bind to trypsin) experimental and theoretical studies of this model system—the "hydrogen atom" of protein dynamics—served to initiate explorations in this field.

Since dynamic studies of biomolecules are now being performed by many workers it is useful to set the field in perspective relative to the more general development of molecular dynamics. Molecular dynamics has followed two pathways which come together in the study of biomolecular dynamics. One of these, usually referred to as trajectory calculations, has been concerned primarily with the study of gas-phase scattering of atoms and molecules. Much has been done in applying the classical trajectory method to a wide range of chemical reactions.[2,3] The classical studies have been supplemented by semiclassical and quantum-mechanical calculations since quantum effects sometimes play an important role.[3,4] Now the focus of trajectory studies is on more complex molecules, their redistribution of internal energy, and the role of this on their reactivity.

The other pathway in molecular dynamics has been concerned with physical rather than chemical interactions and the thermodynamic and average dynamic properties of large numbers of particles at equilibrium, rather than detailed trajectories of a few particles. Although the basic ideas go back to van der Waals and Boltzmann, the modern area began with the work of Alder and Wainright on hard-sphere liquids in the late 1950s.[5] The paper by Rahman[6] in 1964, on a molecular dynamics simulation of liquid argon with a soft-sphere (Lennard-Jones) potential, represented an important next step. Simulations of complex liquids followed; the now classic study of liquid water by Stillinger and Rahman was published in 1974.[7] Since then, there have been many studies on the equilibrium and nonequilibrium behavior of a wide range of systems.[8,9]

This background set the stage for the development of the study of molecular dynamics of biomolecules. The size of an individual molecule, composed of five hundred or more atoms for even a small protein, is such that its simulation in isolation can determine approximate equilibrium properties, as in the molecular dynamics of fluids, although detailed aspects of the atomic motions are of considerable interest, as in trajectory calculations. A basic assumption in initiating such studies was that potential functions could be constructed that were sufficiently accurate to give meaningful results for systems as complex as proteins or nucleic acids. In addition, it was necessary to assume that for such systems of inhomogeneous character, in contrast to the homogeneous character of even "complex" liquids like water, simulations of an attainable time scale (10 psec in the initial studies) could provide a useful sample of the phase space in the neighborhood of the native structure. For neither of these assumptions was there strong supporting evidence. Nevertheless, it seemed worthwhile in 1975 to apply the techniques of molecular dynamics with the available potential functions to the internal atomic motions of proteins with known crystal structures.[1,10]

The most important consequence of the first simulations of biomolecules was that they introduced a conceptual change. Although to chemists and physicists it is self-evident that polymers like proteins and nucleic acids undergo significant fluctuations at room temperature, the classic view of such molecules in their native state had been static in character. This followed from the dominant role of high-resolution X-ray crystallography in providing structural information for these complex systems. The remarkable detail evident in crystal structures led to an image of biomolecules with every atom fixed in place. D. C. Phillips, who determined the first enzyme crystal structure, wrote recently that "the period 1965–75 may be described as the decade of the rigid macromolecule. Brass models of DNA and a variety of proteins dominated the scene and much of the thinking."[11] Molecular-dynamics simulations have been instrumental in changing the static view of the structure of biomolecules to a dynamic picture. It is now recognized that the atoms of which biopolymers are composed are in a state of constant motion at ordinary temperatures. The X-ray structure of a protein provides the average atomic positions, but the atoms exhibit fluid-like motions of sizable amplitudes about these averages. Crystallographers have acceded to this viewpoint and have come so far as to sometimes emphasize the parts of a molecule they do not see in a crystal structure as evidence of motion or disorder.[12] The new understanding of protein dynamics subsumes the static picture in that use of the average positions still allows discussion of many aspects of biomolecule function in the language of structural chemistry. However, the recognition of the importance of fluctuations opens the way for more sophisticated and accurate interpretations.

Simulation studies in this area, as in others, have the possibility of providing the ultimate detail concerning motional phenomena. The primary limitation of simulation methods is that they are approximate. It is here that experiment plays an essential role in validating the simulation methods; that is, comparisons with experimental data can serve to test the accuracy of the calculated results and to provide criteria for improving the methodology. When experimental comparisons indicate that the simulations are meaningful, their capacity for providing detailed results often makes it possible to examine specific aspects of the atomic motions far more easily than by doing measurements.

At the present stage of the molecular dynamics of biomolecules, there is a general understanding of the subnanosecond motions; that is, the types of motion that occur have been demonstrated, their characteristics evaluated, and the important factors determining their properties delineated. Simulation methods have shown that the structural fluctuations are sizable; particularly large fluctuations are found where steric constraints due to molecular packing are small (e.g., in exposed side-chains and external loops), but substantial mobility is also found in the interior of a macromolecule. Local atomic displacements in the interior are correlated in a manner that tends to minimize disturbances of the global structure. This leads to fluctuations larger than would be permitted in a rigid protein matrix.

For motions on a longer timescale, our understanding is more limited. When the motion of interest can be described in terms of a reaction path (e.g., hinge-bending, local activated events), methods exist for examining the nature and the rate of the process. However, for the motions that are slow due to their complexity and involve large-scale structural changes, extensions of the available approaches are required. Harmonic and simplified model dynamics, as well as reaction-path calculations, can provide information on slower motions, such as opening fluctuations and helix-coil transitions.

In applying molecular dynamics to physical studies of biomolecules a number of aspects are important. There is, of course, the direct comparison between the results of calculations and the experimental data. More interesting, however, is the possibility of extending the interpretation of experiments. Also, experimental data can be generated by simulations and analyzed as would real data to test the method used. Finally, new effects can be predicted from the simulation as a stimulus for additional experimental investigations.

In what follows, applications of molecular dynamics that illustrate each of these points are outlined.

A TEST: X-RAY DIFFRACTION

Since atomic fluctuations are the basis of protein dynamics, it is important to have experimental tests of the accuracy of the simulation results concerning them. For the magnitudes of the motions, the most detailed data are provided, in principle, by an analysis of the Debye-Waller or temperature factors obtained in crystallographic refinements of X-ray structures.

It is well known from small-molecule crystallography that the effects of thermal motion must be included in the interpretation of the X-ray data to obtain accurate structural results. Detailed models have been introduced to take account of anisotropic and anharmonic motions of the atoms and these models have been applied to high-resolution data for small molecules.[13] In protein crystallography, the limited data available relative to the large number of parameters that have to be determined, have made it necessary to assume that the atomic motions are isotropic and harmonic. In that case the structure factor, $F(Q)$, which is related to the measured intensity by $I(Q) = |F(Q)|^2$, is given by

$$F(Q) = \sum_{j=1}^{N} f_j(Q) \, e^{iQ \cdot \langle r_j \rangle} \, e^{W_j(Q)} \tag{1}$$

where Q is the scattering vector, $\langle r_j \rangle$ is the average position of atom j with atomic scattering factor $f_j(Q)$ and the sum is over the N atoms in the asymmetric unit of the crystal. The Debye-Waller factor, $W_j(Q)$, is defined by

$$W_j(Q) = -\frac{8}{3} \pi^2 \langle \Delta r_j^2 \rangle s^2 = -B_j s^2 \tag{2}$$

where $s = |Q|/4\pi$. The quantity B_j is usually referred to as the temperature factor, which is directly related to the mean-square atomic fluctuations in the isotropic harmonic model. More generally, if the motion is harmonic but anisotropic, a set of six parameters

$$B_j^{xx} = \langle \Delta x_j^2 \rangle, \, B_j^{xy} = \langle \Delta x_j \Delta y_j \rangle, \, \ldots \, B_j^z = \langle \Delta z_j^2 \rangle)$$

is required to fully characterize the atomic motion. Although in the earlier X-ray studies of proteins, the significance of the temperature factors was ignored (presumably because the data were not at a sufficient level of resolution and accuracy), more recently attempts have been made to relate the observed temperature factors to the atomic motions.

In principle, the temperature factors provide a very detailed measure of these motions because information is available for the mean-square fluctuation of every heavy atom. In practice, there are two types of difficulties in relating the B factors obtained from protein refinements to the atomic motions. The first is that, in addition to thermal fluctuations, any static (lattice) disorder in the crystal contributes to the B factors; i.e., since a crystal is made up of many unit cells, different molecular geometries in the various cells have the same effect on the averge electron density, and therefore the B factor, as atomic motions. In only one case, the iron atom of myoglobin, has there been an experimental attempt to determine the disorder contribution.[14] Since the Mossbauer effect is not altered by static disorder (i.e., each nucleus absorbs independently), but does depend on atomic motions, comparisons of Mossbauer and X-ray data have been used to estimate a disorder contribution for the iron atom; the value obtained is

$$\langle \Delta r_{Fe}^2 \rangle = 0.08 \, \text{Å}^2$$

Although the value is only approximate, it nevertheless indicates that the observed B factors (e.g., on the order of 0.44 Å^2 for backbone atoms and 0.50 Å^2 for side-chain atoms) are dominated by the motional contribution. Most experimental B factor values are compared directly with the molecular dynamics results (i.e., neglecting the disorder contribution) or are rescaled by a constant amount (e.g., by setting the smallest observed B factor to zero) on the assumption that the disorder contribution is the same for all atoms.[15] Second, since simulations have shown that the atomic fluctuations are highly anisotropic and, in some cases, anharmonic, it is important to determine the errors introduced into the refinement process by the assumption of isotropic and harmonic motion. A direct experimental estimate of the errors is difficult because sufficient data are not yet available for protein crystals. Moreover, any data set includes other errors which would obscure the analysis. As an alternative to an experimental analysis of the errors in the refinement of proteins, a purely theoretical approach can be used.[16] The basic idea is to generate X-ray data from a molecular dynamics simulation of a protein and to use these data in a standard refinement procedure. The error in the analysis can then be determined by comparing the refined X-ray structure and temperature factors with the average structure and the mean-square fluctuations from the simulation. Such a comparison, in which no real experimental results are used, avoids problems due to inaccuracies in the measured data (exact calculated intensities are used), to crystal disorder (there is none in the model), and to approximations in the simulation (the simulation is exact for this case). The only question about such a comparison is whether the atomic motions found in the simulation are a meaningful representation of those occurring in proteins. As has been shown,[2,15,17] molecular dynamics simulations provide a reasonable picture of the motions in spite of errors in the potentials, the neglect of the crystal environment, and the finite-time classical trajectories used to obtain the results. However, these inaccuracies do not affect the exactitude of the computer "experiment" for testing the refinement procedure that is described below.

In this study,[16] a 25-psec molecular dynamics trajectory for myoglobin was used.[18] The average structure and the mean-square fluctuations from that structure were calculated directly from the trajectory. To obtain the average electron density, appropriate atomic electron distributions were assigned to the individual atoms and the results for each coordinate set were averaged over the trajectory. Given the symmetry, unit cell dimensions, and position of the myoglobin molecule in the unit cell, average structure factors, $\langle F(Q) \rangle$, and intensities, $I(Q) = |\langle F(Q) \rangle|^2$, were calculated from the Fourier transform of the average electron density, $\langle \rho(r) \rangle$, as a function of position r in the unit cell. Data were generated at 1.5 Å resolution, since this is comparable to the resolution of the best X-ray data currently available for proteins the size of myoglobin.[19,20] The resulting intensities at Bragg reciprocal lattice points were used as input data for the widely applied crystallographic program, PROLSQ.[21] The time-averaged atomic positions obtained from the simulation and a uniform temperature factor provide the initial model for refinement. The positions and an isotropic, harmonic temperature factor for each atom were then refined iteratively against the computer-generated intensities in the standard way. Differences between the refined results for the average atomic positions and their mean-square fluctuations and those obtained from the molecular dynamics trajectory are due to errors introduced by the refinement procedure.

The overall rms error in atomic positions ranged from 0.24 Å to 0.29 Å for slightly different restrained and unrestrained refinement procedures.[16] The errors in backbone positions (0.10–0.20 Å) are generally less than those for side-chain atoms (0.28–0.33 Å); the largest positional errors are on the order of 0.6 Å. The backbone errors, although small, are comparable to the rms deviation of 0.21 Å between the positions of the backbone atoms in the refined experimental structures of oxymyoglobin and carboxy myoglobin.[19,20] Further, the positional errors are not uniform over the whole structure. There is a strong correlation between the positional error and the magnitude of the mean-square fluctuation for an atom, with certain regions of the protein, such as loops and external side-chains, having the largest errors.

The refined mean-square fluctuations are systematically smaller than the fluctuations calculated directly from the simulation. The magnitudes and variation of temperature factors along the backbone are relatively well reproduced, but the refined side-chain fluctuations are almost always significantly smaller than the actual values. The average backbone B factors from different refinements are in the range 11.3 to 11.7 Å^2, as compared with the exact value of 12.4 Å^2; for the side-chains, the refinements yield 16.5 to 17.6 Å^2, relative to the exact value of 26.8 Å^2. Regions of the protein that have high mobility have large errors in temperature factors as well as in positions. Examination of all atoms shows that fluctuations greater than about 0.75 Å^2 (B = 20 Å^2) are almost always underestimated by the refinement. Moreover, while actual mean-square atomic fluctuations have values as large as 5 Å^2, the X-ray refinement leads to an effective upper limit of about 2 Å^2. This arises from the fact that most of the atoms with large fluctuations have multiple conformations and that the refinement procedure picks out one of them.

To do refinements that take some account of anisotropic motions for all but the smallest proteins, it has been necessary to introduce assumptions concerning the nature of the anisotropy. One possibility is to assume anisotropic rigid body motions for side-chains such as tryptophan and phenylalanine.[22,23] An alternative is to introduce a "dictionary" in which the orientation of the anisotropy tensor is related to the stereochemistry around each atom[21]; this reduces the six independent parameters of the anisotropic temperature factor tensor B_j to three parameters per atom. An analysis of a simulation for BPTI[24] has shown that the actual anisotropies in the atomic motions are generally not simply related to the local stereochemistry; an exception is the main-chain carbonyl oxygen, which has its largest motion perpendicular to the $C = O$ bond. Thus, use of stereochemical assumptions in the refinement can yield incorrectly oriented anisotropy tensors and significantly reduced values for the anisotropies. The large-scale motions of atoms are collective and side-chains tend to move as a unit so that the directions of largest motion are not related to the local bond direction and have similar orientations in the different atoms forming a group that is undergoing correlated motions. Consequently, it is necessary to use the full anisotropy tensor to obtain meaningful results. This is possible with proteins that are particularly well ordered, so that the diffraction data extend to better than 1-Å resolution.

AN EXTENSION: NUCLEAR MAGNETIC RESONANCE

Nuclear magnetic resonance (NMR) is an experimental technique that has played an essential role in the analysis of the internal motions of proteins.[10,25] Like X-ray diffraction, it can provide information about individual atoms; unlike X-ray diffraction, NMR is sensitive not only to the magnitude, but also to the timescales of the motions. Nuclear relaxation processes are dependent on atomic motions on the nanosecond to picosecond timescale. Although molecular tumbling is generally the dominant relaxation mechanism for proteins in solution, internal motions contribute as well; for solids, the internal motions are of primary importance. In addition, NMR parameters, such as nuclear spin-spin coupling constants and chemical shifts, depend on the protein environment. In many cases different local conformations exist, but the interconversion is rapid on the NMR timescale, here on the order of milliseconds, so that average values are observed. When the interconversion time is on the order of the NMR timescale or slower, the transition rates can be studied; an example is provided by the reorientation ("flipping") of aromatic rings.[26,27]

In addition to supplying data on the dynamics of proteins, NMR can also be used to obtain structural information. Recent advances in techniques now make it possible to obtain a large number of approximate interproton distances for proteins by the use of nuclear Overhauser effect (NOE) measurements.[28] If the protein is relatively small and has a well-resolved spectrum, a large portion of the protons can be assigned and several hundred distances for these protons can be determined by the use of two-dimensional NMR techniques.[29] Clearly these distances can serve to provide structural information for proteins, analogous to their earlier use for organic molecules.[28,30] Of great interest is the possibility that enough distance information can be measured to actually determine the high-resolution structure of a protein in solution, particularly for proteins that are difficult to crystallize. In what follows we consider two questions related to this possibility. The first concerns the effect of motional averaging on the accuracy of the apparent distances obtained from the NOE studies and the second, whether the number of distances that can be obtained experimentally are sufficient for a structure determination.

For spin-lattice relaxation, such as observed in nuclear Overhauser effect measurements, it is possible to express the behavior of the magnetization of the nuclei being studied by the equation[31,32]

$$\frac{d(I_z(t) - I_o)_i}{dt} = -\rho_i(I_z(t) - I_o)_i - \sum_{i \neq j} \sigma_{ij}(I_z(t) - I_o)_j \tag{3}$$

where $I_z(t)_i$ and I_{oi} are the z components of the magnetization of nucleus i at time t and at equilibrium, ρ_i is the direct relaxation rate of nucleus i, and σ_{ij} is the cross relaxation rate between nuclei i and j. The quantities ρ_i and σ_{ij} can be expressed in terms of spectral densities

$$\rho_i = \frac{6\pi}{5} \gamma_i^2 \gamma_j^2 \hbar^2 \sum_{i \neq j} [\frac{1}{3}J_{ij}(\omega_i - \omega_j) + J_{ij}(\omega_i) + 2J_{ij}(\omega_i + \omega_j)] \tag{4}$$

$$\sigma_{ij} = \frac{6\pi}{5} \gamma_i^2 \gamma_j^2 \hbar^2 [2J_{ij}(\omega_i + \omega_j) - \frac{1}{3}J_{ij}(\omega_i - \omega_j)] \tag{5}$$

where ω_i is the resonance frequency of nucleus i. The spectral density functions can be obtained from the correlation functions for the relative motions of the nuclei with spins i and j,[31,33]

$$J_{ij}^n(\omega) = \int_0^\infty \frac{\langle Y_n^2(\Theta_{lab}(t)\phi_{lab}(t))Y_n^{2*}(\Theta_{lab}(0)\phi_{lab}(0))\rangle}{r_{ij}^3(0)r_{ij}^3(t)} \cos(\omega t) \, dt \tag{6}$$

where $Y_n^2(\Theta(t)\phi(t))$ are second-order spherical harmonics and the angular brackets represent an ensemble average which is approximated by an integral over the molecular dynamics trajectory. The quantities $\Theta_{lab}(t)$ and $\phi_{lab}(t)$ are the polar angles at time t of the internuclear vector between protons i and j with respect to the external magnetic field and r_{ij} is the interproton distance. In the simplest case of a rigid molecule undergoing isotropic tumbling with a correlation time τ_o this reduces to the familiar expression

$$J_{ij}(\omega) = \frac{1}{4\pi r_{ij}^6}\left[\frac{\tau_o}{1 + (\omega\tau_o)^2}\right] \tag{7}$$

The nuclear Overhauser effect corresponds to the selective enhancement of a given resonance by the irradiation of another resonance in a dipolar coupled spin system. Of particular interest for obtaining motional and distance information are measurements that provide time-dependent NOEs from which the cross relaxation rates σ_{ij} (see Eq. 5) can be determined directly or indirectly by solving a set of coupled equations (Eqs. 3–5). Motions on the picosecond timescale are expected to introduce averaging effects that decrease the cross-relaxation rates by a scale factor relative to the rigid model. A lysozyme molecular dynamics simulation[34] has been used to calculate dipole vector correlation functions[31] for proton pairs that have been studied experimentally.[35,36] Four proton pairs on three side-chains (Trp 28, Ile 98, and Met 105) with very different motional properties were examined. Trp 28 is quite rigid, Ile 98 has significant fluctuations, and Met 105 is particularly mobile in that it jumps among different side-chain conformations during the simulation. The rank order of the scale factors (order parameters) is the same in the theoretical and experimental results. However, although the results for the Trp 28 protons agree with the measurements to within the experimental error, for both Ile 98 and Met 105 the motional averaging found from the NOEs is significantly greater than the calculated value. This suggests that these residues are undergoing rare fluctuations involving transitions that are not adequately sampled by the simulation.

If nuclear Overhauser effects are measured between pairs of protons whose distance is not fixed by the structure of a residue, the strong distance-dependence of the cross-relaxation rates $(1/r^6)$ can be used to obtain estimates of the interproton distances.[29,35-37] The simplest application of this approach is to assume that proteins are rigid and tumble isotropically. The lysozyme molecular dynamics simulation was used to determine whether picosecond fluctuations are likely to introduce important errors into such an analysis.[31] The results show that the presence of the motions will cause a general decrease in most NOE effects observed in a protein. However, because the distance depends on the sixth root of the observed NOE, motional errors of factor of two in the latter lead to only a 12% uncertainty in the distance. Thus, the decrease is usually too small to produce a significant change in the distance estimated from the measured NOE value. This is consistent with the excellent correlation found between experimental NOE values and those calculated using distances from a crystal structure.[36] Specific NOEs can, however, be altered by the internal motions to such a degree that the effective distances obtained are considerably different from those predicted for a static structure. Such possibilities must, therefore, be considered in any structure determination based on NOE data. This is true particularly for cases involving averaging over large fluctuations.

Because of the inverse sixth power of the NOE distance-dependence, experimental data so far are limited to protons that are separated by less than 5 Å. Thus, the long-range information required for a direct protein structure determination is not available. To overcome this limitation it is possible to introduce additional information provided by empirical energy functions.[38] One way of proceeding is to do molecular dynamics simulations with the approximate interproton distances introduced as restraints in the form of skewed biharmonic potentials[37,39] and the force constants chosen to correspond to the experimental uncertainties in the distance.

A model study of the small protein crambin[39] was made with realistic NOE restraints. Two hundred forty approximate interproton distances less than 4 Å were used, including 184 short-range distances (i.e., connecting protons in two residues that were less than five residues apart in the sequence) and 56 long-range distances. The molecular dynamics simulations converged to the known crambin structure from different initial extended structures. The average structure obtained from the simulations with a series of different protocols had rms deviations of 1.3 Å for the backbone atoms, and 1.9 Å for the side-chain atoms. Individual converged simulations had rms deviations in the range 1.5 to 2.1 Å and 2.1 to 2.8 Å for the backbone and side-chain atoms, respectively. Further, it was shown that a dynamics structure with significantly larger deviations (5.7 Å) could be characterized as incorrect, independent of a knowledge of the crystal structure because of its higher energy and the fact that the NOE restraints were not satisfied within the limits of error. The incorrect structure resulted when all NOE restraints were introduced simultaneously, rather than allowing the dynamics to proceed first in the presence of only the short-range restraints followed by introduction of the long-range restraints. Although crambin has three disulfide bridges it was not necessary to introduce information concerning them to obtain an accurate structure.

The folding process as simulated by the restrained dynamics is very rapid. At the end of the first 2 psec the secondary structure is essentially established while the molecule is still in an extended conformation. Some tertiary folding occurs even in the absence of long-range restraints. When they are introduced, it takes about 5 psec to obtain a tertiary structure that is approximately correct and another 6 psec to introduce the small adjustments required to converge to the final structure. It is of interest to consider whether the results obtained in the restrained dynamics simulation have any relation to actual protein folding. That correctly folded structures are achieved only when the secondary structural elements are at least partly formed before the tertiary restraints are introduced is suggestive of the diffusion-collision model of protein folding.[40] Clearly, the specific pathway has no physical meaning since it is dominated by the NOE restraints. Also, the time scale of the simulated folding process is 12 orders of magnitude faster than experimental estimates. About 6 to 8 orders of magnitude of the rate increase are due to the fact that the secondary structure is stable once it is formed, in contrast to a real protein, where the secondary structural elements spend only a small fraction of time in the native conformation until coalescence has occurred. The remainder of the artificial rate increase presumably arises from the fact that the protein follows a single direct path to the folded state in the presence of the NOE restraints, instead of having to go through a complex search process.

A PREDICTION: STRUCTURAL ROLE OF ACTIVE-SITE WATERS IN RIBONUCLEASE A

To achieve a realistic treatment of solvent-accessible active sites, a molecular dynamics simulation method, called the stochastic boundary method, has been implemented.[41-43] It makes possible the simulation of a localized region, approximately spherical in shape, that is composed of the active site with or without ligands, the essential portions of the protein in the neighborhood of the active site, and the surrounding solvent. The approach provides a simple and convenient method for reducing the total number of atoms included in the simulation, while avoiding spurious edge effects.

The stochastic boundary method for solvated proteins starts with a known X-ray structure; for the present problem the refined high-resolution (1.5 to 2 Å) X-ray structures provided by G. Petsko and coworkers was used.[44,45] The region of interest (here the active site of ribonuclease A) was defined by choosing a reference point (which was taken at the position of the phosphorus atom in the CpA inhibitor complex) and constructing a sphere of 12 Å radius around this point. Space within the sphere not occupied by crystallographically determined atoms was filled by water molecules, introduced from an equilibrated sample of liquid water. The 12-Å sphere was further subdivided into a reaction region (10 Å radius) treated by full molecular dynamics and a buffer region (the volume between 10 and 12 Å) treated by Langevin dynamics, in which Newton's equations of motion for the nonhydrogen atoms are augmented by a frictional term and a random-force term; these additional terms approximate the effects of the neglected parts of the system and permit energy transfer in and out of the reaction region. Water molecules diffuse freely between the reaction and buffer regions, but are prevented from escaping by an average boundary force.[42] The protein atoms in the buffer region are constrained by harmonic forces derived from crystallographic temperature factors.[43] The forces on the atoms and their dynamics were calculated with the CHARMM program[38]; the water molecules were represented by the ST2 model.[46]

One of the striking aspects of the active site of ribonuclease is the presence of a large number of positively charged groups, some of which may be involved in guiding and/or binding the subtrate.[47] The simulation demonstrated that these residues are stabilized in the absence of ligands by well-defined water networks. A particular example includes Lys-7, Lys-41, Lys-66, Arg-39 and the doubly protonated His-119. Bridging waters, some of which are organized into trigonal bipyramidal structures, were found to stabilize the otherwise very unfavorable configuration of near-neighbor positive groups because the interaction energy between water and the charged $C\text{-}NH_n^+$ ($n = 1$, 2, or 3) moieties is very large; e.g., at a donor-acceptor distance of 2.8 Å, the $C\text{-}NH_3^+\text{-}H_2O$ energy is -19 kcal/mol with the empirical potential used for the simulation,[38] in approximate agreement with accurate quantum-mechanical calculations[48] and gas-phase ion-molecule data.[49] The average stabilization energy of the charged groups (Lys-7, Lys-41, Lys-66, Arg-39, and His-119) and the 106 water molecules included in the simulation is -376.6 kcal/mol. This energy is calculated as the difference between the simulated system and a system composed of separate protein and bulk water. Unfavorable protein-protein charged-group interactions are balanced by favorable water-protein and water-water interactions. The average energy per molecule of pure water from an equivalent stochastic boundary simulation[42] was -9.0 kcal/mol, whereas that of the waters included in the active-site simulation was -10.2 kcal/mol; in the latter a large contribution to the energy came from the interactions between the water molecules and the protein atoms. It is such energy differences that are essential to a correct evaluation of binding equilibria and the changes introduced by site-specific mutagenesis.[50]

During the simulation, the water molecules involved in the charged-group interactions oscillated around their average positions, generally without performing exchange. On a longer time scale, it is expected that the waters would exchange and that the side-chains would undergo larger-scale displacements. This is in accord with the disorder found in the X-ray results for lysine and arginine residues (e.g., Lys-41 and Arg-39),[44,51] a fact that makes difficult a crystallographic determination of the water structure in this case. It is also of interest that Lys-7 and Lys-41 have an average separation of only 4 Å in the simulation, less than that found in the X-ray structure. That this like-charged pair can exist in such a configuration is corroborated by experiments that have shown that the two lysines can be cross-linked[52]; the structure of this compound has been reported recently[53] and is similar to that found in the native protein.

In addition to the role of water in stabilizing the charged groups that span the active site and participate in catalysis, water molecules make hydrogen bonds to protein polar groups that become involved in ligand binding. A particularly clear example is provided by the adenine-binding site in the CpA simulation. The NH_2 group of adenine acted as a donor, making hydrogen bonds to the carbonyl of Asn-67, and the ring N^{1A} of adenine acted as an acceptor for a hydrogen bond from the amide group of Glu-69. Corresponding hydrogen bonds were present in the free ribonuclease simulation, with appropriately bound water molecules replacing the substrate. These waters and those that interact with the pyrimidine-site residues Thr-45 and Ser-123 help to preserve the protein structure in the optimal arrangement for binding. Similar substrate "mimicry" has been observed in X-ray structures of lysozyme[54] and of penicillopepsin,[55] but has not yet been seen in ribonuclease.

ACKNOWLEDGEMENTS

I gratefully acknowledge the work done by many collaborators in the research described in this review. Their essential contributions are made clear by the citations in the reference list.

NOTES AND REFERENCES

1. McCAMMON, J. A., B. R. GELIN & M. KARPLUS. 1977. 267: 585.
2. PORTER, R. N. 1974. Ann. Rev. Phys. Chem. 25: 371.
3. WALKER, R. B. & J. C. LIGHT. 1980. Ann. Rev. Phys. Chem. 31: 401.
4. SCHATZ, G. C., & A. KUPPERMANN. 1980. J. Chem. Phys. 62: 2502.
5. ALDER, B. J. & T. E. WAINRIGHT. 1959. J. Chem. Phys. 31: 459.
6. RAHMAN, A. 1964. Phys. Rev. A136: 405.
7. STILLINGER, F. H. & A. RAHMAN. 1974. J. Chem. Phys. 60: 1545.
8. WOOD, W. W. & J. J. ERPENBECK. 1976. Ann. Rev. Phys. Chem. 27: 319.
9. HOOVER, W. G. 1983. Ann. Rev. Phys. Chem. 34: 103.
10. For early reviews of experimental and theoretical developments see GURD, F. R. N. & J. M. ROTHGEB. 1979. Adv. Prot. Chem. 33: 73 and KARPLUS, M. & J. A. McCAMMON. 1986. CRC Crit. Rev. Biochem. 9: 293.
11. PHILLIPS, D. C. 1981. In Biomolecular Stereodynamics. R. H. Sarma, Ed.: 497. Adenine. New York, NY.
12. MARQUART, M., J. DEISENDORFER, R. HUBER & W. PALM. 1980. J. Mol. Biol. 141: 369.
13. ZUCKER, U. H. & H. SCHULZ. 1982. Acta Crystallogr A38: 563.
14. HARTMANN, H., F. PARAK, W. STEIGEMANN, G. A. PETSKO, D. R. PONZI & H. FRAUENFELDER. 1982. Proc. Natl. Acad. Sci. USA 79: 4967.
15. PETSKO, G. A. & D. RINGE. 1984. Ann. Rev. Biophys. Bioeng. 13: 331.
16. KURIYAN, J., G. A. PETSKO, R. M. LEVY & M. KARPLUS. 1986. J. Mol. Biol. 190: 227.
17. KARPLUS, M. & J. A. McCAMMON. 1983. Ann. Rev. Biochem. 52: 263.
18. LEVY, R. M., R. P. SHERIDAN, J. W. KEEPERS, G. S. DUBEY, S. SWAMINATHAN & M. KARPLUS. 1985. Biophys. J. 48: 509.
19. KURIYAN, J., G. A. PETSKO & M. KARPLUS. To be published.
20. PHILLIPS, S. E. V. 1980. J. Mol. Biol. 142: 531.
21. KONNERT, J. H. & W. A. HENDRICKSON. 1980. Acta Cryst. A36: 344.
22. GLOVER, I., I. HANEEF, J. PITTS, S. WOOD, D. MOSS, I. TICKLE & T. BLUNDELL. 1983. Biopolymers 22: 293.
23. ARTYMIUK, P. J., C. C. F. BLAKE, D. E. P. GRACE, S. J. OATLEY, D. C. PHILLIPS & J. J. E. STERNBERG. 1979. Nature 280: 563.
24. YU, H., M. KARPLUS & W. A. HENDRICKSON. 1985. Acta Cryst. B41: 191.
25. CAMPBELL, I. D., C. M. DOBSON & R. J. P. WILLIAMS. 1978. Adv. Chem. Phys. 39: 55.
26. CAMPBELL, I. D., C. M. DOBSON, G. R. MOORE, S. J. PERKINS & R. J. P. WILLIAMS. 1976. FEBS Lett. 70: 96.

27. WAGNER, G., A. DeMARCO & K. WUTHRICH. 1976. Biophys. Struct. Mech. **2:** 139.
28. NOGGLE, J. H. & R. E. SCHIRMER. 1971. The Nuclear Overhauser Effect. Academic Press. New York, NY.
29. WAGNER, G. & K. WUTHRICH. 1982. J. Mol. Biol. **160:** 343.
30. HONIG, B., HUDSON, B. D. SYKES & M. KARPLUS. 1971. Proc. Natl. Acad. Sci. USA **68:** 1289.
31. OLEJNICZAK, E. T., C. M. DOBSON, M. KARPLUS & R. M. LEVY. 1984. J. Am. Chem. Soc. **106:** 1923.
32. SOLOMON, I. 1955. Phys. Rev. **99:** 559.
33. LEVY, R. M., M. KARPLUS & P. G. WOLYNES. 1981. J. Am. Chem. Soc. **103:** 5998.
34. ICHIYE, T., B. OLAFSON, S. SWAMINATHAN & M. KARPLUS. Biopolymers. In press.
35. OLEJNICZAK, E. T., F. M. POULSEN & D. M. DOBSON. 1981. J. Am. Chem. Soc. **103:** 6574.
36. POULSEN, F. M., J. C. HOCH & C. M. DOBSON. 1980. Biochemistry **19:** 2597.
37. CLORE, G. M., A. M. GRONENBORN, A. T. BRUNGER & M. KARPLUS. 1985. J. Mol. Biol. **186:** 435.
38. BROOKS, B. R., R. E. BRUCCOLERI, B. D. OLAFSON, D. J. STATES, S. SWAMINATHAN & M. KARPLUS. 1983. J. Comp. Chem. **4:** 187.
39. BRUNGER, A. T., G. M. CLORE, A. M. GRONENBORN & M. KARPLUS. 1986. Proc. Natl. Acad. Sci. USA **83:** 380.
40. BASHFORD, D., D. L. WEAVER & M. KARPLUS. 1984. J. Biomol. Struct. Dyn. **1:** 1243.
41. BROOKS, C. L. & M. KARPLUS. 1983. J. Chem. Phys. **79:** 6312.
42. BRUNGER, A., C. L. BROOKS & M. KARPLUS. 1984. Chem. Phys. Lett. **105:** 495.
43. BROOKS, C. L., A. BRUNGER & M. KARPLUS. 1985. Biopolymers **24:** 843.
44. GILBERT, W. A., A. L. FINK & G. A. PETSKO. Biochemistry. In press.
45. CAMPBELL, R. L. & G. A. PETSKO. Biochemistry. In press.
46. STILLINGER, F. H. & A. RAHMAN. 1974. J. Chem. Phys. **60:** 1545.
47. MATTHEW, J. B. & F. M. RICHARDS. 1982. Biochemistry **21:** 4989.
48. DESMEULES, D. J. & L. C. ALLEN. 1980. J. Chem. Phys. **72:** 4731.
49. KEBARLE, P. 1977. Annu. Rev. Phys. Chem. **28:** 445.
50. FERSHT, A. R., J.-P. SHI, J. KNILL-JONES, D. M. LOWE, A. J. WILKINSON, D. M. BLOWQ, P. BRICK, P. CARTER, M. M. Y. WAYE & G. WINTER. 1985. Nature (London) **314:** 235.
51. WLODAWER, A. 1985. *In* Biological Macromolecules and Assemblies: Vol. 2, Nucleic Acids and Interactive Proteins. F. A. Jurnak and A. McPherson, Eds.: 394. Wiley. New York, NY.
52. MARFEY, P. S., M. UZIEL & J. LITTLE. 1965. J. Biol. Chem. **240:** 3270.
53. WEBER, P. C., F. R. SALEMME, S. H. LIN, Y. KONISHI & H. A. SCHERAGA. 1985. J. Mol. Biol. **181:** 453.
54. BLAKE, C. C. F., W. C. A. PULFORD & P. J. ARTYMIUK. 1983. J. Mol. Biol. **167:** 693.
55. JAMES, M. N. G. & A. R. SIELECKI. 1983. J. Mol. Biol. **163:** 299.

A Lysozyme Molecular Dynamics Simulation

CAROL B. POST,[a] MARTIN KARPLUS,[a]
AND CHRISTOPHER DOBSON[b]

[a]*Department of Chemistry*
Harvard University
Boston, Massachusetts 02138

[b]*Oxford University*
Oxford, England

The most detailed theoretical method for studying the internal motions of proteins is by molecular dynamics simulation. Starting with the refined crystallographic coordinates,[1] we have calculated a 100-psec trajectory of lysozyme with a version of the CHARMM program[2] modified to run on a CRAY-1S supercomputer. Crystallographic waters of structural importance, as determined by hydrogen bonding, were included in the simulation.

The nature of the dynamic average structures has been examined by comparisons with the X-ray results for the atomic positions, the main-chain and side-chain dihedral angles and the hydrogen-bonding geometry.[3] There is good overall agreement between the X-ray and the MD results. Of particular structural importance is the hydrogen-bonding geometry. Distributions of the heavy-atom distance and the acceptor angle and the donor angle were calculated from the X-ray structure and from the dynamics 1-psec average structures. The X-ray geometry is generally reproduced in the simulation. For reverse-turn hydrogen bonds between residues i and $i + 3$, however, there was a general tendency for the CO—HN geometry to become more linear.

Convergence of the atomic fluctuations for the different secondary structural elements of lysozyme is shown in the figure. The fluctuations in the beta-sheet and helices B and D have reached a plateau value and it appears that configuration space has been well sampled for these elements of the protein. The fluctuations for many of the secondary structures have not converged either because of longer time motional processes or rare events. For example, the kink in the curve for the C to D loop is due to main-chain dihedral angle transitions which occur only twice during the simulation and give rise to a flip in the main chain, resulting in a concerted motion of residues 101 to 105.

The dynamic behavior of the 53 structural waters was analyzed by examining the correlation between the fluctuations of the water oxygens and the lysozyme heavy-atoms to which they are hydrogen-bonded. Comparison was made of two constraining potentials on the waters: (1) quartic constraints to the initial oxygen positions and (2) quartic constraints to the protein center of mass. The fluctuation cross-correlations between the waters and the hydrogen-bonded protein atoms were strong for trajectories calculated with either constraining potential. However, the initial position constraint gave a relationship in the fluctuation magnitude of the water and the hydrogen-bonded protein atom more similar to the crystallographic results than the center of mass constraint.

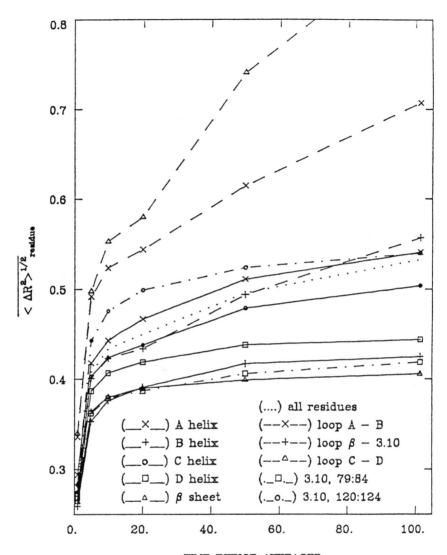

TIME PERIOD AVERAGED

FIGURE 1. Atomic fluctuations as a function of the length of the time period of averaging. For times less than the full simulation (100 psec), the results for the individual time periods were ensemble-averaged. The curves show fluctuations averaged over the main-chain atoms for the residues of the secondary structural elements indicated on the plot.

REFERENCES

1. HANDOLL, H., P. ARTYMIUK & D. C. PHILLIPS. Personal communication.
2. BROOKS, B. R., R. E. BRUCCOLERI, B. D. OLAFSON, D. J. STATES, S. SWAMINATHAN & M. KARPLUS. 1983. J. Comput. Chem **4**: 187–217.
3. POST, C. B., B. R. BROOKS, M. KARPLUS, C. M. DOBSON, P. J. ARTYMIUK, CHEETHAM & D. C. PHILLIPS. 1986. J. Mol. Biol. **190**: 455–479.

Simulations of Proteins in Water[a]

H. J. C. BERENDSEN, W. F. VAN GUNSTEREN,[b]
H. R. J. ZWINDERMAN, AND R. G. GEURTSEN

Laboratory of Physical Chemistry
University of Groningen
9747 AG Groningen, the Netherlands

INTRODUCTION

Computer simulation techniques have been shown to be a powerful tool for probing the properties of molecular liquids or solutions of flexible molecules. They give access to details of the atomic motion which are often inaccessible to direct experimental observation. Only a few proteins have been studied by computer simulation, in which various approximations have been made in order to decrease the complexity of a simulation including all degrees of freedom of both protein and solvent. Molecular dynamics (MD) simulations *in vacuo,* ignoring the effect of the solvent or crystalline environment on the dynamics of the protein, have been performed for pancreatic trypsin inhibitor (PTI),[1] cytochrome c,[2] rubredoxin,[3] lysozyme,[4] myoglobin,[5] and the C-terminal fragment of the L_7/L_{12} ribosomal protein.[6] MD simulation of protein *in solution* has just started: PTI in a Lennard-Jones solvent[7] and in water[8] and avian pancreatic polypeptide (aPP) in water.[9] A few simulation studies of *protein crystals* are known. In a Monte Carlo (MC) simulation of the unit cell of lysozyme all protein atoms were kept in fixed positions and only the water and counterions were allowed to move.[10] The next step was to allow the protein side-chain to move as well.[11] Only in two studies—one of PTI[12] and the other of aPP[13]—were both all protein atoms and all water molecules in the unit cell allowed to move.

In order to assess the validity of computer simulations of proteins the simulated properties must be compared with the available experimental data. The main source of experimental data on atomic structure and motion in proteins is X-ray diffraction of protein crystals. High-resolution data, beyond 1 Å, have recently become available for three small proteins: aPP (0.98 Å),[14] crambin (0.945 Å, Hendrickson, personal communication),[31] and PTI (0.94 Å).[15,16] A preliminary comparison of the simulated and observed atomic properties of the aPP unit cell has been reported.[13] Here we report a molecular dynamics simulation of the full unit cell of a PTI crystal (crystal form II[15,16]), involving 4 protein molecules, 552 water molecules, and 24 counterions.

The simulation reported here differs from a previous one[12] in a number of aspects: (1) Crystal form II,[15,16] rather than crystal form I,[17] is simulated. (2) The protein force field has been slightly changed with respect to van der Waals interaction between united atoms that are separated by three covalent bonds. (3) The range of the Coulombic force is extended from 0.8 nm to 1.1 nm. (4) In order to obtain a neutral

[a]This work was supported by SON, the Foundation of Chemical Research, under the auspices of the Netherlands Organization for the Advancement of Pure Research (ZWO).
[b]To whom correspondence should be addressed.

system, 24 chlorine counterions have been included in the simulation. (5) Finally, the length of 40 psec of the present simulation is twice that of the previous one.

In the next section we shall briefly describe the model and computational procedure, which we subsequently assess in the following section by comparison of the simulated protein properties with experimentally observed ones. Finally, we deal with protein hydration and water dynamics and describe the behavior of the counterions.

MODEL AND COMPUTATIONAL PROCEDURE

The protein PTI consists of 454 heavy atoms. Hydrogen atoms attached to carbon atoms are incorporated into the latter, forming united atoms, whereas the other 114 hydrogen atoms, which may form hydrogen bonds, are explicitly treated. The potential energy function describing the interaction between protein atoms, water molecules, and counterions is identical to the one used in previous studies,[6,8,9,12,13] except for an improved treatment of third-neighbor interactions. It is composed of terms representing bond-angle bending, harmonic (out-of-plane, out-of-tetrahedral configuration) dihedral bending, sinusoidal dihedral torsion, van der Waals and electrostatic (Coulombic) interactions. As in the other simulations that explicitly include solvent molecules,[8,9,12,13] atomic partial charges[18] are used without modification; a dielectric constant of 1 is applied. The van der Waals parameters of the force field have been published.[18] These are considerably larger than the ones that are generally used,[3,20–23] but reproduce the correct crystal densities of small molecules when tested by applying MD at constant pressure.[32] These van der Waals parameters also avoid the contraction of a protein *in vacuo*, as observed for insulin.[24] However, when applied to united atoms that are separated by three covalent bonds (third neighbors) they induce too large a repulsion in *gauche* conformations.

In order to avoid this effect, the (smaller) van der Waals parameters of Ref. 22 are used for united atoms that are separated by three covalent bonds in the improved force field. The nonbonded interaction should be calculated for all atom pairs that are separated by more than two covalent bonds. A cut-off radius $R_2^c = 1.1$ nm is applied, beyond which no interactions are included. This value is chosen slightly smaller than half the value of the smallest lattice parameter of the PTI(II) crystal, viz., $b = 2.34$ nm. In this manner an atom cannot interact with both another atom and its periodic image. In order to reduce computing costs, another cut-off radius $R_1^c = 0.8$ nm is applied beyond which only Coulombic forces are calculated and kept fixed during $N_i^c = 10$ MD time-steps of $\Delta t = 0.002$ psec. The nonbonded interactions within R_1^c are calculated at each MD time-step from a pair list that is updated only after every N_i^c time-steps. In this way the long-range Coulombic forces can be approximately taken into account, without increasing the computing effort, at the expense of neglecting the fluctuation of the forces beyond R_1^c during N_i^c time-steps. We note that the cut-off radii R_1^c and R_2^c are both applied to centers of geometry of neutral atom groups in the protein and to the oxygen atoms of the water molecules, in order to avoid the breaking of the charge neutrality of a group or water molecule in case an atom–atom cut-off is applied. No switching functions are used to smooth the cut-off effects.

As in the previous simulations[8,9,12,13] all bond lengths are kept rigid during the simulation by using the SHAKE method[25,26] and the water molecules are modeled by a simple rigid three-point charge (SPC) model.[27] Interactions between water, protein,

TABLE 1. Potential Energy of PTI (II) Unit Cell

			Potential Energy (kJ mol⁻¹)						
	Total	Bonded	Nonbonded						
		Internal	PTI-PTI	PTI-Cl	Cl-Cl	PTI-H_2O	Cl-H_2O	H_2O-H_2O	
4 PTI, exp.	1900	3054	-1154						
4 PTI + 576 H_2O	$3.6*10^8$	3054	-1154			$3.4*10^8$		$2.0*10^7$	
EM (100 steps)	-29781	3453	-9354			-16133		-7747	
4 PTI + 552H_2O + 24Cl	-36337	3473	-9737	-16925	4507	-15160	4580	-7074	
EM (100 steps)	-43441	3751	-9497	-20626	4340	-12730	-1554	-7125	
MD									
10 psec	-60853	6576	-14970	-20479	3785	-21142	-5042	-9585	
20 psec	-60466	6440	-15284	-20070	4086	-20314	-6005	-9359	
30 psec	-61573	6449	-15574	-20196	3963	-20811	-6033	-9370	
40 psec	-59551	6751	-13403	-19979	3558	-21423	-6175	-8886	

NOTE: The experimental structure has been deduced from joint X-ray and neutron diffraction data refinement.[15,16]

and Cl^- atoms are obtained from combination rules, using $\sqrt{C_{12}} = 10.34 \ 10^{-3}$ (kJ mol^{-1}/nm^{12})$^{1/2}$ and $\sqrt{C_6} = 117.5 \ 10^{-3}$ (kJ mol^{-1}/nm^6)$^{1/2}$ for the latter.

The initial configuration was obtained as follows (TABLE 1). The $P_{2_1 2_1 2_1}$ symmetry transformation was applied to the PTI(II) atomic coordinates[15,16] together with those of the four internal water molecules in order to obtain a unit cell (II, a = 7.41 nm, b = 2.34 nm, c = 2.89 nm) configuration. Water molecules were inserted in the crystal by immersing the whole unit cell in an equilibrium configuration of bulk SPC water and subsequently removing all water molecules that are outside the unit cell or of which the oxygen atom lies within $R^i = 0.23$ nm of a nonhydrogen protein atom. In this way 560 water molecules were inserted in the crystal, yielding a density of $\rho = 1.22$ gcm^{-3}. The high nonbonded energy of this configuration, which is due to the small value of R^i (PTI–H_2O) or to the unit cell periodicity (H_2O–H_2O), was relaxed by performing steepest-descents energy minimization (EM). At pH 8.2 a PTI molecule has a net charge of +6e, due to the presence of the following charged residues: 1 Arg, 3 Asp, 7 Glu, 15 Lys, 17 Arg, 20 Arg, 26 Lys, 39 Arg, 41 Lys, 42 Arg, 46 Lys, 49 Glu, 50 Asp, 53 Arg, and the amino- and carboxy-termini of the polypeptide chain.

In order to neutralize the unit cell, 24 negative ions had to be inserted. For reasons of simplicity we chose these to be Cl^- ions, although the PTI(II) crystal has been grown from a K_2HPO_4 solution. The Cl ions were inserted as follows: For the configuration of 4 PTI + 576 waters, the electric potential at all water oxygen positions was calculated and the water molecule at the highest potential was replaced by a Cl ion. This procedure was repeated 24 times and then followed by EM of the whole system, which allowed for a relaxation of the Cl-PTI and Cl-H_2O interactions. Thus, the system that is simulated contains 3952 atoms, involving 11856 degrees of freedom.

Initial velocities for the atoms were taken from a maxwellian distribution at 300 K, independently for each of the four molecules. In order to avoid slow temperature drift, the system was weakly coupled to a thermal bath of $T_0 = 300$ K, when integrating the equations of motion with time-step $\Delta t = 2$ fsec. This was done by applying the algorithm of Ref. 28 with temperature relaxation time $\tau = 0.1$ psec. This value makes the temperature coupling weak enough to avoid any significant effect on the atomic properties of the system.[28] Periodic boundary conditions corresponding to the crystal translational symmetry were applied.

The MD run covered a time span of 40 psec, which took about 16 central processing unit (CPU) hours on a one-pipe Cyber 205 supercomputer. FIGURE 1 shows that the potential energy drops rapidly during the first few picoseconds and stabilizes after 10 psec. The root mean square (rms) deviation of the actual atomic positions from the starting (experimental) ones shows only a very small drift after this equilibration period. Therefore, the final 30 psec of the run has been used, with a time resolution of 0.05 psec, for calculating averages of various quantities.

The simulations were carried out using the program package GROMOS.[c]

PROTEIN STRUCTURE AND DYNAMICS

TABLE 2 shows the rms differences between simulated structures at various times and the experimental structures. Although the PTI(II) experimental structure has

[c]GROMOS: Groningen Molecular Simulation package, developed at the University of Groningen, can be requested from the second author.

been derived by joint refinement of X-ray and neutron diffraction data,[15,16] we will refer to it as X-ray in the tables. The deviation of the time-averaged structures from the PTI(II) experimental structure is comparable to the previously found deviation from the PTI(I) one; 0.10 nm for the C^α atoms and 0.15 nm for all atoms. Averaging over the four molecules in the unit cell reduces these numbers to 0.09 nm for the C^α atoms and 0.13 nm for all atoms. FIGURE 2 demonstrates that the largest deviations from the experimental structure occur at the ends of the polypeptide chain. If we exclude the first and last two residues, the rms deviation averaged over the α-carbon drops to 0.067 nm, which value is smaller than the experimental rms positional fluctuation for the α-carbons of 0.070 nm, as it can be derived from the B-factors. The rms differences between the different molecules in the unit cell are significantly smaller than in the PTI(I) unit cell. Yet the four time-averaged structures differ by 0.14 nm for all atoms or by 0.08 nm for the α-carbons. This is reflected in the fact that when the rms positional fluctuations are calculated by averaging over time and over all four molecules in the unit cell (FIG. 3a), much larger values are found than when the fluctuations are obtained per molecule (FIG. 3b).

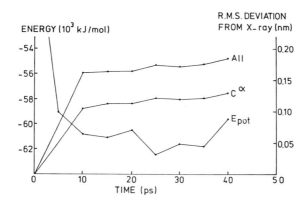

FIGURE 1. Total potential energy of the PT(II) unit cell and root mean square deviation of the simulated structures from the X-ray structure of PTI(II) as a function of time and averaged over the backbone α-carbons and over all atoms of the four molecules.

The correspondence between the MD fluctuations per molecule and the experimental values is satisfactory, given the approximations that are usually made in isotropic B-factor refinement procedures. We note that the largest differences occur for residues 1, 2 and 46–54 (FIG. 3b). These residues also show the largest differences in experimental B-factors between the PTI(I) and PTI(II) crystal forms, as can be observed from Figure 12 of Ref. 16.

A protein structure may also be characterized by its hydrogen bonds. In TABLE 3 the intra- and intermolecular hydrogen bonds of PTI(II) are listed. Three backbone–backbone hydrogen bonds out of 23 are not reproduced by the simulation; the X-ray hydrogen bonds 16 Ala H–36 Gly O and 17 Arg H–25 Ala O are also not observed in the PTI(I) structure when a slightly different hydrogen bond angle criterion of 145° is

TABLE 2. Root Mean Square Differences for All (C$^\alpha$) Atoms between Various Structures

Time (psec)	1/X-ray	2/X-ray	3/X-ray	4/X-ray	⟨1–4⟩/X-ray	1/2	2/3	3/4
10	0.176(0.109)	0.145(0.094)	0.172(0.117)	0.149(0.095)		0.181(0.104)	0.177(0.091)	0.183(0.111)
20	0.185(0.116)	0.145(0.092)	0.160(0.111)	0.162(0.128)		0.177(0.102)	0.151(0.090)	0.170(0.119)
30	0.181(0.118)	0.166(0.119)	0.162(0.104)	0.174(0.133)		0.187(0.116)	0.174(0.118)	0.172(0.112)
40	0.198(0.120)	0.179(0.136)	0.188(0.127)	0.176(0.133)		0.216(0.135)	0.186(0.126)	0.188(0.110)
Average 10–40	0.166(0.101)	0.140(0.098)	0.152(0.101)	0.150(0.115)	0.126(0.091)	0.143(0.072)	0.125(0.071)	0.140(0.086)
Average 8–20	0.166(0.113)	0.157(0.112)	0.162(0.106)	0.163(0.094)	0.119(0.082)	0.157(0.118)	0.169(0.099)	0.208(0.094)

NOTE: The rms difference $\langle(\bar{r}_i - \bar{r}_j)^2\rangle^{1/2}$ between two structures i and j is given in nm, ⟨ ⟩ denotes averaging over all atoms or all C$^\alpha$ atoms (between parentheses). The four molecules in the unit cell are denoted by 1, 2, 3 and 4 and their average structure by ⟨1–4⟩. The first five lines of this table give results for the PTI(II) unit cell[15,16] and the last line gives those of an earlier MD simulation[12] of the PTI(I) unit cell.[17]

applied[8]; the hydrogen bond from 52 Met O to 56 Gly H is shifted to 57 Gly H in the simulation. The backbone–side-chain hydrogen bonds show a comparable picture; the hydrogen bond 13 Pro O–15 Lys HZ is not observed in the PTI(I) structure in Ref. 8 and weakly reproduced in the simulation; the hydrogen bond from 55 Cys O to 1 Arg HH in PTI(II) is shifted to 1 Arg NE in PTI(I) and not observed in the simulation. The PTI(I) structure yields two side-chain–side-chain hydrogen bonds: 20 Arg HH–44 Asn OD and 24 Asn HD–31 Gln OE; the PTI(II) only yields the former, whereas the simulation predicts a third one between 7 Glu OE and 42 Arg HH. We note that in the PTI(II) crystal the 7 Glu side-chain is disordered.[16] None of the intermolecular hydrogen bonds in the PTI(II) crystal mentioned in Ref. 16 (17 Arg H–25 Ala, O, 21 Tyr HH–27 Ala O, 39 Arg HH–42 Arg O, 39 Arg HH–44 Asn O) satisfies the hydrogen bond criterion that is used here. For the latter three pairs the simulation yields sizeable percentages of hydrogen bonding.

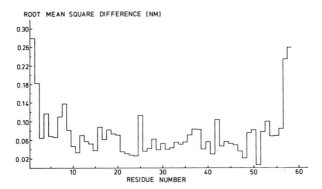

FIGURE 2. Difference between the PTI simulated time- and molecule-averaged atomic positions ($< $ 1-4 $>$) and the X-ray values (PTI(II)) for the C^α atoms.

TABLE 4 lists the rms difference in backbone torsion angles φ, ψ and ω between various structures. Averaged over each of the four molecules in the unit cell the difference in φ, ψ angle between the time-averaged and the experimental structures is considerable, although better than that found for a simulation of one PTI molecule in a rigid crystalline (I) environment without water (LCC).[7] Notwithstanding the difference in rigidity of the environment in both simulations, the torsion angle fluctuations are of the same size; in the LCC simulation the constraining effect of the rigid protein environment is compensated for by the omission of water molecules in the crystal, which allows space for the protein to move. FIGURE 4 illustrates that the overall difference in φ, ψ angle between simulated and experimental structures is due to a few peptide planes which have a different orientation in the two structures; these are the planes 15 Lys–16 Ala, 36 Gly–37 Gly, 37 Gly–38 Cys, 39 Arg–40 Ala and the ones at the disordered carboxy terminal. Eighty percent of the simulated φ, ψ angles show a deviation from the experimental structure which is of the same size as the fluctuation of the φ, ψ angles during the simulation.

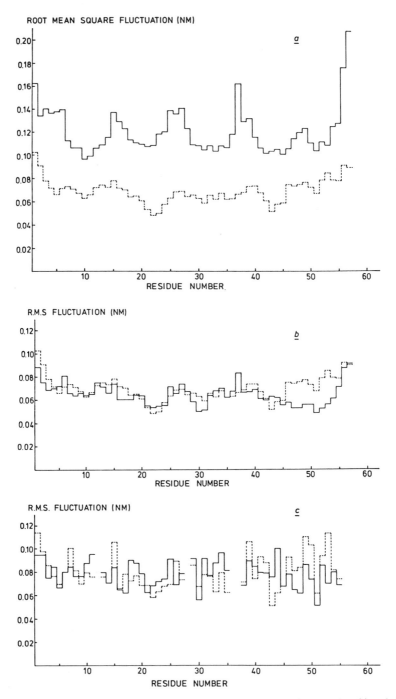

FIGURE 3. The rms positional fluctuations for the PTI atoms averaged over each residue. *Solid line:* MD results. *Broken line:* fluctuations from a set of X-ray temperature factors.[15,16] (***a***) Fluctuations of backbone atoms around the average structure obtained by averaging over time *and* over all four molecules in the unit cell; (***b***) mean (over four molecules) fluctuations of backbone atoms *per molecule;* (***c***) mean (over four molecules) fluctuations of side-chain atoms *per molecule.*

TABLE 3. Percentage of Observed and Simulated Hydrogen Bonds in the PTI(II) Unit Cell

Intramolecular Hydrogen Bonds		X-ray	MD, Molecule			
			1	2	3	4
Backbone	Backbone					
1 Arg H	37 Gly O	0	0	34	1	0
2 Pro O	5 Cys H	100	69	67	83	42
2 Pro O	6 Leu H	0	7	20	0	0
3 Asp O	6 Leu H	100	73	39	82	84
3 Asp O	7 Glu H	0	0	25	6	2
4 Phe O	7 Glu H	100	14	28	12	69
5 Cys O	7 Glu H	0	12	3	14	1
11 Thr O	36 Gly H	100	25	67	87	58
14 Cys O	16 Ala H	0	8	19	33	10
14 Cys O	37 Gly H	0	37	39	0	50
16 Ala H	36 Gly O	100	0	0	2	0
17 Arg H	25 Ala O	100	0	3	12	1
17 Arg H	26 Lys O	0	71	65	13	72
18 Ile H	35 Tyr O	100	92	94	94	93
18 Ile O	35 Tyr H	100	94	90	94	77
20 Arg H	33 Phe O	100	72	94	87	87
20 Arg O	33 Phe H	100	96	99	96	97
21 Tyr H	45 Phe O	100	82	87	73	81
21 Tyr O	45 Phe H	100	93	84	84	95
22 Phe H	31 Gln O	100	97	99	98	99
22 Phe O	31 Gln H	100	90	77	88	89
24 Asn H	29 Leu O	100	82	90	83	89
24 Asn O	28 Gly H	100	35	32	22	29
24 Asn O	29 Leu H	0	57	19	32	22
25 Ala O	28 Gly H	0	9	21	18	11
41 Lys O	44 Asn H	0	0	1	1	33
42 Arg O	44 Asn H	100	50	46	59	33
47 Ser O	50 Asp H	0	1	1	2	20
47 Ser O	51 Cys H	100	94	91	70	26
48 Ala O	52 Met H	100	98	85	94	97
49 Glu O	53 Arg H	100	72	60	29	58
50 Asp O	53 Arg H	0	4	24	13	4
50 Asp O	54 Thr H	100	32	47	10	23
51 Cys O	55 Cys H	100	89	54	88	22
51 Cys O	56 Gly H	0	0	0	2	66
52 Met O	56 Gly H	100	0	0	5	7
52 Met O	57 Gly H	0	55	55	74	0
55 Cys O	57 Gly H	0	0	14	0	12
Backbone	Side-Chain					
3 Asp H	3 Asp OD	0	12	27	17	11
4 Phe O	42 Arg HH	0	12	0	0	0
5 Cys O	43 Asn HD	0	15	9	10	14
7 Glu O	43 Asn HD	100	84	55	61	41
10 Tyr O	44 Asn HD	0	0	0	0	23
13 Pro O	15 Lys HZ	100	2	34	38	0
23 Tyr H	43 Asn OD	100	80	87	90	83
23 Tyr O	43 Asn HD	100	75	86	67	51
27 Ala H	24 Asn OD	100	92	74	65	78
34 Val O	11 Thr HG	0	19	54	26	55
37 Gly O	35 Tyr HH	0	51	4	21	8
39 Arg O	35 Tyr HH	0	17	66	3	76

TABLE 3. *Continued*

Intramolecular Hydrogen Bonds		X-ray	MD, Molecule			
			1	2	3	4
41 Lys H	44 Asn OD	0	0	0	0	30
42 Arg H	7 Glu OE	0	38	63	58	54
43 Asn H	7 Glu OE	100	92	3	37	11
46 Lys H	50 Asp OD	0	29	92	4	7
47 Ser H	50 Asp OD	0	44	0	94	76
49 Glu H	49 Glu OE	100	33	16	52	80
50 Asp H	47 Ser OG	100	81	84	78	2
50 Asp O	54 Thr HG	0	0	25	0	6
52 Met O	26 Lys HZ	0	0	0	0	40
55 Cys H	54 Thr OG	0	0	4	0	18
55 Cys O	1 Arg HH	100	0	0	2	18
56 Gly O	1 Arg HH	0	2	8	52	0
57 Gly O	26 Lys NZ	0	2	0	0	11
58 Ala O	17 Arg NH	0	10	35	0	0
Side-Chain	Side-Chain					
1 Arg HH	23 Tyr OH	0	3	5	6	41
7 Glu OE	42 Arg HH	0	67	100	33	67
20 Arg HH	44 Asn OD	100	0	15	78	2
24 Asn HD	31 Gln OE	0	91	37	49	54
35 Tyr OH	44 Asn HD	0	0	0	14	0
47 Ser HG	49 Glu OE	0	0	0	0	100
47 Ser HG	50 Asp OD	0	39	0	98	0
50 Asp OD	53 Arg HH	0	11	1	28	5
Intermolecular Hydrogen Bonds						
17 Arg HH	49 Glu OE	0	54(2)	25(1)	3(4)	14(3)
21 Tyr HH	27 Ala O	0	0	40(1)	41(4)	84(3)
21 Tyr HH	28 Gly O	0	0	52(1)	0	4(3)
26 Lys HZ	49 Glu OE	0	0	0	13(4)	0
39 Arg HH	41 Lys O	0	0	9(3)	0	40(1)
39 Arg HH	42 Arg O	0	0	45(3)	2(2)	26(1)
39 Arg HH	44 Asn O	0	0	26(3)	0	49(1)
40 Ala H	10 Tyr OH	0	29(4)	0	41(2)	0
41 Lys HZ	1 Arg O	0	0	2(3)	0	21(1)
54 Thr HG	3 Asp OD	0	0	0	80(1)	0

NOTE: The protein-protein hydrogen bonds are listed when the percentage of the configurations that satisfy the hydrogen bond criterion is larger than 10%. The criterion is: the donor-hydrogen-acceptor angle must be larger than 135° and the hydrogen-acceptor distance smaller than 0.24 nm. For the intermolecular hydrogen bonds the number between parentheses denotes the molecule in the unit cell to which the hydrogen bond is made.

PROTEIN HYDRATION AND WATER DYNAMICS

Since the atoms of the protein molecules in the unit cell have shifted away from the experimental positions by an average 0.15 nm, it is not meaningful to compare the time-averaged water positions with the experimental ones, unless this is done in a local coordinate frame defined in terms of neighboring protein atoms. This type of analysis of water positions is still to be completed.

Here we only give some examples of the hydration of protein atoms. FIGURE 5 shows the radial distribution of water oxygens around 46 Lys_1 NZ, which clearly displays a hydration shell. Defining a hydration shell radius $R^{sh} = 0.4$ nm, one may

TABLE 4. Root Mean Square Differences in Torsion Angles between Various Structures and Fluctuations of Torsion Angles

Torsion Angle q	RMS Difference $\langle \overline{(q_i - q_j)^2} \rangle^{1/2}$				
	1/X-ray	2/X-ray	3/X-ray	4/X-ray	LCC/X-ray
C-N-CA-C (φ)	34.5	39.4	31.5	33.0	39.5
N-CA-C-N (ψ)	36.1	35.0	28.4	31.9	41.6
CA-C-N-CA (ω)	4.8	4.9	4.3	3.9	9.7
	RMS Fluctuation $\langle \overline{(q_i - q_i)^2} \rangle^{1/2}$				
	1	2	3	4	LCC
C-N-CA-C (φ)	15.8	16.6	17.3	18.0	15.9
N-CA-C-N (ψ)	15.2	15.0	16.9	17.0	15.5
CA-C-N-CA (ω)	7.2	7.4	7.5	7.3	7.8

NOTE: The rms difference $\langle \overline{(q_i - q_j)^2} \rangle^{1/2}$ between two structures i and j is given in degrees. The time average (MD) is denoted by the overbar and $\langle \ \rangle$ denotes averaging over residues. The four molecules in the unit cell are denoted by 1, 2, 3 and 4. The last column contains results of an earlier MD simulation of one PTI molecule in a rigid crystalline (I) environment without water.[7]

calculate the residence distribution of water molecules in the hydration shell (FIG. 6), from which the half-life $\tau_{1/2}$ of water molecules in the hydration shell can be deduced. Values of $\tau_{1/2}$ for different atoms of the four protein molecules in the unit cell are given in TABLE 5. There one also finds the averaged (over 30 psec) number of protein-water hydrogen bonds N^{hb} in which a specific protein atom is involved. From the number of

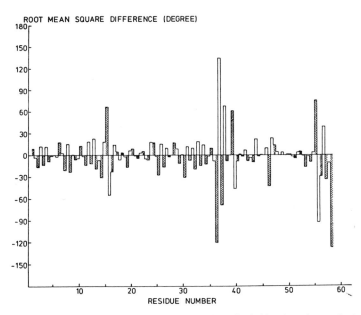

FIGURE 4. Difference in backbone φ (*blank bars*) and ψ (*shaded bars*) torsion angles between the PTI simulated time- and molecule-averaged structure ($< 1-4 >$) and the X-ray (PTI(II)) structure.

different water molecules N^ω that is forming the hydrogen bond an estimate for the protein-water hydrogen bond life-time τ^{hb} can be obtained. Values for τ^{hb}, which are of course shorter than $\tau_{1/2}$, center around 3–4 psec.

The distribution of the mobility of simulated waters is plotted in FIGURE 7. The diffusion constant of SPC water[27] is $D = 0.0036$ nm^2 psec^{-1}, which yields an rms fluctuation of 0.8 nm for bulk water over a period of 30 psec. In the PTI(II) crystal simulation no water molecules exhibit such bulk water mobility. The distribution gives no clear indication of the occurrence of distinct mobility classes for crystalline water. When FIGURE 3b is compared with FIGURE 3c it can be estimated that the protein backbone atoms show a mobility of 0.05–0.07 nm compared with 0.07–0.09 nm for the side-chain atoms in the MD simulation. It is estimated from FIGURE 7 that the number of water molecules in these two mobility classes are 8 and 18, respectively. Two hundred fifty-two ordered water positions are experimentally observed in the PTI(II) crystal. The results of the simulation indicate that waters do reside in these positions for periods of the order of ten picoseconds.

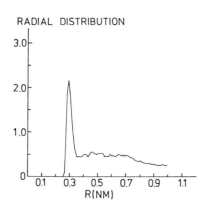

FIGURE 5. Radial (R) distribution of water oxygens around the NZ atom of 46 Lys in molecule 1 from the PTI(II) unit cell molecular dynamics simulation.

BEHAVIOR OF COUNTERIONS

The 24 Cl$^-$ ions in the PTI(II) unit cell show considerable differences in diffusional behavior. Two extreme cases are shown in FIGURE 8, where the mean square displacement of two different Cl$^-$ ions is given as a function of time. The curves for the other 22 Cl$^-$ ions fall in between the ones that are displayed. From the slope of the curves values for the diffusion constant, D, can be calculated. We find (TABLE 6) $D_{Cl}^{min} = 3.2 \ 10^{-5}$ nm^2/psec^{-1} and $D_{Cl}^{max} = 1.5 \ 10^{-3}$ nm^2/psec^{-1}, which are, as expected, lower than the experimental value for Cl$^-$ diffusion in bulk water at $T = 300$ K: $D_{Cl}^{exp} = 2.0 \ 10^{-3}$ nm^{2}psec^{-1}.

The radial distribution function for Cl$^-$-water oxygen pairs is given in FIGURE 9. It may be compared to MD simulation results for Cl$^-$ in solution.[29,30] The structure of the first coordination shell around Cl$^-$ in the PTI(II) crystal corresponds very well with that in bulk water, as can be seen from TABLE 6. However, the second-neighbor maximum in the radial distribution function of PTI(II) is much lower (0.5 versus 1.5) and broader than in bulk.[29,30] The presence of the protein molecules in the PTI(II) unit

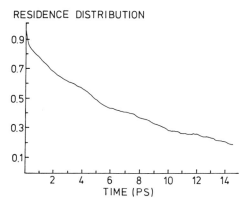

RESIDENCE DISTRIBUTION

FIGURE 6. Residence time distribution of water molecules (oxygen atoms) in the hydration shell (R < 0.4 nm) of the NZ atom of 46 Lys in molecule 1 from the PTI(II) unit cell molecular dynamics simulation.

cell does not allow the formation of a regular second hydration shell around the counterions. The half-life of water molecules in the first hydration shell of the Cl⁻ ions found here in PTI(II) is much longer than the one found for Cl⁻ in bulk water. The surrounding protein seems to stabilize the Cl⁻ hydration shell, since ⅔ of the Cl ions show $\tau_{1/2}$ values for water molecules that are longer than 15 psec. (TABLE 6).

TABLE 7 gives the percentage of simulated protein-Cl hydrogen bonds. About 40% of the Cl ions shows no hydrogen bonding at all, and the others are only weakly bound to the protein molecules. This corresponds well with the presence of a strong hydration shell around Cl ions in the PTI(II) crystal. Because protein-Cl hydrogen bonding is

TABLE 5. Protein Hydration

Protein		Water Residence Distribution		Protein-Water Hydrogen Bonding		
Atom	Molecule	R^{sh}	$\tau_{1/2}$	N^{hb}	N^{ω}	τ^{hb}
1 Arg NH1	1		>15	1.1	8	4.1
	2		9.0	0.8	9	2.6
	3	0.45	11.0	0.8	7	3.5
	4		8.9	0.9	9	2.9
46 Lys NZ	1		4.8	2.2	16	4.1
	2		5.6	2.2	18	3.6
	3	0.40	7.8	2.0	16	3.7
	4		9.7	2.1	15	4.3
52 Met CE	1		12.2			
	2		7.8			
	3	0.55	12.0			
	4		>15			

NOTE: The radius (nm) of the hydration shell of a protein atom is denoted by R^{sh}. The residence half-life $\tau_{1/2}$ (psec) yields the period of time after which only half of the water molecules originally present are still within R^{sh} of the protein atom. The average number of protein-water hydrogen bonds, using the same hydrogen bond criterion as in TABLE 3, is denoted by N^{hb}. The symbol N^{ω} denotes the number of different water molecules that are forming the protein-water hydrogen bond. The hydrogen bond life time τ^{hb} is obtained from $(N^{hb}/N^{\omega})* 30$ psec.

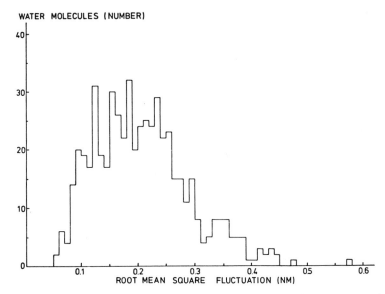

FIGURE 7. Distribution of the water positional fluctuations (MD) of the 552 waters in the PTI(II) unit cell.

weak, no clear correlation between protein atom and Cl^- mobility is observed in TABLE 7, except for the few relative most stable hydrogen bonds.

DISCUSSION

The results of the 40-psec MD simulation of the PT(II) crystal unit cell confirm the picture obtained from shorter simulations of the PTI(I)[12] and of the aPP[13] unit cell.

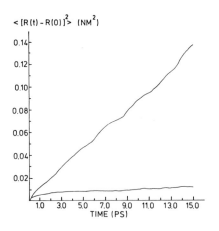

FIGURE 8. Mean square displacement of two different Cl ions in the PTI(II) unit cell molecular dynamics simulation.

TABLE 6. Chlorine Diffusion and Structure and Dynamics of Chlorine Hydration

	PTI (II) MD	Bulk MD[29]	Bulk MD[30]	Bulk Exp.[30]
T (K)	300	287	279	294
D_{Cl} (10^{-3} nm^{2}psec^{-1})	0.032 (min) / 1.5 (max)	2.3	—	2.0
R^{max} (1) (nm)	0.320	0.329	0.323	0.320 ± 0.04
R^{sh} (1) (nm)	0.4	0.4	0.4	
Coordination number	6.0	7.2	5.6–6.2	5.7 ± 0.2
$\tau_{1/2}$ (psec)	9.1 (min) / >15 (max)	3.1	—	
R^{max} (2) (nm)	0.50	0.52	0.49	

NOTE: The temperature T is given in K, the chlorine diffusion constant D_{Cl} in 10^{-3} nm^2 psec^{-1}. The values of R at which the Cl-water oxygen radial distribution function shows maxima and minima are denoted by $R^{max}(i)$ and $R^{sh}(i)$, for i = 1, 2, ... The half-life $\tau_{1/2}$ is defined as in TABLE 5.

The experimental atomic positions are reproduced within 0.1–0.15 nm and the atomic mobilities agree with X-ray temperature factors. Although smaller than in the previous unit cell simulations, the rms positional differences between the time-averaged structures of the four molecules in the unit cell are still larger (0.8–1.4 nm) than the experimental rms fluctuations (0.6–0.8). This might be due to the procedure by which the water molecules and counterions have been inserted into the unit cell. When the unfavorable interactions of water and counterions (TABLE 1) are relaxed by performing energy minimization, *all* atoms in the unit cell are allowed to move. In this way protein atoms can be pushed from their initial positions by the water molecules or

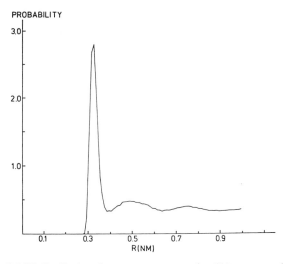

FIGURE 9. Radial (R) distribution of water oxygens around a Cl ion, averaged over 24 Cl ions in the PTI(II) unit cell molecular dynamics simulation.

counterions. It would be appropriate to restrain the protein atoms to their initial positions when the relaxation (EM) of the unfavorable water and counterion interactions is performed.

SUMMARY

Molecular dynamics simulations of hydrated protein crystals have been carried out in only three cases: two simulations of pancreatic trypsin inhibitor (PTI) and one of

TABLE 7. Percentage of Simulated Protein-Cl Hydrogen Bonds for the 24 Cl$^-$ Ions in the PTI(II) Unit Cell

Protein		Cl	Hydrogen Bond (%)
Atom	rms Fluctuation	rms Fluctuation	
17 Arg$_3$ NH	0.104 ⎱	0.070	35 ⎱
24 Asn$_4$ ND	0.094 ⎰		11 ⎰
—	—	0.071	—
17 Arg$_4$ NH	0.079	0.074	40
24 Asn$_2$ ND	0.088	0.078	30
15 Lys$_2$ N	0.051 ⎱	0.085	17 ⎱
42 Arg$_3$ NH	0.127 ⎰		16 ⎰
26 Lys$_1$ NZ	0.071	0.087	22
—	—	0.087	—
41 Lys$_1$ NZ	0.115	0.091	15
1 Arg$_1$ N	0.093	0.095	12
—	—	0.100	—
—	—	0.101	—
42 Arg$_2$ NH	0.087	0.102	17
—	—	0.106	—
—	—	0.114	—
31 Gln$_2$ NE	0.127	0.116	15
41 Lys$_1$ NZ	0.115	0.117	10
—	—	0.124	—
35 Tyr$_3$ OH	0.110	0.140	19
20 Arg$_3$ NH	0.085	0.140	15
—	—	0.143	—
15 Lys$_2$ NZ	0.103	0.146	20
39 Arg$_2$ NE	0.111	0.157	30
—	—	0.172	—
—	—	0.206	—

NOTE: The hydrogen bond criterion is identical to the one quoted in TABLE 3. The rms fluctuations are given in nm.

avian pancreatic polypeptide (aPP). The purpose of such simulations is to evaluate the accuracy and reliability of molecular dynamics simulations of hydrated proteins. A 40-psec simulation on crystalline PTI, involving 4 protein molecules (of 58 amino acids each), 552 water molecules, and 24 Cl$^-$ ions is described. Considerations are energetic and structural stability, division of interaction energy among water and protein, and precision of average structure and structural fluctuations compared to X-ray data from a new 0.94-Å resolution study of Wlodawer *et al.*[16] Water and ion dynamics are considered by analysis of diffusional motions and of residence times of water molecules in specific sites.

ACKNOWLEDGMENTS

We thank Dr. R. Huber for supplying us with the refined PTI(II) coordinates prior to publication. The molecular dynamics simulation has been performed on the one-pipe Cyber 205 supercomputer at SARA in Amsterdam. We thank Dr. A. Rozendaal of SARA for his generous assistance.

REFERENCES

1. McCAMMON, J. A., B. R. GELIN & M. KARPLUS. 1977. Nature (London) **267:** 585–590.
2. NORTHRUP, S. H., M. R. PEAR, J. A. McCAMMON & M. KARPLUS. 1980. Nature (London) **286:** 304–305.
3. LEVITT, M. 1980. *In* Protein Folding. R. Jaenicke, Ed.: 17–39. Elsevier/North Holland. Amsterdam.
4. ICHIYE, T. & M. KARPLUS. 1983. Biochemistry **22:** 2884–2893.
5. LEVY, R. M. 1985. NMR relaxation and protein dynamics. *In* Proceedings of the Workshop on Molecular Dynamics and Protein Structure. J. Hermans, Ed.: 62–64. Polycrystal Book Service. Western Springs, IL.
6. AQVIST, J., W. F. VAN GUNSTEREN, M. LEIJONMARCK & O. TAPIA. 1985. J. Mol. Biol. **183:** 461–477.
7. VAN GUNSTEREN, W. F. & M. KARPLUS. 1982. Biochemistry **21:** 2259–2274.
8. VAN GUNSTEREN, W. F. & H. J. C. BERENDSEN. 1984. J. Mol. Biol. **176:** 559–564.
9. KRUEGER, P., W. STRASSBURGER, A. WOLLMER & W. F. VAN GUNSTEREN. 1985. Eur. Biophys. J. **13:** 77–88.
10. HAGLER, A. & J. MOULT. 1978. Nature (London) **272:** 222–226.
11. HERMANS, J. & M. VACATELLO. 1980. *In* Water in Polymers. S.P. Rowland, Ed.: 199–214. American Chemical Society. Washington, DC.
12. VAN GUNSTEREN, W. F., H. J. C. BERENDSEN, J. HERMANS, W. G. J. HOL & J. P. M. POSTMA. 1983. Proc. Natl. Acad. Sci. USA **80:** 4315–4319.
13. HANEEF, I., I. D. GLOVER, I. J. TICKLE, D. S. MOSS, S. P. WOOD, T. L. BLUNDELL & W. F. VAN GUNSTEREN. 1985. The dynamics of pancreatic polypeptide: A comparison of X-ray anisotropic refinement at 0.98 Å resolution, molecular dynamics and normal mode analysis. *In* Proceedings of the Workshop on Molecular Dynamics and Protein Structure. J. Hermans, Ed.: 85–91. Polycrystal Book Service. Western Springs, IL.
14. GLOVER, I., I. HANEEF, J. PITTS, S. WOOD, D. MOSS, I. TICKLE & T. BLUNDELL. 1983. Biopolymers **22:** 293–304.
15. WALTER, J. & R. HUBER. 1983. J. Mol. Biol. **167:** 911–917.
16. WLODAWER, A., J. WALTER, R. HUBER & L. SJOLIN. 1984. J. Mol. Biol. **180:** 301–329.
17. DEISENHOFER, J. & W. STEIGEMANN. 1975. Acta Crystallogr. Sect. B **31:** 238–250.
18. HERMANS, J., H. J. C. BERENDSEN, W. F. VAN GUNSTEREN & J. P. M. POSTMA. 1984. Biopolymers **23:** 1513–1518.
19. DUNFIELD, L. G., A. W. BURGESS & H. A. SCHERAGA. 1978. J. Phys. Chem. **82:** 2609–2616.
20. BROOKS, B. R., R. E. BRUCCOLERI, B. D. OLAFSON, D. J. STATES, S. SWAMINATHAN & M. KARPLUS 1983. J. Comput. Chem. **4:** 187–217.
21. KOLLMAN, P. A., P. K. WEINER & A. DEARING. 1981. Biopolymers **20:** 2583–2621.
22. VAN GUNSTEREN, W. F. & M. KARPLUS. 1982. Macromolecules **15:** 1528–1544.
23. LEVITT, M. 1983. J. Mol. Biol. **168:** 595–620.
24. WODAK, S. J., P. ALARD, P. DELHAISE & C. RENNEBOOG-SQUILBIN. 1984. J. Mol. Biol. **181:** 317–322.

25. RYCKAERT, J. P., G. CICCOTTI & H. J. C. BERENDSEN. 1977. J. Comput. Phys. **23:** 327–341.
26. VAN GUNSTEREN, W. F. & H. J. C. BERENDSEN. 1977. Mol. Phys. **34:** 1311–1327.
27. BERENDSEN, H. J. C., J. P. M. POSTMA, W. F. VAN GUNSTEREN & J. HERMANS. 1981. *In* Intermolecular Forces. B. Pullman, Ed.: 331–342. Reidel. Dordrecht, the Netherlands.
28. BERENDSEN, H. J. C., J. P. M. POSTMA, A. DINOLA, W. F. VAN GUNSTEREN & J. R. HAAK. 1984. J. Chem. Phys. **81:** 3684–3690.
29. IMPEY, R. W., P. A. MADDEN & I. R. MCDONALD. 1983. J. Phys. Chem. **87:** 5071–5083.
30. BOUNDS, D. G.. 1985. Mol. Phys. **54:** 1335–1355.
31. TEETER, M.. 1984. Proc. Natl. Acad. Sci. USA **81:** 6014–6018.
32. VAN GUNSTEREN, W. F. & H. J. C. BERENDSEN. 1985. Molecular dynamics simulations: Techniques and applications to proteins. *In* Proceedings of the Workshop on Molecular Dynamics and Protein Structure. J. Hermans, Ed.: 5–14. Polycrystal Book Service. Western Springs, IL.

A Molecular Dynamics Computer Simulation of an Eight-Base-Pair DNA Fragment in Aqueous Solution: Comparison with Experimental Two-Dimensional NMR Data[a]

W. F. VAN GUNSTEREN, H. J. C. BERENDSEN,
R. G. GEURTSEN, AND H. R. J. ZWINDERMAN

Laboratory of Physical Chemistry
University of Groningen
9747 AG Groningen, the Netherlands

INTRODUCTION

Computer simulation of biological macromolecules is a rapidly growing field. Since 1977 a number of proteins have been simulated by the molecular dynamics (MD) method. In most cases the simulations were performed *in vacuo,* that is, the effect of the solvent environment on the protein dynamics was neglected. Only in a few studies were the solvating water molecules explicitly simulated.[1] When a more or less spherical protein is simulated, the distortion of the structure due to the vacuum boundary condition will be limited to residues at the surface of the molecule or loops protruding out of the protein core. However, when long extended structures like DNA are simulated, the vacuum boundary condition will severely distort the structure, which lacks a compact inner core that supplies some stability. Moreover, the omission of counterions and the shielding effect of the surrounding water will profoundly change the energy balance of a DNA fragment with its many negatively charged phosphate groups. Therefore, only a few attempts have been made to simulate DNA fragments *in vacuo,* and the mentioned problems in these cases were partly avoided by removing all charges entirely,[2] by reducing the phosphate charges in combination with the use of a distance-dependent dielectric constant,[3] or by introducing large implicitly·hydrated counterions and a distance-dependent dielectric constant.[4] In a molecular dynamics simulation of tRNA *in vacuo* similar approximations were made.[5]

Another line of development is the simulation of crystal hydrates, in which the nucleic acid is kept fixed and only water and counterions are allowed to move.[6–9] However, this type of study is overshadowed by the question to what extent the considerable flexibility of DNA fragments will change the results when taken into account.

Recently a 114-psec MD simulation of a 5-base-pair DNA fragment in a liquid drop consisting of 830 water molecules and 8 Na^+ counterions has been reported.[10] The

[a]This work was supported by SON, the Foundation for Chemical Research, under the auspices of the Netherlands Organization for the Advancement of Pure Research (ZWO).

analysis is focused on the influence of the presence of water and counterions on the DNA flexibility by comparing the results with those of a previous simulation without water.[4] Because of the lack of structural experimental data on this fragment, no direct experimental test of the simulation could be performed.

Structural data are available for only a few DNA fragments from X-ray diffraction studies[11] or from two-dimensional nuclear magnetic resonance (2D NMR) nuclear Overhauser enhancement (NOE) experiments.[12] Preliminary data have been published[13] in which the results of a molecular dynamics simulation of a crystal unit cell containing four 12-base-pair DNA fragments, 1220 water molecules, and 88 Na^+ ions are compared with X-ray data.[11] Here, we report an 80-psec MD simulation of an 8-base-pair DNA fragment solvated in 1231 water molecules plus 14 Na^+ ions in a rectangular box using periodic boundary conditions. For this fragment a set of 174 proton-proton distances is available, which has been derived from 2D NOE experiments (R. Boelens and R. Kaptein, private communication). As with the Dickerson 12-base-pair fragment simulation, this availability of accurate NOE distances allows for a direct structural test of the simulation results.

In the next two sections we shall briefly describe the model and computational procedure, which is subsequently assessed by comparing the simulated NOE proton–proton distances with the experimentally observed ones. The subsequent sections will deal with the DNA properties, DNA hydration and water dynamics and the behavior of the counterions.

We note that this study differs from the ones in which the experimental NOE distance data are used in an MD simulation in order to find a three-dimensional structure of a protein or DNA fragment that satisfies the experimentally observed NOE distances.[14-16] Here, the NOE data are not used in the MD simulation, but serve as a test of the simulated DNA structure.

MODEL AND COMPUTATIONAL PROCEDURE

The DNA fragment (dCGCAACGC/dGCGTTGCG) consists of 322 heavy atoms. No phosphate groups are placed at the ends of the strands, which have O5* and O3* as terminal atoms. Hydrogen atoms attached to carbon atoms are incorporated into the latter, forming united atoms, whereas the other 40 hydrogen atoms, which may form hydrogen bonds, are explicitly treated. The potential-energy function describing the interaction between DNA atoms, water molecules, and counterions is identical to the one used in Ref. 1. It is composed of terms representing bond-angle bending, harmonic (out-of-plane, out-of-tetrahedral configuration) dihedral bending, sinusoidal dihedral torsion, and van der Waals and electrostatic Coulomb interactions. Atomic partial charges[17] are used without modification; a dielectric constant of 1 is applied. In order to evaluate electrostatic interactions with sufficient accuracy a long cut-off range has to be used; a value of at least 1.5 nm seems necessary. But such a range is very expensive when pair interactions are evaluated; the number of neighbor atoms within 1.5 nm will exceed 300. We have adopted a twin-range method: the total nonbonded interactions are evaluated every MD time-step of $\Delta t = 0.002$ psec within a short cut-off range $R_1^c = 0.8$ nm, while the longer range (Coulomb) interactions within $R_2^c = 1.5$ nm are evaluated less frequently (only every $N_1^c - 10$ time-steps) and are kept fixed between updates. As in the protein simulations[1] all bond lengths are kept rigid during the simulation by using the SHAKE method[18,19] and the water molecules are modeled by a simple rigid three-point charge (SPC) model.[20] Interactions between water, DNA and Na$^+$ions are obtained from combination rules, using $\sqrt{C_{12}} = 14.5 \ 10^{-5}$ (kJ mol^{-1} nm^{12})$^{1/2}$ and $\sqrt{C_6} = 8.49 \ 10^{-3}$ (kJ mol^{-1} nm^6)$^{1/2}$ for the latter.

The B-DNA structure as given by Arnott and Hukins[21] was taken as initial configuration for the 8-base-pair fragment (TABLE 1). This structure was placed in the center of a box of which the dimensions (a = 3.040 nm, b = 3.398 nm, c = 4.012 nm) were chosen such that the minimum distance of any DNA atom to the wall of the box was 0.5 nm. Water molecules were inserted in the box by immersing it in an equilibrium configuration of bulk SPC water and subsequently removing all water molecules that are outside the box or of which the oxygen atom lies within $R^i = 0.20$ nm of a nonhydrogen DNA atom. In this way 1245 water molecules were inserted in the box, yielding a density of $\rho = 1.09$ g cm^{-3}. The high nonbonded energy of this configuration, which is due to the small value of R^i (DNA-H$_2$O) or to the periodic boundary condition (H$_2$O–H$_2$O), was relaxed by performing steepest-descents energy minimization (EM). In order to obtain a neutral system 14 Na$^+$ ions were inserted as follows: For the configuration of DNA plus 1245 waters, the electric potential at all water oxygen positions was calculated and the water molecule at the lowest potential was replaced by a Na$^+$ ion. This procedure was repeated 14 times and then followed by EM of the whole system, which allowed for relaxation of the Na–DNA and Na–H$_2$O interactions. Thus, the system that is simulated contains 4069 atoms, involving 12,207 degrees of freedom.

Initial velocities for the atoms were taken from a maxwellian distribution at 300 K. The simulation was done at constant temperature and pressure, that is, the system was weakly coupled to a thermal bath of $T_0 = 300$ K and to a pressure bath of $P_0 = 1$ atm = 0.06102 kJ mol^{-1} nm^{-3}, when integrating the equations of motion with time step $\Delta t = 2$ fsec. This was done by applying the algorithm of Ref. 22 with temperature relaxation time $\tau_T = 0.1$ psec (0.01 psec during the first 5 psec of the run) and pressure relaxation time $\tau_P = 0.5$ psec. These values make the coupling to the temperature and pressure baths weak enough to avoid any significant effect on the atomic properties of the system.[22]

The value for the compressibility, β, was chosen in the following way: the compressibility of water at various T_0 and P_0 is known.[23] The value for biological macromolecules like proteins is about five times smaller than that for water.[24] Thus, from the ratio of the amount of DNA and water in the box a rough estimate of the compressibility of the whole system can be made. We found $\beta = 6.19 \ 10^{-4}$ (kJ mol^{-1} nm^{-3})$^{-1}$. It is not crucial to use the physically correct β in the MD run, since β only influences the timescale of the coupling of the system to the pressure bath. We note that the pressure scaling of the periodic box was done independently along the three edges of the box.

The MD run covered a time span of 80 psec, which took about 40 central processing unit (CPU) hours on a one-pipe Cyber 205 supercomputer. FIGURE 1 shows that the potential energy drops rapidly during the first few picoseconds and stabilizes after 20 psec. The root mean square (rms) deviation of the actual atomic positions from the starting ones (B-DNA) shows a small drift after this equilibration period. The final 60 psec of the run have been used, with a time resolution of 0.05 psec, for analysis.

TABLE 1. Potential Energy of DNA (8 BP) in Solution at Constant Temperature and Pressure

| | | | Potential Energy (kJ/mol) | | | | | | | |
	Total	Bonded (Internal)	DNA-DNA	DNA-Na	Na-Na	DNA-H_2O	Na-H_2O	H_2O-H_2O	Volume (nm^3)
						Nonbonded				
DNA (B)	540	1443	-903						
DNA + 1245 H_2O	$7.9*10^{12}$	1443	-903			18186		$7.9*10^{12}$	
EM (88 steps)	-42223	1510	-1457			-2726		-39550	
DNA + 1231 H_2O + 14 Na	-52556	1521	-1455	-13936	3637	-2892	-801	-38630	
EM (36 steps)	-53767	1519	-1474	-14218	3742	-2746	-2835	-37755	41.44
MD									
10 psec	-66497	2027	-2513	-12709	3743	-6553	-7985	-42507	42.39
20 psec	-66809	1908	-2704	-12452	4311	-6529	-8338	-43005	41.98
30 psec	-67187	1938	-2154	-12867	4380	-6870	-8891	-42723	42.18
40 psec	-68054	1934	-2601	-12306	3951	-7275	-9158	-42599	41.66
50 psec	-67565	2008	-2463	-11865	3552	-7277	-8856	-42664	41.71
60 psec	-67331	1858	-1841	-11714	3549	-8014	-8430	-42839	41.92
70 psec	-68229	1883	-2685	-11244	2994	-7992	-8931	-42254	41.67
80 psec	-68124	1931	-2685	-11825	3534	-7930	-8993	-42156	41.96
20–50 psec (time average)	-67404	1947	-2480	-12372	4048	-6988	-8811	-42748	41.88
50–80 psec (time average)	-67812	1920	-2418	-11662	3407	-7803	-8802	-42478	41.82

Averages of various quantities are in most cases taken over two separate periods of 30 psec, in order to assess whether equilibration during 20 psec was sufficient.

TABLE 1 shows that the average volume of the box differs by only 1% from its initial value. This means that the minimum DNA-water distance of $R^i = 0.20$ nm for eliminating water molecules has been chosen appropriately.

COMPARISON OF MOLECULAR DYNAMICS WITH TWO-DIMENSIONAL NMR DATA

TABLE 2 compares proton–proton distances in different DNA structures with the experimental set of 174 NOE proton–proton distances derived from 2D NMR NOE experiments (R. Boelens and R. Kaptein, private communication). The average size of the experimental NOE distances is 0.297 nm and the experimental error is estimated to

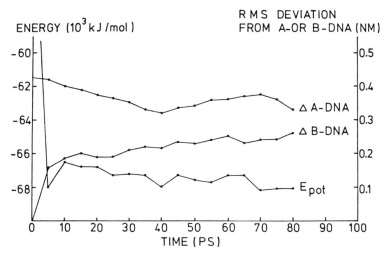

FIGURE 1. Total potential energy of the system and root mean square deviation of the simulated DNA structures from the A-DNA and B-DNA structures[21] as a function of time and averaged over all DNA atoms.

be on the order of 10% of the NOE distance itself. Therefore, we consider an NOE distance to be violated when the simulated distance is larger than 1.1 times the NOE distance itself. This is quite a stringent test on the simulated structures.

None of the two ideal (A- and B-) DNA structures satisfies all experimental NOE distances. In A-DNA 52% of the base-ribose NOEs is violated, whereas in B-DNA 20% of the internucleotide intra strand NOEs is violated. As expected, energy minimization does not change the structure much. From the configurations at 20 psec, 50 psec, and 80 psec it can be seen that the number and size of the violations increase gradually compared with those of the initial B-DNA structure. However, there are two

TABLE 2. Percentage (V) and Size (ΔR) of Violations of Experimental NOE Distances in DNA (8 BP) in Solution

	Intranucleotide						Internucleotide						Total		
	Base-Base		Base-Ribose		Rib-Rib		Intrastrand				Interstrand				
							Base-Base		Base-Rib		Base-Base				
	8		27		76		5		50		8		174		
Number of NOEs	ΔR	V	ΔR	V	ΔR	V	ΔR	V	ΔR	V	ΔR	V	ΔR	V	ΔR_{max}
A-DNA	.006	13	.074	52	.006	8	.009	0	.040	52	.002	0	.026	28	.156
B-DNA	.006	13	.002	4	.007	7	.015	20	.015	18	.001	0	.008	10	.111
EM	.008	13	.002	4	.003	1	.018	20	.018	20	.001	0	.008	8	.120
MD															
20 psec	.006	0	.011	15	.002	1	.067	60	.073	62	.023	50	.027	24	.305
50 psec	.006	13	.011	22	.003	1	.020	20	.070	58	.024	38	.025	23	.296
80 psec	.007	0	.017	30	.002	4	.056	40	.085	68	.106	75	.035	30	.275
20–50 psec (time-averaged)	.004	0	.006	7	.001	1	.027	20	.050	54	.032	25	.018	20	.202
50–80 psec (time-averaged)	.004	0	.007	11	.002	1	.052	20	.052	50	.084	75	.022	20	.294

NOTE: The size, ΔR, of the violation of the experimental NOE distances is given in nanometers and averaged over different types of proton-proton pairs. The percentage of violations, V, is obtained by counting violations that are larger than 10% of the experimental NOE distance. ΔR_{max} denotes the maximum observed ΔR.

exceptions: the intranucleotide ribose–ribose NOEs are better satisfied than in either A- or B-DNA, and the maximum violation is a little reduced compared to the value after equilibration. The next two lines in TABLE 2 show that the time-averaged structures satisfy the NOEs better than do the individual configurations at specific time points. In view of the fact that no fit to the experimental NOEs has been performed the simulated DNA structure agrees surprisingly well with the 174 NOE distances: 80% of these NOE distances is satisfied within experimental error, the mean violation is 0.022 nm, and the maximum violation amounts to 0.294 nm.

From the data in TABLE 2, but also from FIGURE 1, it may be concluded that the simulated DNA structure is a mixture of the idealized A- and B-DNA structures. This is illustrated in TABLE 3, where it is shown that both time-averaged structures are closer, either measured in Cartesian coordinates or in internal torsion coordinates, to both the A- and B-DNA structures than these latter two are to each other. The values suggest that the MD structure is a mixture of 40% A-DNA and 60% B-DNA. We note that also Dickerson's 12-base-pair X-ray structure[11] deviates from the ideal B-DNA

TABLE 3. Root Mean Square Difference between Various DNA Structures

	A-DNA	B-DNA	20–50 psec (time-averaged)	50–80 psec (time-averaged)
A-DNA	.0	50.2	46.9	48.1
B-DNA	.425	.0	36.3	38.5
20–50 psec (time-averaged)	.332	.195	.0	11.1
50–80 psec (time-averaged)	.351	.223	.123	.0

NOTE: *Upper right triangle:* rms difference (in degrees) between pairs of structures averaged over all torsion angles. *Lower left triangle:* rms difference (in nm) between pairs of structures averaged over all atom positions.

structure for this fragment; the difference is 0.133 nm averaged over all atom positions, or 33° when averaged over all torsion angles.

The difference between the 20–50-psec and 50–80-psec time-averaged structures is only 0.123 nm averaged over all atoms or, alternatively, 11.1° when averaged over all torsion angles. This difference is smaller than the torsion angle fluctuations over both time spans, as these are given in TABLE 6. This indicates the absence of a possible slow drift, with a relaxation time longer than 30 psec of the structure during the simulation.

DNA STRUCTURE AND DYNAMICS

TABLE 4 lists the hydrogen bonds between base pairs. During the MD run hydrogen bonds are broken and formed again. From the percentage hydrogen bonding it is observed that the ends of the helix are less stable than the central part, notwithstanding the A-T type of the central base pairs.

TABLE 5 contains some helix parameters for various DNA structures. In the two time-averaged 8-base-pair solution structures the angle between the bases of a pair is

TABLE 4. Percentage of Hydrogen Bonds between Base Pairs

		A-DNA	B-DNA	20 psec	50 psec	80 psec	20–50 psec	50–80 psec
C-G	O2-H21	100	100	0	100	0	15	16
	N3-H1	100	100	0	100	0	13	21
	H42-O6	100	100	0	100	0	65	8
G-C	H21-O2	100	100	100	100	100	83	90
	H1-N3	100	100	100	100	100	88	92
	O6-H42	100	100	0	100	100	87	83
C-G	O2-H21	100	100	100	100	100	56	86
	N3-H1	100	100	100	100	100	68	84
	H42-O6	100	100	100	100	100	94	81
A-T	N1-H3	100	100	100	100	100	82	60
	H61-O4	100	100	100	100	100	94	77
A-T	N1-H3	100	100	0	100	0	90	32
	H61-O4	100	100	100	100	100	76	90
C-G	O2-H21	100	100	0	100	0	60	1
	N3-H1	100	100	100	100	0	81	14
	H42-O6	100	100	100	100	100	96	85
G-C	H21-O2	100	100	100	100	0	89	16
	H1-N3	100	100	100	100	0	82	51
	O6-H42	100	100	100	100	100	77	81
C-G	O2-H21	100	100	0	0	0	1	0
	N3-H1	100	100	100	0	0	15	0
	H42-O6	100	100	100	0	0	84	10

NOTE: The hydrogen bond criterion is: the donor-hydrogen-acceptor angle must be larger than 135° and the hydrogen-acceptor distance smaller than 0.25 nm.

TABLE 5. Absolute Value of Helix Parameters for Various Structures Averaged over the Base Pairs

	Angle between Bases of a Pair	Base-Pair Roll	Base-Pair Tilt	Helix Repeat
8 base pairs				
A-DNA	11.7	2.0	19.9	32.7
B-DNA	4.2	0.7	5.7	36.0
EM	4.5 ± 2.7	1.2 ± 0.9	5.7 ± 0.4	36.0 ± 0.6
MD				
20 psec	30.1 ± 17.1	6.8 ± 5.9	11.8 ± 3.4	37.4 ± 5.0
50 psec	27.7 ± 11.6	5.6 ± 3.3	14.3 ± 6.8	36.3 ± 5.3
80 psec	34.2 ± 16.6	9.4 ± 4.5	11.0 ± 3.7	36.6 ± 11.2
20–50 psec (time-averaged)	20.7 ± 5.2	2.3 ± 1.8	13.4 ± 2.6	36.1 ± 3.6
50–80 psec (time-averaged)	29.6 ± 16.2	7.1 ± 5.3	11.2 ± 3.5	34.5 ± 5.0
12 base pairs				
B-DNA	4.2	0.7	5.7	36.0
X-ray	14.2 ± 4.8	4.9 ± 3.1	4.7 ± 4.0	36.1 ± 3.9

NOTE: The X-ray values have been obtained from the CGCGAATTCGCG structure of Ref 11. In each column the second value denotes the variation over the base pairs.

TABLE 6. Average Torsion Angles and Fluctuations for Various DNA Structures

	Backbone						Ribose								Base
	α	β	γ	δ	ϵ	ζ	ν_0	ν_1	ν_2	ν_3	ν_4	P	ν_{max}	Conformation	χ
8 base pairs															
A-DNA	-85	-152	46	83	178	-46	4	-26	37	-36	20	13	4	C_3-endo	-154
B-DNA	-47	-146	36	156	155	-95	-4	25	-35	33	-18	192	4	C_3-exo	-98
EM	-62	-149	49	145	154	-88	-12	30	-35	31	-12	181	12	{ C_2-endo / C_3-exo	-97
20–50 psec (time-averaged)	-78	161	72	118	-163	-107	-37	37	-23	2	22	127	62	C_1-exo	-115
fluctuation	22	22	11	17	18	25	11	10	13	16	14				16
50–80 psec (time-averaged)	-56	164	75	116	-160	-107	-39	38	-22	0	24	124	70	C_1-exo	-117
fluctuation	19	19	20	17	17	20	10	12	16	16	14				16
12 base pairs															
B-DNA	-47	-146	36	156	155	-95	-4	25	-35	33	-18	192	4	C_3-exo	-98
X-ray	-63	171	59	123	-169	-108	-39	39	-26	4	22	130	61	C_1-exo	-117

NOTE: The torsion angles have been averaged over the nucleotides, and defined following Ref. 25; they are given in degrees. The central two atoms of the different angles are: P-O5* (α), O5*-C5* (β), C5*-C4* (γ), C4*-C3* (δ), C3*-O3* (ϵ), O3*-P (ζ), C4*-C1* (ν_0), C1*-C2* (ν_1), C2*-C3* (ν_2), C3*-C4* (ν_3), C4*-O4* (ν_4). P is the pseudorotation phase angle[25] and the value of ν_{max} follows from the relation $\nu_{max} = \nu_0/\cos$ P. The X-ray values have been obtained from Ref. 11.

larger than in A- or B-DNA. The value in the crystalline 12-base-pair fragment[11] is also larger than in A- or B-DNA, but only half as large as in the 50–80-psec solution structure. The base-pair roll in solution is comparable to the X-ray value, but the tilt is again much larger, reflecting the presence of A- and B-DNA type structures during the simulation. The values for the helix repeat in solution lie between the A- and B-DNA values, with a preference for the latter.

TABLE 6 gives the torsion angles averaged over the 16 nucleotides. The MD time-averaged values for the two periods, 20–50 psec and 50–80 psec, differ by at most 3°, except for the α-angle. The fluctuations of the backbone torsion angles are slightly

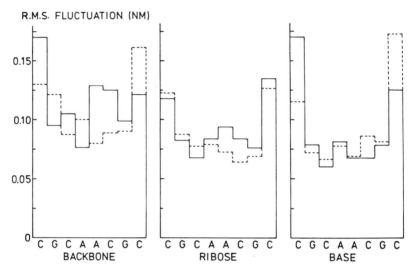

FIGURE 2. Rms positional fluctuations for the period 20–80 psec of the DNA atoms, averaged over the indicated parts of each nucleotide. *Solid line:* strand dCGCAACGC; *broken line:* strand dGCGTTGCG.

larger than those observed in proteins[1]; those of the ribose torsion angles are slightly smaller. The torsion angles in solution are rather different from both those in A-DNA and in B-DNA. The same observation holds for Dickerson's crystalline 12-base-pair fragment. Interestingly, the MD solution torsion angles resemble very much the X-ray ones: the rms difference is only 7°. In both cases the sugar pucker is $C_{1'}$-exo. When the 16 individual sugar puckers are analyzed, it turns out that they range from $C_{3'}$-endo (A) to $C_{3'}$-exo (B), yielding a characterization of the solution conformation at 75% B-DNA and 25% A-DNA.

FIGURE 2 illustrates the rms positional fluctuations over the 20–80-psec period for the different types of atoms. As expected, the nucleotides at the ends of the helix show a larger mobility than do the central ones. The backbone atoms show a bit larger mobility than do those in the riboses or bases.

RADIAL DISTRIBUTION

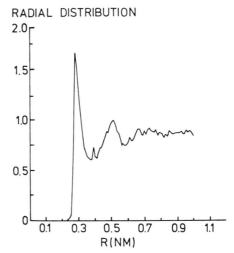

FIGURE 3. Radial distribution of water oxygens around the OP1 atom of 4 Ade calculated from the 20–80 psec part of the MD simulation.

DNA HYDRATION AND WATER DYNAMICS

The hydration of the 8-base-pair fragment has not yet been completely analyzed. Therefore, we only give some examples of the hydration of a few DNA atoms. In FIGURE 3 the radial distribution of water oxygens around 4 Ade OP1 is shown, which clearly displays the presence of a hydration shell. Defining a hydration shell radius $R^{sh} = 0.40$ nm, one may calculate the residence distribution of water molecules in the

RESIDENCE DISTRIBUTION

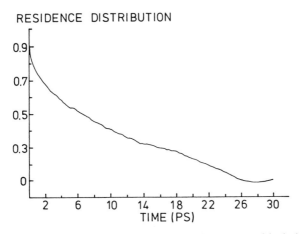

FIGURE 4. Residence time distribution of water molecules (oxygen atoms) in the hydration shell ($R < 0.4$ nm) of the OP1 atom of 4 Ade calculated from the 20–80 psec part of the MD simulation.

TABLE 7. DNA Hydration

DNA		Water Residence Distribution		DNA-Water Hydrogen Bonding			
Atom	Location	R^{sh}	$\tau_{1/2}$	N^{hb}	N^{ω}	τ^{hb}	τ^{hb}_{max}
4 Ade OP1	Edge	0.40	6.5	2.3	25	5.5	21.7
4 Ade O4*	Minor	0.45	12.2	0.7	9	4.4	35.9
4 Ade C2*	Major	0.45	6.8	—	—	—	—
3 Cyt O2	Minor	0.45	>30	0.7	4	10.6	34.3
14 Gua N2	Minor	0.55	16.3	0.8	12	3.9	10.5

NOTE: The residence half-life $\tau_{1/2}$ (psec) yields the period of time after which only half of the water molecules originally present within the hydration shell of radius R^{sh} are still within distance R^{sh} of the DNA atom. The average number of DNA-water hydrogen bonds, using the same hydrogen bond criterion as in TABLE 4, is denoted by N^{hb}. The symbol N^{ω} denotes the number of different water molecules that are forming the DNA-water hydrogen bond. The hydrogen bond life-time τ^{hb} is obtained from $(N^{hb}/N^{\omega}) * 60$ psec. The symbol τ^{hb}_{max} denotes the longest time during which a specific water molecule is bound to the DNA atom.

hydration shell (FIG. 4), from which the half-life $\tau_{1/2}$ of water molecules in the hydration shell can be deduced. Values of $\tau_{1/2}$ for different atoms in the minor and major grooves and on the outer edge of the DNA helix are given in TABLE 7. There one also finds the averaged (over 60 psec) number of DNA-water hyrdogen bonds N^{hb} in which a specific DNA atom is involved. From the number of different water molecules N^{ω} that is forming the hydrogen bond, an estimate for the DNA-water hydrogen bond life-time τ^{hb} can be obtained. Values for $\tau_{1/2}$ and τ^{hb} show a considerable spread. The phosphate oxygen shows three times more hydrogen bonding than the other oxygen atoms in DNA.

FIGURE 5 plots the distribution of the mobility of simulated waters. The two periods, 20–50 psec and 50–80 psec, display about the same patterns, except that the

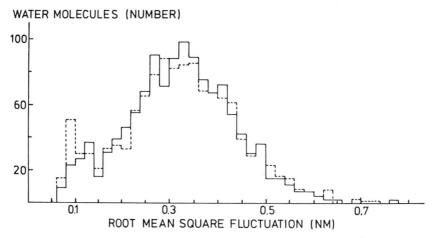

FIGURE 5. Distribution of the water positional fluctuations (MD) of the 1231 waters in the computational box. *Solid line:* fluctuation over the period of 20–50 psec; *broken line:* fluctuation over the period 50–80 psec.

latter period yields more waters having a mobility comparable to that of the DNA atoms: the number of waters with a rms fluctuation lower than 0.1 nm is 32 (20–50 psec) or 66 (50–80 psec). The diffusion constant of SPC water[20] is $D = 0.0036$ $nm^2\,psec^{-1}$, which yields an rms fluctuation of 0.8 nm for bulk water over a period of 30 psec. In the DNA simulation no water molecules exhibit such bulk water mobility.

A more extensive analysis of DNA hydration and water dynamics will be given elsewhere.

BEHAVIOR OF SODIUM COUNTERIONS

The 14 Na^+ ions in the computational box show considerable differences in diffusional behavior. Two extreme cases are shown in FIGURE 6, where the mean

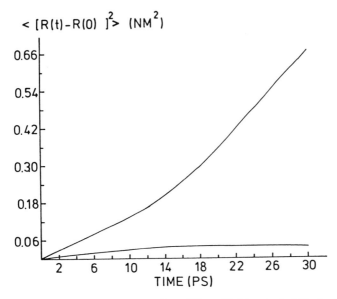

FIGURE 6. Mean square displacement of two different Na^+ ions in the MD simulation.

square displacement of two different Na^+ ions is given as a function of time. The curves for the other 12 Na^+ ions fall between the ones that are displayed. From the slope of the curves values for the diffusion constant D can be calculated. We find (TABLE 8) values between virtually zero and $D = 5.0\ 10^{-3}\ nm^2\,psec^{-1}$. The latter represents an extreme case. The other 13 values are smaller than $1.0\ 10^{-3}\ nm^2\,psec^{-1}$.

The distribution of Na-phosphate oxygen distances is shown in FIGURE 7 for two periods of the MD simulation which follow the equilibration period of 20 psec. No Na^+ ions are found within 0.3 nm of any phosphate oxygen. However, at the start of the MD simulation most of the Na ions were near the DNA: nine within 0.3 nm, two within 0.34 nm, and the remaining three in solution. This means that during the equilibration

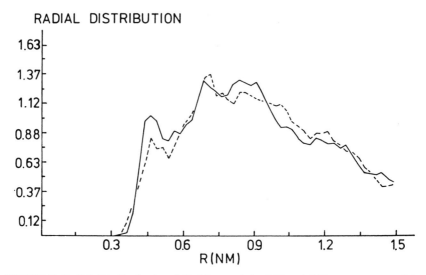

FIGURE 7. Radial (R) distribution of Na$^+$ ions and the O1P and O2P oxygen atoms of the phosphate groups in DNA. *Solid line:* distribution for the period 20–50 psec; *broken line:* distribution for the period 50–80 psec.

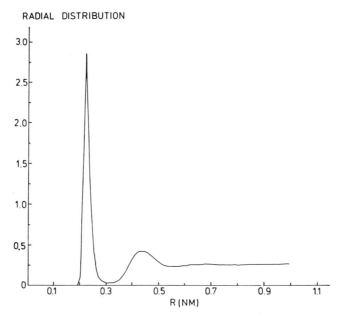

FIGURE 8. Radial (R) distribution of water oxygens around a Na$^+$ ion, averaged over 14 Na ions and 20–80 psec.

period of the simulation the sodium ions went into solution. They prefer a distance to the phosphate oxygens of about 0.46 nm or 0.7 nm. So, both phosphate oxygens and Na ions are almost completely hydrated.

The radial distribution function for Na–water oxygen pairs is given in FIGURE 8. It may be compared to MD simulation results for Na^+ in solution.[26,27] The structure of the first coordination shell around Na^+ ions near a DNA fragment corresponds well with that in bulk water, as can be seen from TABLE 8. The coordination number is in the presence of DNA a little lower than in bulk, which may be due to the influence of phosphate oxygens in the second coordination shell.

The half-life of water molecules in the first hydration shell of the Na^+ ions that hydrate DNA is found to be longer than for a pure Na solution (bulk). For the Na^+ ion that displays the large diffusion constant of $D = 5.10^{-3}$ nm^2 $psec^{-1}$ the half-life of the water molecules in the first hydration shell is $\tau_{1/2} = 11.7$ psec, whereas for the other

TABLE 8. Sodium Diffusion and Structure and Dynamics of Sodium Hydration

	DNA	Bulk	
	MD	MD[26]	MD[27]
T(K)	300	282	287
D_{Na} $(10^{-3}$ $nm^2psec^{-1})$	0.0 (min) $\left.\right\}$ 5.0 (max)	1.0	—
R^{max} (1) (nm)	0.225	0.229	0.235
R^{sh} (1) (nm)	0.32	0.31	0.31
Coordination number	5.6	6.0	6.2
$\tau_{1/2}$ (psec)	11.7 (min) $\left.\right\}$ >30 (max)	6.9	—
R^{max} (2) (nm)	0.43	0.44	0.44

NOTE: The temperature, T, is given in K, the sodium diffusion constant, D_{Na}, in 10^{-3} nm^2 $psec^{-1}$. The values of R at which the Na-water oxygen radial distribution function shows maxima and minima are denoted by $R^{max}(i)$ and $R^{sh}(i)$, for i = 1, 2, ... The half-life, $\tau_{1/2}$, is defined as in TABLE 7.

Na^+ ions the values of $\tau_{1/2}$ are larger than 20 psec. It seems that the presence of the DNA, which is hydrated by hydrated Na^+ counterions, influences the properties of the Na^+ hydration shells itself.

CONCLUSIONS

The results of the 80-psec MD simulation of the 8-base-pair DNA fragment in a rectangular box with 1231 H_2O molecules and 14 Na^+ counterions, using periodic boundary conditions, yield a consistent picture of the structure and dynamics of a small DNA fragment in solution.

Eighty percent of a set of 174 experimental NOE distances is reproduced within the experimental error. The solution structure can be roughly characterized as a mixture of A-DNA (30%) and B-DNA (70%) features. In terms of mean torsion angles the solution structure resembles much the Dickerson 12-base-pair crystal structure.[11]

The sodium ions show a clear tendency to solvate at a distance of at least 0.46 nm from the phosphate oxygens.

SUMMARY

The structure and dynamics of an 8-base-pair DNA fragment (dCGCAACGC/dGCGTTGCG) in aqueous solution (14 Na^+ ions, 1231 water molecules) have been simulated by using the molecular-dynamics method. Interproton distances have been calculated for various structures and are compared with a set of 174 distances which have been derived from 2D NOE experiments. The averaged MD structures are compared with ideal A-DNA and B-DNA structures in terms of helix parameters, dihedral angles, and so forth. The hydration of various atoms of the DNA fragment and of the Na^+ ions is analyzed by calculating coordination numbers and first-neighbor shell residence times.

ACKNOWLEDGMENTS

We thank Rolf Boelens for supplying us with the set of experimental NOE distances prior to publication, and Thea Koning for generating the Arnott and Hukins A-DNA structure. The MD simulation has been performed on the one-pipe Cyber 205 supercomputer at SARA in Amsterdam. We thank Alon Rozendaal of SARA for his invaluable help.

REFERENCES

1. BERENDSEN, H. J. C., W. F. VAN GUNSTEREN, H. R. J. ZWINDERMAN & R. G. GEURTSEN. 1986. Simulations of proteins in water. This volume.
2. LEVITT, M. 1983. Cold Spring Harbor Symp. Quant. Biol. **47**: 251–275.
3. TIDOR, B., K. K. IRIKURA, B. R. BROOKS & M. KARPLUS. 1983. J. Biomol. Struct. Dyn. **1**: 231–252.
4. SINGH, U. C., S. J. WEINER & P. KOLLMAN. 1985. Proc. Natl. Acad. Sci. USA **82**: 755–759.
5. PRABHAKARAN, M., S. C. HARVEY, B. MAO & J. A. MCCAMMON. 1983. J. Biomol. Struct. Dyn. **1**: 357–369.
6. CLEMENTI, E. & G. CORONGIU. 1982. Biopolymers **21**: 763–777.
7. MEZEI, M., D. L. BEVERIDGE, H. M. BERMAN, J. M. GOODFELLOW, J. FINNEY & S. NEIDLE. 1983. J. Biomol. Struct. Dyn. **1**: 287–297.
8. LEE, W. K., Y. GAO & E. W. PROHOFSKY. 1984. Biopolymers **23**: 257–270.
9. BEVERIDGE, D. L., P. V. MAYE, B. JAYARAM, G. RAVISHANKER & M. MEZEI. 1984. J. Biomol. Struct. Dyn. **2**: 261–270.
10. SEIBEL, G. L., U. C. SINGH & P. A. KOLLMAN. 1985. Proc. Natl. Acad. Sci. USA. **82**: 6537–6540.
11. DREW, H. R., R. M. WING, T. TAKANO, C. BROKA, S. TANAKA, K. ITAKURA & R. E. DICKERSON. 1981. Proc. Natl. Acad. Sci. USA **78**: 2179–2183.
12. SCHEEK, R. M., N. RUSSO, R. BOELENS, R. KAPTEIN & J. H. VAN BOOM. 1983. J. Am. Chem. Soc. **105**: 2914–2916.
13. VAN GUNSTEREN, W. F. & H. J. C. BERENDSEN. 1985. Molecular dynamics simulations:

techniques and applications to proteins. *In* Proceedings of the Workshop on Molecular Dynamics and Protein Structure. J. Hermans, Ed.: 5–14. Polycrystal Book Service. Western Springs, IL.

14. VAN GUNSTEREN, W. F., R. KAPTEIN & E. R. P. ZUIDERWEG. 1983. Use of the molecular dynamics computer simulation method when determining protein structure by 2D-NMR. *In* Nucleic Acid Conformation and Dynamics, Report of NATO/CECAM Workshop. W. K. Olson, Ed.: 79–92. Orsay, France.

15. KAPTEIN, R., E. R. P. ZUIDERWEG, R. M. SCHEEK, R. BOELENS & W. F. VAN GUNSTEREN. 1985. J. Mol. Biol **182**: 179–182.

16. VAN GUNSTEREN, W. F., R. BOELENS, R. KAPTEIN, R. M. SCHEEK & E. R. P. ZUIDERWEG. 1984. An improved restrained molecular dynamics technique to obtain protein tertiary structure from nuclear magnetic resonance data. *In* Proceedings of the Workshop on Molecular Dynamics and Protein Structure. J. Hermans, Ed.: 92–99. Polycrystal Book Service. Western Springs, IL.

17. HERMANS, J., H. J. C. BERENDSEN, W. F. VAN GUNSTEREN & J. P. M. POSTMA. 1984. Biopolymers **23**: 1513–1518.

18. RYCKAERT, J. P., G. CICCOTTI & H. J. C. BERENDSEN. 1977. J. Comput. Phys. **23**: 327–341.

19. VAN GUNSTEREN, W. F. & H. J. C. BERENDSEN. 1977. Mol. Phys. **34**: 1311–1327.

20. BERENDSEN, H. J. C., J. P. M. POSTMA, W. F. VAN GUNSTEREN & J. HERMANS. 1981. *In* Intermolecular Forces. B. Pullman, Ed.: 331–342. Reidel. Dordecht, the Netherlands.

21. ARNOTT, S. & D. W. L. HUKINS. 1972. Biochem. Biophys. Res. Commun. **47**: 1504–1509.

22. BERENDSEN, H. J. C., J. P. M. POSTMA, A. DI NOLA, W. F. VAN GUNSTEREN & J. R. HAAK. 1984. J. Chem. Phys. **81**: 3684–3690.

23. KELL, G. S. 1967. J. Chem. Eng. Data **12**: 66–69.

24. GAVISH, B., E. GRATTON & C. J. HARDY. 1983. Proc. Natl. Acad. Sci. USA **81**: 750–754.

25. SAENGER, W. 1984. Principles of Nucleic Acid Structure. Springer. New York. NY.

26. IMPEY, R. W., P. A. MADDEN & I. R. McDONALD. 1983. J. Phys. Chem. **87**: 5071–5083.

27. BOUNDS, D. G. 1985. Mol. Phys. **54**: 1335–1355.

Tryptophan Structure and Dynamics Using GROMOS

RICHARD A. ENGH, LIN X. CHEN,
AND GRAHAM R. FLEMING

The University of Chicago
Chicago, Illinois 60637

The desire to use tryptophan as a fluorescent probe to study protein motion has motivated studies on tryptophan in aqueous solutions. Tryptophan's biexponential fluorescence decay in solution has given rise to models which account for this behavior

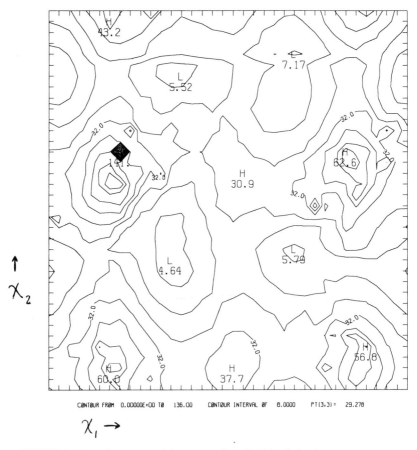

CONTOUR FROM 0.00000E+00 TO 136.00   CONTOUR INTERVAL OF 8.0000   PT(3,3) = 29.278

FIGURE 1. Approximate potential energy surface (in kJ/mol) for the χ_1, χ_2 rotamers.

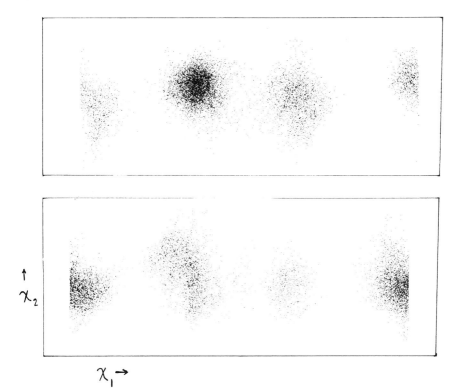

FIGURE 2. Two typical MD trajectories plotted as a function of χ_1 and χ_2 angles. All χ_1 rotamers are sampled during each 100-psec trajectory, while no χ_2 transitions occur. (*Top*) χ_2 = perpendicular; (*bottom*) χ_2 = antiperpendicular.

by postulating the existence of rotamers about the C_α-C_β bond. These models require that the rotamers do not interconvert during the fluorescence lifetime; this assumption has not been verified. Further, there are certain peculiarities in assigning particular fluorescence lifetimes to these rotamers on the basis of proximity of the charged quenching groups. To address these issues we have used energy minimization and classical molecular dynamics as bases for transition rate and population estimates. These suggest that the C_β-C_γ dihedral (χ_2) determines the fluorescence lifetime for equilibrating groups of C_α-C_β rotamers.

The energy minimization was similar to that already reported in the literature,[1,2] but with GROMOS charges (for tryptophan in solution) and GROMOS interaction potentials. A potential surface for the energies of χ_1, χ_2 conformations was generated by constraining the dihedrals and minimizing the potential energy of the remaining degrees of freedom. Energy barriers between conformers range from about 3 kcal/mol to about 8 kcal/mol. The smaller barriers correspond to χ_1 rotations. Seven 100-psec classical dynamics trajectories were calculated for the system coupled to a heat bath[3] with a time constant of 0.1 psec (solvent molecules are not explicitly included). During these trajectories, only one χ_2 transition was observed (and this was during the equilibration period), whereas χ_1 transitions occurred several times in each simulation. In light of these results, a re-evaluation of the χ_1 rotamer models is required. It may be that the χ_2 dihedral specifies the conformers associated with a particular lifetime, while χ_1 rotamers remain in equilibrium during the fluorescence decay. Also suggestive, but perhaps fortuitous, is the close correspondence between NMR estimates of χ_2 population distributions[4] and the fluorescence pre-exponential factors.[5]

ACKNOWLEDGMENTS

We are grateful to Professor van Gunsteren and Dr. Al Cross for providing us with the GROMOS package and assisting us in using it.

REFERENCES

1. ANDERSON, J. S., G. S. BOWITCH & R. L. BREWSTER. 1983. Biopolymers 22: 2459–2476.
2. VASQUEZ, M., G. NEMETHY & H. A. SCHERAGA. 1983. Macromolecules 16: 1043–1049.
3. BERENDSEN, H. J. C., et al. 1984. J. Chem. Phys. 81 (8): 3684–3690.
4. DEZUBE, B., C. M. DOBSON & C. E. TEAGUE. 1981. J. Chem. Soc. Perkin Trans. 2: 730–735.
5. PETRICH, J. W., M. C. CHANG, D. B. McDONALD & G. R. FLEMING. 1983. J. Am. Chem. Soc. 105: 3824–3832.

Index of Contributors